Proceedings of the Probabilistic Safety Assessment and Management (PSAM) 12 Conference - Volume 2

Foreword

It is was our honor to welcome you to Honolulu, Hawaii, for the twelfth rendition of the Probabilistic Safety Assessment and Management (PSAM) Conference. The planning for PSAM Honolulu began back in 2007 (before PSAM 9 in Hong Kong), when we looked at several locations around the United States, included Arizona, California, Boston, and even considered locations in Oceania. Based upon the feedback both during and after the conference, PSAM 12 proved to be a great suc-cess.

We would like to thank all of the volunteers, those that served before, during, and after the Conference. Members of the Technical Program Committee, the Organizing Committee, the session chairs, and the presenters have our gratitude for making PSAM 12 the most memorable PSAM yet.

This publication represents the technical proceedings for the Conference. Due to the large number of published papers (a total of 391), we have subdivided the technical content (papers) into multiple volumes.

On behalf of the International Association for Probabilistic Safety Assessment and Management Board of Directors, we hope that this publication will provide a valuable technical resource in addition to a reminder of the memorable stay in the Hawaiian Islands.

Dr. Curtis Smith
Technical Program Chairs

Dr. Todd Paulos
General Chair

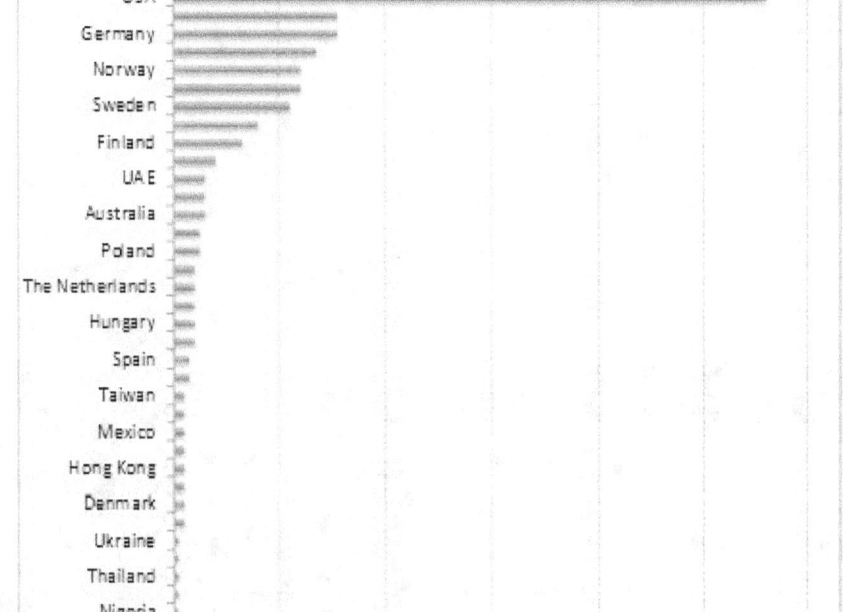

Number of Papers Presented at PSAM 12 (by country)

Sponsors

Sponsors

EPRI Assesses Seismic Resistance of Electronic Components

Together... Shaping the Future of Electricity

Technical Program Committee

Technical Program Chair: Curtis Smith, INL USA

Assistant Technical Program Chairs: Steve Epstein, Lloyd's Register Japan
Vinh Dang, PSI Switzerland
Ted Steinberg, QUT Australia

We would like to thank the members of the PSAM 12 Technical Program Committee. These individuals helped to make PSAM 12 a success by reviewing abstracts, technical papers, organizing sessions, and providing technical leadership for the conference.

Technical Committee Members:

Roland Akselsson	Vyacheslav S. Kharchenko
S. Massoud (Mike) Azizi	James Knudsen
Tito Bonano	Zoltan Kovacs
Ronald Boring	Ping Li
Roger Boyer	Harry Liao
Mario Brito	Francois van Loggerenberg
Kaushik Chatterjee	Jerome Lonchampt
Vinh Dang	Soliman A. Mahmoud
Claver Diallo	Diego Mandelli
Nsimah Ekanem	Donoval Mathias
Steve Epstein	Zahra Mohaghegh
Fernando Ferrante	Thor Myklebust
Federico Gabriele	Cen Nan
Ray Gallucci	Mohammad Pourgolmohammad
S. Tina Ghosh	Marina Roewekamp
David Grabaskas	Clayton Smith
Katrina Groth	Shawn St. Germain
Seth Guikema	Ted Steinberg
Steve Hess	Kurt Vedros
Christopher J. Jablonowski	Smain Yalaoui
Moosung Jae	Robert Youngblood
Jeffrey Joe	Enrico Zio

Organizing Committee

General Chair: Dr. Todd Paulos

General Vice Chair: Prof. Stephen Hora, USC

Technical Program Chair: Curtis Smith, INL USA

Webmaster, Registration,
Support for Papers/Abstracts
Submission and Review: Hanna Shapira, TICS

Table of Content

Table of Content

Page   Paper

Table of Content

Life Analysis for the Main bearing of Aircraft Engines

Peng Qin[a], Xiaoling Zhang[a], Liping He[a], Liangliang Ding[a]

[a] School of Mechanics, Electronic, and Industrial Engineering, University of Electronic Science and Technology of China, Chengdu, China

Email: lipinghe@uestc.edu.cn

ABSTRACT: Life of main bearings in a aircraft engine directly affects the reliability, safety and feasibility of aircraft engines. In order to optimize and improve the performance of the existing aircraft engines, as well meet the needs of a new generation of aircraft engines, this paper analyses and estimate of main bearing's life of aircraft engines by taking the deep groove ball bearing as an example. Firstly, the 3D model of deep groove ball bearings is established by using Pro-E software and then converted into a finite element model. Secondly, such features as stiffness, strength and fatigue life of the deep groove ball bearing are investigated by ANSYS software, which result in some theoretical are discussion and some relevant measures that enable improving the service life of main bearings of the aircraft engine.

Keywords: aircraft engines, deep groove ball bearing, service life, reliability, finite element analysis

1. INTRODUCTION

The bearing of main shaft is a major component of aircraft engine. With the development of aircraft engines, the demand of working condition has been increasing rapidly. Since life and reliability of main bearing directly affect the life and reliability of the engine, failures of main bearing would lead to disastrous consequences.

As the loading support and movement connecting component, bearing can transform the sliding friction into the rolling friction. So main bearing plays a vital role in the aircraft engine, it's operating temperature is $300°C$, the value of DN (bearing diameter (mm) × speed (r/min)) is larger than $2.3×10^6$. Meanwhile, as the engine bearing unit, aircraft engine main bearing also bear the static load, the axial concentricity caused by static and dynamic load, maneuvering load, steady vibration load, rotor hot bending, the compressor dynamic load, temperature load and so on.

As the key component of the aircraft engine, its advanced research and development has been concerned in home and abroad. Such as NASA of America, it began a "speed bearing research and development program" in the 1959[2]. But we have a large gap in the research of main bearing of aircraft engine, especially in the design. In order to meet the needs of aircraft development, we should develop and apply advanced design and analysis techniques of bearing.

According to study of Chen[2],Tang[3] and Zhao[4], we take a in-depth research about the recent progress of the current finite element analysis method for fatigue life prediction in this paper.

Lot of study has been done about the life prediction of aircraft engine in the world. In the present paper, we take the deep groove ball bearings as an example, which mainly composed of outer ring, cage, rolling elements and inner ring. Since fatigue failure generally does not occur on the retainer, the bearing life is mainly decided by the life of the outer ring, inner ring and rolling elements. According to the Lundberg-Palmgren bearing life theory, the life of the bearing outer ring, inner ring and rolling elements are a series of random numbers subjected to Weibull distribution. So the life of bearing can be estimated through the life of the bearing outer ring, inner ring and a rolling element .The main work of the present paper shown as follows:

(1) In section 2, we introduced the basic theory and models for fatigue life prediction of the deep groove ball bearing, and analyzed the basic procedure of fatigue life prediction.

(2) Established parametric model and loaded the appropriate boundary conditions by using the static analysis module of ANSYS, then obtained the bearing stress and deformation under static load.

(3) Got the fatigue life of the deep groove ball bearing with fatigue module of ANSYS, and compared it with the theory value in the end, and proposed a series of measures to improve the bearing's life.

2.   COMPUTATIONAL BASIS OF MAIN BEARING MODELING

Fatigue life of rolling bearings is when the rear bearing starts running, including any part on the inner and outer ring or rolling, before material fatigue caused by the damage, the total number of bearing rotation or a few hours. Contact fatigue is the major failure mode of rolling bearing, so fatigue life is an important indicator of bearing's design and application. The experience shows that the L-P model is simple and has a sufficient accuracy for the bearing life assessment. Thus we use it to calculate the fatigue life in this paper. For the particular bearing materials, the greater space of stress and more cycles can make the material fatigue failure probability increasing, therefore gives the following empirical formula:

$$\ln \frac{1}{S} \propto \frac{\tau_0^c N^e V}{Z_0^h} \tag{1}$$

Where τ_0 is the maximum alternating shear stress;

Z_0 is the maximum depth of alternating shear stress is located;

c, e, h are determined by the index data bearing test

V is the volume by stress.

Assume that the width of contact ellipse stress value 2a, depth Z_0, the raceway length is l, then we obtain

$$V \propto aZ_0 l \tag{2}$$

Meanwhile, the number of stress cycles is proportional to the fatigue life, i.e. $N \propto L$

In the case of point contact, with the Hertz theory of elastic rolling contact, there exists

$$a \propto Q^{\frac{1}{3}} \quad (3) \qquad \tau_0 \propto Q^{\frac{1}{3}} \quad (4) \qquad Z_0 \propto Q^{\frac{1}{3}} \quad (5)$$

Take Eq. (3), (4), (5) into Eq.

$$\ln\frac{1}{S} \propto Q^{\frac{c-h+2}{3}} L^e \tag{6}$$

Where Q is rolling load

When a certain bearing is given, S is a constant probability of survival, we can get:

$$Q^{\frac{c-h+2}{3}} L^e = \text{constant} \tag{7}$$

If Q_c described as the rolling loads (i.e., rolling dynamic load rating) when $L = 1$ (i.e., running one million revolutions), then get:

$$Q^{\frac{c-h+2}{3}} L^e = Q_c^{\frac{c-h+2}{3}} \tag{8} \qquad L = \left(\frac{Q_c}{Q}\right)^{\frac{c-h+2}{3e}} \tag{9}$$

In case of line contact, we get

$$\tau_0 \propto Q^{\frac{1}{2}} \tag{10} \qquad Z_0 \propto Q^{\frac{1}{2}} \tag{11}$$

Considering that the roller's length is equal to a, we get

$$L = \left(\frac{Q_c}{Q}\right)^{\frac{c-h+2}{2e}} \tag{12}$$

Take Eq. (10), (11) into the Eq. (12)

$$L = \left(\frac{Q_c}{Q}\right)^P \tag{13}$$

Where for the point contact $P = (c-h+2)/3e$

for the line contact $P = (c-h+2)/2e$

Where P is determined by the experimental data. and $P = 3$, so

$$L = \left(\frac{Q_c}{Q}\right)^3 \tag{14}$$

Eq. (14) is the basic formula for fatigue life calculation of the deep groove ball bearing
The fatigue life of inner and outer rings can be described as

$$L_i = \left(\frac{Q_{ci}}{Q_i}\right)^3 \tag{15} \qquad L_e = \left(\frac{Q_{ce}}{Q_e}\right)^3 \tag{16}$$

Where Q_{ci} Q_{ce} are inner and outer rings of the dynamic load rating

Q_i Q_e are inner and outer rings of the contact load rating

The fatigue life L of the bearing is the intersection of the fatigue life of inner and outer ring.

$$L = \left[\sum_{j=1}^{N_b} \left(L_i^{-e_w} + L_e^{-e_w}\right)_j\right]^{-1/e_w}, (j=1,2,\cdots,N_b) \tag{17}$$

Where e_w is the Weibull slope and $e_w = 10/9$

In the Eq. (17), we only consider the fatigue damage of the inner and outer rings. In the actual project, rolling fatigue spall also accounted for a considerable proportion. Therefore, it is necessary to consider the impact of fatigue life for rolling element bearing.

Rolling elements with the inner and outer contact fatigue life can be described as

$$L_{bi} = \left(\frac{Q_{bi}}{Q_i} \right)^3 \qquad (18) \qquad L_{be} = \left(\frac{Q_{be}}{Q_e} \right)^3 \qquad (19)$$

When considering the impact of rolling elements, bearing fatigue life can be described as L

$$L = \left[\sum_{j=1}^{N_b} \left(L_i^{-e_w} + L_e^{-e_w} + L_{bi}^{-e_w} + L_{be}^{-e_w} \right)_j \right]^{-1/e_w}, (j = 1, 2, \cdots, N_b) \qquad (20)$$

Where L_i, L_e are the fatigue life of the inner and outer ring.

L_{bi}, L_{be} are the fatigue life of the rolling elements contact with the inner and outer ring. The calculation process for fatigue life, is shown as Figure 1.

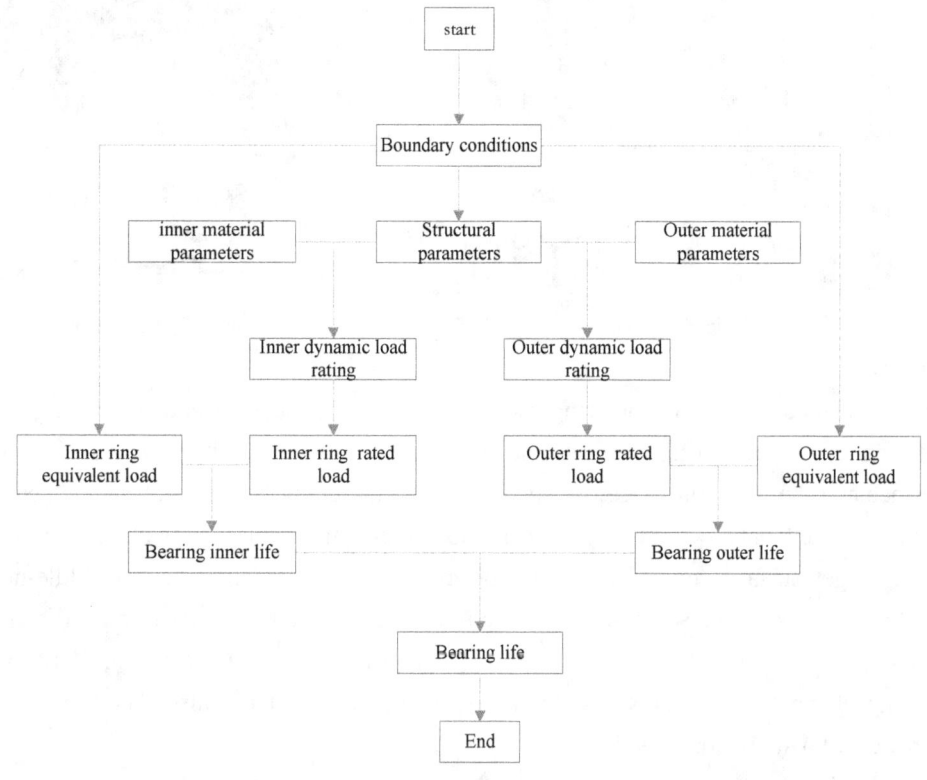

Figure 1 The calculation process of main bearing's fatigue life

3. CONSTRUCTION OF FINITE ELEMENT MODEL

Finite element method is a way that takes the continuous geometry into finite number units, which set a finite number of nodes in each unit. As the successive body is regarded as a set of units aggregates which connected only in node, we can convert continuous infinite degrees of freedom into a finite discrete-DOF domain, this way will greatly simplify the problem.

3.1. Descriptions of finite element models

1. A three-dimensional model

Using Pro/E establish the deep groove ball bearing 6217 3D model, the outer diameter D is 150mm, diameter d is 85mm, the thickness is 28mm. And import it into ANSYS workbench software modules. We can see Figure 2.

2. Definition of material properties

After importing the solid models of deep groove ball bearing successfully, we define material properties and choose the elastic modulus (2.07×1011Pa)and Poisson's ratio (0.3).

3. Meshing

Using free mesh, inner and outer rings of the grid cell size of 2mm, refining the grid of different levels to the contact area, the grid rolling contact area is more refined than the inner, outer ring raceway grid. We can see Figure 3.

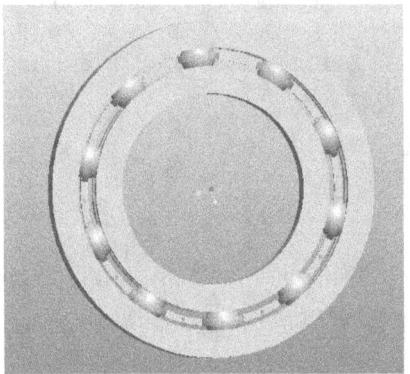 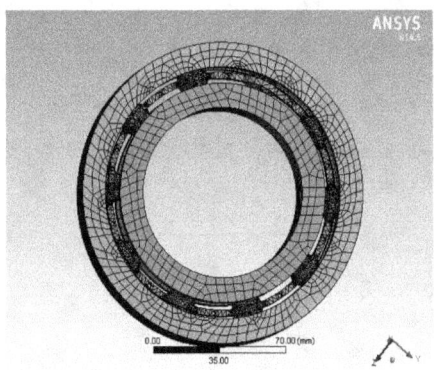

Figure 2 3D model   Figure 3 Finite element model

4. Contact pairs

To establish the contact between inner ring raceway, outer ring raceway and the rolling elements, determine the inner and outer ring raceway and rolling elements are flexible contact, the outer ring raceway surface and the inner ring raceway surface are the target surface, the rolling element surface as the contact surface. Contact element type as the node containing eight nodes quadrilateral element CONTA 174. Target surface contact element type as three-node which have no-middle-node unit TARGEl70, In each pick up on, set rolling with the inner and outer raceway friction coefficient MU = 0.003, normal contact stiffness factor FKN = 0.1, the initial near factor ICONT = 0.01. Each contact with the rolling element and inner and outer rings, create two contact pairs. There are 10 rolling elements, created 20 pairs of contacts right.

In order to ensure the accuracy of the calculation and reduce the amount of calculation, the occurrence of the contact area may use a denser meshing style; the model consists of 13,423 elements and 26,971 nodes, shown as Figure 3.

3.2. Determination of boundary condition

1. Impose boundary conditions

(1) Simulation bearing seat: Constraint all freedom degrees of the outer surface of the bearing outer

ring;

(2) Simulation the retainer: Constraint the circumferential (UY) and axial (UZ) displacement of all nodes in the neutral plane equator of rolling.

(3) Freedom of the inner surface of coupling load: The coupling of all nodes on the inner circle of the radial surface (UX) and circumferential (UY) degrees of freedom.

(4) Simulation flanges: Constraint the UZ displacement of the side of the bearing ring body.

(5) Apply radial load: Applying a radial node radial force is applied in the form of the average radial force to each node on a straight line, can be directly loaded in the direction UX.

(6) Applying a centrifugal force: Bearing rotary inertia force in ANSYS structural static analysis can be used to simulate the dynamic conditions imposed by way of inertial load

1) Add "Rotational Velocity", the "Define By" option into "Component".

2) The actual angular velocity is applied in the global Cartesian Z-axis.

2. Solving set

(1) Time-step control: "Analysis Settings" under "Static Structural", the "Number of Steps" to step 8.

(2) Analysis of control: Contact analysis due to the large local strain, we choose "large displacement static", and open the linear search.

4.   APPLICATION OF LIFE PREDICTION FOR MAIN BEARINGS

4.1. Finite element analysis

Considering the impact of the number of rolling elements, the rolling element diameter and groove curvature coefficient, to get result of the contact stress before the finite element analysis of bearing life. Then analysis the various factors on the bearing contact stress and deformation, study for the next step various factors affecting bearing life and improve bearing life measures provide an important basis for the analysis and the analytical results more reasonable.

We get the conclusion as following:

(1) The maximum equivalent stress and equivalent strain occurred in the radial force at the bottom line is contact point, that is mean bearing the risk of fatigue damage part is the rolling element and raceway contact points. And the shape of the contact area is approximately ellipse.

(2) Maximum rolling contact stress occurs in the radial force beneath line where rolling contact with the raceway.

(3) Overall radial displacement occurs where the inner ring bear the radial load.

4.1.1 Results and Its Analysis

Figure 4 shows the effect of the number of rolling element bearings for bearing contact stress and deformation curves. The diameter of the rolling element bearings can be seen $D_w = 16.5$, the inner and outer groove curvature coefficients were $f_i = 0.505$ and $f_e = 0.51$.

Figure 5 shows the effect of the diameter of rolling element bearings for bearing contact stress and deformation curves. As the number of its rolling is 10, the inner and outer groove the curvature coefficients were $f_i = 0.505$ and $f_e = 0.51$.

Figure 6 shows the effect of the inner and outer groove curvature coefficient of rolling bearings for bearing contact stress and deformation curves. The number of rolling bearings n = 10, the diameter of the rolling element bearing D_w = 16.5mm, difference in coefficient of curvature of the inner and outer groove $f_i - f_e = 0.005$.

According to the figures and data we obtained above, we can know the results as follows:

1. After getting a deep groove ball bearings, we know that the outer contact area is approximately elliptical, the contact area of the contact stress distribution; the inner and outer race on contact stress will decrease when the number of rolling increase, the inner and outer race on contact stress will decrease when the rolling diameter increase, the inner and outer race on contact stress decrease when inner and outer curvature coefficient is $f_i \leq 0.51$, $f_e \leq 0.515$;and its will increase when $f_i \geq 0.51$,

$f_e \geq 0.515$.

2. The radial deformation will decrease when the number of rolling increase and the rolling diameter increase, but it will increase when the inner and outer curvature coefficient increase.

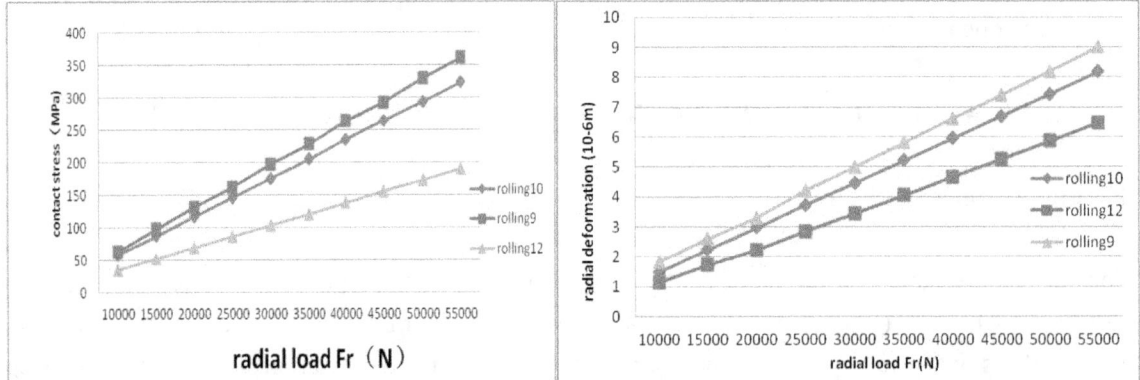

Figure 4 The impact under the change of the number of rolling elements

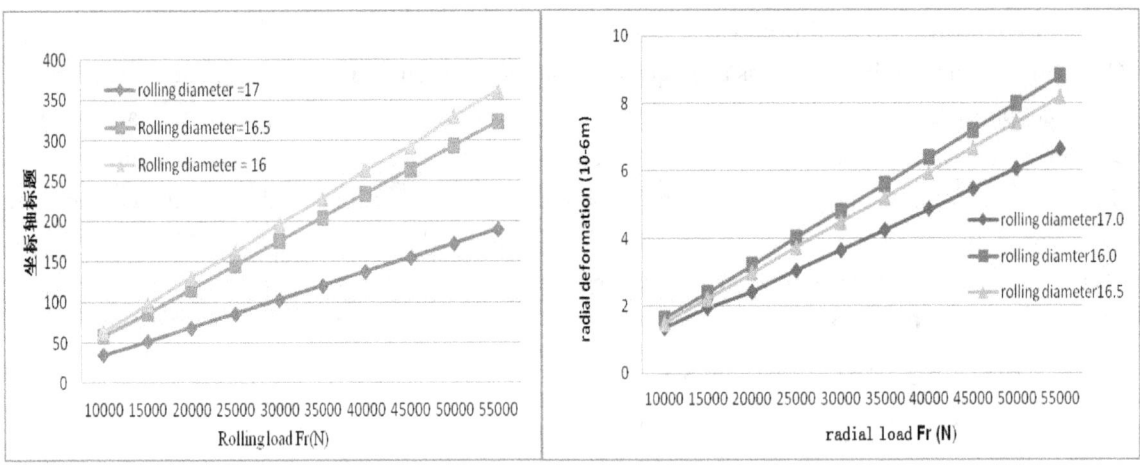

Figure 5 The impact of the chang of the rolling diameter

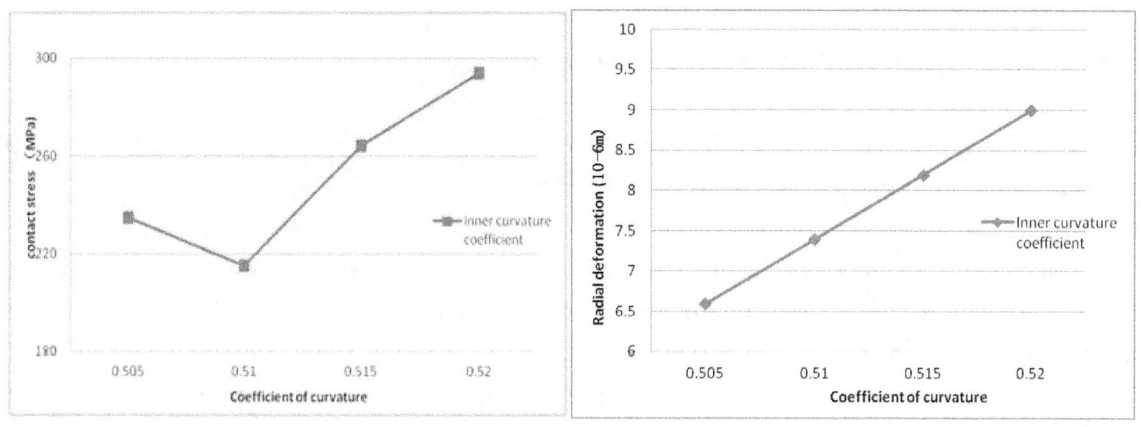

Figure 6 The change of the curvature coefficient of inner and outer race

4.2. Fatigue life analysis

In terms of numerical simulation of the fatigue life，the article will consider the effect of the number of rolling of bearings, rolling diameter and curvature coefficient of inner and outer race and other factors, and use the nominal stress method for fatigue life of deep groove ball beating through numerical simulation. With the help of ANSYS software fatigue analysis module, the fatigue life of bearing is simulated, and the influence discipline of various factors on the beating fatigue life is obtained, the results with the theoretical fatigue life of L-P results are compared, and discuss difference reason between the two results；finally it propose the technical measures to improve bearing life．We also get the diagram which describes the changes of the fatigue life of the bearing.

The results show that the fatigue dangerous parts of deep groove ball bearings appears at the contact point of between rolling units and raceway；bearings fatigue life will increase when the number of rolling increase, fatigue life will increase when rolling diameter increase, fatigue life will increase when inner and outer curvature coefficient is $f_i \leq 0.51$, $f_e \leq 0.515$, and it will decrease when $f_i \geq 0.51$, $f_e \geq 0.515$. Simulation method of fatigue life for deep groove ball bearing can calculate the complex working conditions of bearing and the fatigue life prediction has a high reliability.

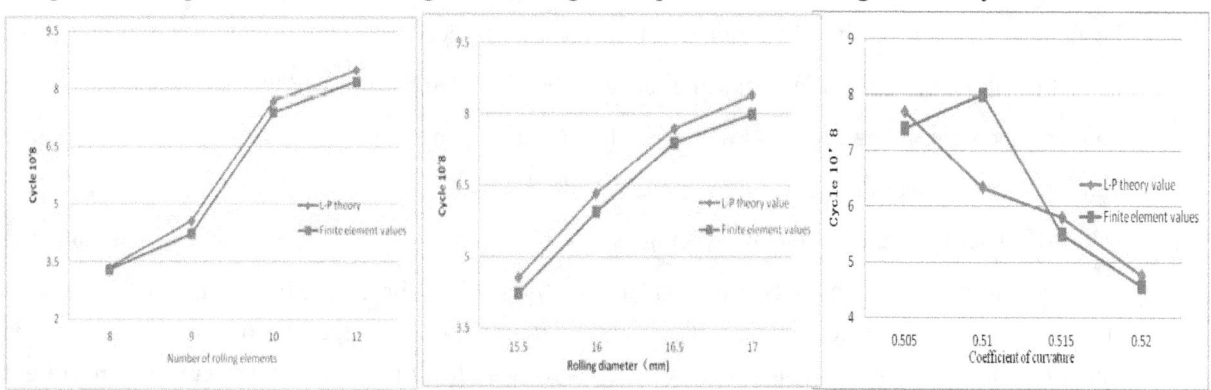

Figure 7 the change of the fatigue life

4.3. Measures to improve the fatigue life for bearings

Based on the analysis above and references [8]-[12], in order to improve the fatigue life of deep groove ball bearings, the following steps can be used in the design.

1. An increase of the number of rolling elements

When increase the number of rolling elements, each rolling element exposed the radial load is decrease, the contact load for each rolling element is also smaller, so fatigue life of improve.

2. An increase of the diameter of the rolling

Rolling diameter increases, the equivalent bearing structure increases bearing capacity increases, so the rolling element bearing life increases with the diameter increased.

3. Reasonable assignment of rolling element and raceway contact parameters.

For deep groove ball bearings, inner groove curvature coefficient $f_i \leq 0.52$, outer groove curvature coefficient of the bearing $f_e \leq 0.53$. It should also be noted between f_i and f_e matches.

In addition, you can also reduce the pitch diameter of the bearing, the bearing does not slip under the premise of guaranteed minimize preload, at high speed, rolling ceramic material instead of steel materials.

5. CONCLUSION

1. The results indicate that the outer contact area of the deep groove ball bearings is approximately elliptical, also the contact stress distribution of the contact area is obtained. Besides, we also get the conclusion that the inner and outer race on contact stress will decrease when the number of rolling increase, the inner and outer race on contact stress will decrease when the rolling diameter increase, the inner and outer race on contact stress decrease when inner and outer curvature coefficient is $f_i \leq 0.51$, $f_e \leq 0.515$, and it will increase when $f_i \geq 0.51$, $f_e \geq 0.515$.

2. The radial deformation will decrease when the rolling number and the rolling diameter increase, but it will increase when the inner and outer curvature coefficient increase. The results of the simulation of the fatigue life indicate that the fatigue dangerous parts of deep groove ball bearings appears at the contact point of between rolling units and raceway, bearings fatigue life will increase when the rolling number increase, fatigue life will increase when rolling diameter increase, fatigue life will increase when inner and outer curvature coefficient is $f_i \leq 0.51$, $f_e \leq 0.515$, and it will decrease when $f_i \geq 0.51$, $f_e \geq 0.515$. The simulation method of fatigue life for deep groove ball bearing can simulate the complex working conditions of bearing and has a high reliability for fatigue life prediction.

In order to shorten the test cycle, reduce cost and improve the accuracy of life prediction, the numerical simulation method used to simulate the fatigue life of the bearing calculation in present paper is very effective.

Acknowledgements

The authors would like to acknowledge the partial supports provided by the National Natural Science Foundation of China under the contract number 51275077, and the Oversea Academic Training Fund, University of Electronic Science and Technology of China.

References

[1]. Miao Xuewen, Hong Jie. *"A Comprehensive Prognostics Approach for Prediction Aero-engine Bearing Life"* [R]. School of Jet Propulsion, Beihang University, Beijing, China, 2000.

[2]. Chen Guang. *"Aero-engine structural design analysis"*[M]. Beijing, beihang university, 2006.

[3]. TangYunbing，Gao Deping. *"Research of Aero—Engine High—Speed Ball Bearing"*. Journal of Aerospace Power, 2006, 2, (12)354-360.

[4]. Zhao Shaobian. *"Fatigue design"*[M].Beijing, Machinery Industry Press,1994.01.

[5]. Yang Xiaowen. *"Study on Three Parameter Weibull Distribution of the Rolling Bearing fatigue life"*[D].Anhui, HeFei University of Technology,Master thesis.2003.

[6]. Jia Qunyi. *"Rolling bearing design principles and application of technology"*[M]. Beijing, High Education Press, 1993.

[7]. Yao Wenxing. *"Fatigue Life Prediction of Structure"*[M].Beijing：National Defense Industry Press, 2003.

[8]. Li Shun Ming. *"Mechanical fatigue and reliability design"*[M].Beijing, Science Press, 2006

[9]. Wang Dali, Sun Liming, Shan Fubing, Xu Hao. *"Application of ANSYS Software in Solution of Contact Problem of Bearing"*. Bearing, 2002, (9), 1-4.

[10].   Lin Hua, Xing Hui Han, Song Huang. *"Finite element analysis of fatigue life for deep groove ball bearing"* [D], Proceedings of the Institution of Mechanical Engineers, Part L: Journal of Materials: Design and Applications January 2013 vol. 227 no. 1 70-81.

[11].   Sriram Pattabhiraman, George Levesque. *"Uncertainty analysis for rolling contact fatigue failure probability of silicon nitride ball bearings"* [D], International Journal of Solids and Structures Volume 47, Issues 18–19, September 2010, Pages 2543–2553.

[12].Liangliang Ding, Li-Ping He, Shunpeng Zhu, Hong-Zhong Huang. *"Probabilistic fatigue life prediction of turbine disc considering model parameters uncertainty"*. Proceedings of 2013 International Conference on Quality, Reliability, Risk, Maintenance, and Safety Engineering, e'mei, China, July 15-18th, 2013, Pages 1089-1091.

Development of a Dynamic, Plant Condition-Dependent Probabilistic Safety Assessment

Radoslaw Lewandowski[a], Richard Denning[a], Tunc Aldemir[a] and Jinsuo Zhang[a]**

[a]The Ohio State University, Columbus, Ohio, USA

Abstract: Although each nuclear power plant has a plant-specific probabilistic risk assessment (PRA) that reflects design differences from other plants, the condition of each plant changes uniquely with time. A great deal of surveillance data are collected for the plant that reflect the changing condition of the plant. In some instances, plant staff use these data to guide the plant's preventative maintenance and surveillance programs. In general, however, these data are not used to characterize the evolving risk of the plant. Our understanding of the underlying mechanisms for the degradation of systems, although far from perfect, is improving with time. The possibility of developing a condition-dependent PRA is explored that would take a first principles approach to modeling the progression of degradation mechanisms, periodically adapting the model to account for surveillance results, and using the model as a basis for a time-dependent characterization of plant specific risk. Because surveillance data would be used to periodically assess the consistency of the observed behavior with model predictions, it might be possible to provide early identification of unanticipated degradation mechanisms. A case study is described involving a potential bypass accident sequence involving the progression of flow-accelerated corrosion in secondary system piping and stress corrosion cracking of steam generator tubes.

Keywords: Probabilistic safety assessment, plant condition monitoring, dynamic risk assessment

1. INTRODUCTION

In recent years, with support from the nuclear industry, the U.S. Nuclear Regulatory Commission has fostered a proactive approach to the management of plant aging processes [1]. The objective of proactive materials degradation assessment (PMDA) is to focus surveillance and preventive maintenance on areas of the plant that are known to be susceptible to degradation processes. In many respects, while PRAs are intended to be plant-specific, they rely extensively on generic data bases. Although plant-specific data are often used to update generic data, a PRA is more representative of a population of similar designs than it is a plant-specific assessment based on that plant's observed conditions. Effectively, a PRA only goes skin deep rather than reflecting the observable measures of the condition of the plant such as the results of eddy current testing, ultrasonic testing, valve monitoring, and other monitoring techniques. A PRA that projects the condition-dependent risk of a plant could be an effective tool that would improve management of age-driven risk. In particular, it would provide strong support to PMDA. At this point, it should be indicated that while existing risk monitors somewhat reflect the current plant condition, the information that they display is restickted to only active components and does not include the condition of passive components.

Section 2 presents a process by which a dynamic assessment of the progression of degradation mechanisms allowing quantifying of time-dependent risk better reflects the evolving plant conditions by using the results of surveillance activities on passive components to continuously update the degradation models.The condition-dependent PRA of the plant could then be used as a tool for the management of surveillance and maintenance activities.

The case study described in Section 3 of this paper involves the mechanistic assessment of multiple degradation mechanisms. To determine their contribution to risk of the scenario, the frequency of the accident initiating event (a steam-line break), the failure on demand of an active system (main steam-

---

* denning.8@osu.edu

line isolation valve), and the induced failure of degraded steam generator (SG) tubes due to the depressurization of the secondary side of the affected steam generator are all under consideration. The modes of degradation modeled are flow-accelerated corrosion (FAC) of the steam-line and primary water stress corrosion cracking (SCC) of the SG tubes (Alloy 600). The analysis is performed in a dynamic manner, in which the time-dependent evolution of degradation mechanisms is followed over the plant lifetime. The analysis includes assumed periodic surveillance activities with some probability of detection and repair depending on the extent of degradation and the reliability of the surveillance tools. If the secondary side of the steam generator depressurizes as the result of an unisolated steam line break, degraded steam generator tubes could fail. During each operating cycle, the probabilities are assessed of a spontaneous steam generator tube rupture (SGTR), a steam-line break resulting from FAC, the conditional probability of a MSIV failure to close, and the conditional probability of steam generator tube failure resulting from depressurization of the secondary side of the steam generator. The number of tubes plugged at the end of a cycle is also assessed.

Section 4 describes the approach and assumptions in a MATLAB model developed to project the time-dependent progression of component degradation and Section 5 presents the results of those analyses. Finally, in Section 6 conclusions are drawn regarding the feasibility and potential benefits of the concept of a living, condition-dependent probabilistic safety assessment.

2. DEVELOPMENT OF A CONDITION-DEPENDENT RISK MODEL

One of the limitations of PRA methodology as it has been applied since the mid-1970s is the static nature of the event tree/fault tree approach. Considerable research has been performed to explore the benefits of dynamic event trees (DETs) [2] to improve PRA predictions. Although some of the events analyzed in traditional static event trees naturally occur in an established order, that is not necessarily true for all cases. Particularly within the context of uncertainty analysis, the order of events can vary based on different sets of inputs [3]. DETs are not restricted by a priori assumptions about the order of events. They also enable analyses of accident progression to be performed in a manner that is mechanistically consistent with computer codes designed for transient analysis (e.g. RELAP [4], MELCOR [5]). DETs are an essential element of advanced human reliability analysis tools that attempt to simulate the cognitive response of human operators to the changing environment of an accident.

DET approaches have been examined for Level 1, 2 and 3 PRAs and dynamic tools, such as ADS [6], ADAPT [7], MCDET [8] have been developed to implement these approaches. In the Risk Informed Safety Margins Characterization Program [9], the U.S. Department of Energy (DOE) is extending the concept of dynamic analysis to plant performance assessment in a New Generation Systems Code (NGSC), also known as RELAP 7, capable of spanning the complete spectrum of time scales and multi-physics aspects of the time-dependent evolution of events in a nuclear power plant [9]. RELAP 7, for example, will not only address the plant's dynamic behavior within the time-scale of a loss of coolant accident, but should be able to assess slowly evolving changes in plant conditions during the fuel cycle leading up to the loss of coolant accident. It is within the conceptual framework of NGSC that we are considering a dynamic condition-dependent PRA. The other aspect of the dynamic condition-dependent PRA that is different from current practice is the very tight coupling between the evolution of the condition-dependent PRA and the co-evolution of surveillance and maintenance practices at the plant.

Clearly, there are practical limits today in the extent to which this type of dynamic condition-dependent PRA could be performed. The analyst would limit the scope of the effort to those mechanisms that are most likely to lead to failure, are sufficiently known to be able to make reliable predictions of degradation, and for which failures have risk significance. This paper describes a conceptual approach to converting an existing PRA to a dynamic, condition-dependent PRA that reflects component degradation. The approach in this study relies to a large extent on the results of expert elicitations documented in NUREG/CR-6923, "Expert Panel Report on Proactive Materials Degradation Assessment" by Brookhaven National Laboratory [1].

Degradation mechanisms affect the performance of both active and passive systems. In this paper, we focus on the impact of degradation of passive systems. In general, passive system failures can either be the initiator in the chain of events of an accident scenario or can affect the performance of the plant's systems, structures and components (SSCs) in responding to the event. Much of the focus of nuclear reactor regulation has been on the potential for pipe breaks in the reactor coolant system and the ability of safety systems to prevent core damage. The degradation mechanisms, particularly stress corrosion cracking of stainless steel, that can lead to the failure of primary system piping have been well studied [1]. The degradation mechanisms that affect the carbon steel portions of the power conversion system are different from those in the primary system. The rupture of secondary system piping not only could potentially lead to severe core damage but also represents a serious threat to plant operating personnel. Historically, FAC has been a major problem in the performance of components of the power conversion system including events leading to fatalities. As a result, all plants now have a FAC management program to address situations wherein vulnerable areas are identified and piping is replaced with steels that are less susceptible to FAC or surveillance is periodically undertaken.

Pipe breaks in either the primary or secondary system arising from corrosion mechanisms are found to be important initiating events in plant risk assessments. Degradation mechanisms could also affect the performance of safety systems that are intended to mitigate the severity of the scenarios that develop from the initiating event. For example, the containment structure of a plant provides the final barrier to the release of radioactive material to the environment in a severe accident. In a PRA, the failure probability and mode of failure of a containment are typically treated in a probabilistic manner as a function of internal pressure only. However, degradation of the containment shell can lead to failure at a lower pressure than for a pristine containment shell [10].

One measure of importance of a degradation mechanism is the extent to which it impacts core damage frequency (CDF). However, even if consideration were only given to degradation mechanisms that affect the frequency of initiating events, CDF does not fully characterize the risk significance of an event. For example, an initiating event leading to containment bypass is of much greater concern than a primary system LOCA within an intact containment. For this reason, a measure of the importance of a degradation mechanism is the potential impact on the consequences of a scenario in addition to its frequency. Very few full PRAs (Level 3 PRAs), which include the assessment of offsite consequence, have been performed for U.S. nuclear power plants. On the other hand, all nuclear power plants have limited Level 2 risk analyses that address the probability of different modes of containment failure, in particular the probability of a large early release of radioactive material to the environment. The results of the existing plant PRAs are used in risk informed regulatory activities through the risk meassures of CDF and of large early release frequency (LERF), which are considered surrogates for the NRC's probabilistic safety goals [12].

The modes and rates of degradation processes typically depend on the time-dependent thermal-hydraulic and stress environment to which the SSC is exposed. Figure 1 illustrates a process by which the condition-dependent behavior of the plant risk would be assessed. The time-dependent environment would be calculated using the *Plant Simulator*, for example the RELAP 7 code (or a legacy code such as RELAP 5 or MELCOR). At time $t=0$, plant condition, plant configuration and state of process variables (Initial Conditions) are fed into the *Plant Simulator*. Plant configuration and component failure rates/probabilities are also fed into the *PRA Code* for the prediction of *Risk Metrics and Importances* at $t=kT$. The variable T is a user specified time interval, possibly chosen to represent the duration of an operating cycle or a surveillance interval as well as to model degradation dynamics adequately and k is the number of the time interval. The *Plant Simulator* produces the thermal-hydraulic/neutronic/stress data which are fed into the *Component Aging Progression Model* which is assumed to stay constant within the time interval T to predict failure rates/probabilities at $t=kT$. The *Plant Simulator* in Figure 1 may be required to operate over distinctly different time scales such as: 1)

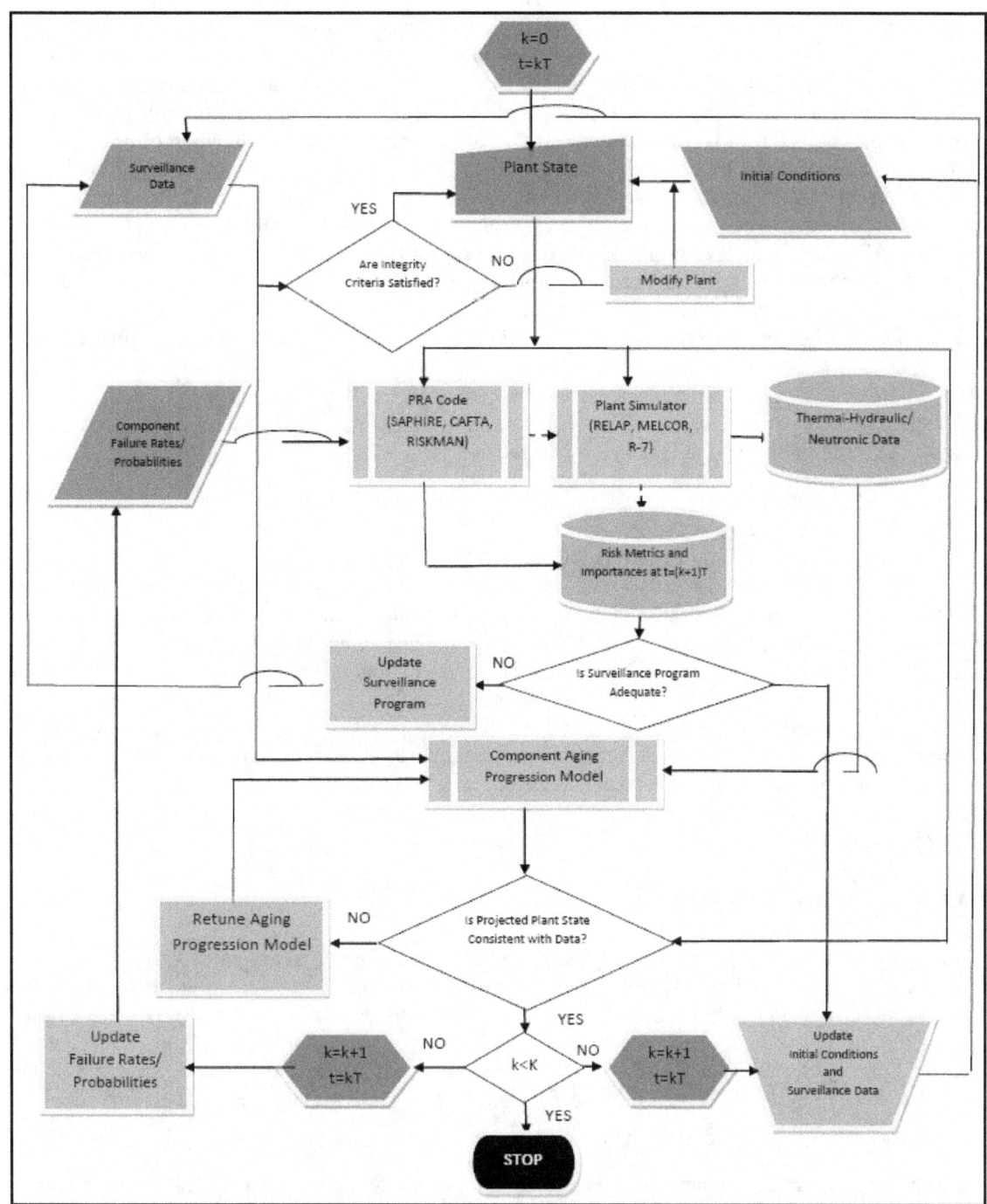

Figure 1: Evolution of Dynamic, Condition-Dependent PRA

the quasi-steady state condition while the plant is at power during a fuel cycle, and, 2) the dynamic time frame of a reactor shutdown and startup or the transient response of the plant to an accident. For the quasi-steady state condition, it is likely that the thermal-hydraulic conditions will be maintained constant based on the results of offline steady-state calculations performed with the *Plant Simulator*. *Initial conditions/surveillance data*, maintenance and repair actions affecting the plant state and *Component Failure Rates/Probabilities* inferred from these data are also updated at each time *kT*.

At the end of each time interval *T*, the plant condition is re-evaluated as it impacts the determination of the plant risk for that time interval. If the predicted *Risk Metrics and Importances* are found inadequate, the surveillance program is updated. As degradation processes continue over the time period, the potential of some kind of an initiating event such as a leak or rupture of a pipe at a weld will grow. Based on the condition, the likelihood of an initiating event of this nature will be determined, which will affect the plant risk for that time interval. Similarly, degradation will occur in components that need to operate in response to an initiating event, again affecting the outcome of the risk assessment. Thus, it is necessary not only to project degradation as a function of time but also to interpret the impact of a level of degradation on the probability of the occurrence of an initiating event or the impact of a level of degradation on the performance of a component. If the predicted *Plant Condition* is found to be inconsistent with surveillance data, then the aging progression model would be retuned. Similarly, the results of surveillance performed within a particular time interval could indicate the need to repair or replace a component or structure. Thus, components or structures can be returned to some initial state at which degradation mechanisms will again continue to degrade their performance. The time is incremented by T and the process is repeated until the target time horizon kT is reached.

For the update of *Initial conditions/ surveillance data*, the proposed approach would employ a model similar to that employed by [13] in that repair rate is

$$\omega = \frac{P_I P_D}{T_F + T_R} \tag{1}$$

where T_F is the mean time between inspections for flaws (i.e., the inspection interval), T_R is the mean time to repair once the flaw is detected, P_I represents the probability of inclusion or the probability that a piping element with a flaw will be inspected per inspection interval given that a certain percentage of elements are inspected. This concept is explored in detail by [14] and [15]. P_D is the probability of detection. The concept of modeling probability of detection for a specific testing method (e.g., UT) and as a function of the dominant physical parameter (e.g., crack length) has been developed in the literature [16-18]. Human factors are implicitly considered; however, some detection models provide for more explicit consideration. With the probability of detection often developed by testing performed on artificial defects, a Bayesian approach may be employed [19,20] to update "generic" data to reflect plant-specific surveillance testing techniques and results.

3. CASE STUDY

3.1. Scenario Description

The accident scenario selected for analysis involves failures of passive components due to degradation mechanisms for which there has been extensive operational experience. More specifically, in the case studied here, a steam line break outside the containment and downstream of a main steam isolation valve (MSIV) serves as the initiating event. Following the steam line break, the accident continues with a failure of a MSIV and the rupture of a flawed SG tube in the steam generator upstream of the MSIV. FAC of piping in the power conversion system and SCC of steam generator tubes were selected as the acting degradation mechanisms.

Steam lines penetrate the containment wallmaking it necessary for each line to contain an isolation valve external to the containment. The containment building becomes pressurized if the steam line breaks inside the building. This is a design basis accident analyzed in safety analysis reports (SARs). The containment integrity is not lost if the MSIVs close properly. If the depressurization of the SG leads to rupture of a tube, the reactor coolant system (RCS) begins to depressurize and it becomes necessary to inject emergency core cooling water into the RCS to prevent core damage. The conditional probability of core meltdown should be similar to that seen in small break LOCA scenarios. However, if the emergency core cooling system (ECCS) works in both injection mode and recirculation mode, core meltdown would be avoided.

If the steam line breaks downstream of the isolation valve, the situation is different. If the MSIV fails to actuate and a tube rupture is induced by the increased pressure differential, there is the potential for core meltdown with the containment bypassed. In this case, the ECCS is effective in the injection phase but fails due to lack of water in the sump when the switch to recirculation occurs. Although the likelihood of this scenario is expected to be very small, it is of interest because of the potential for containment bypass and a large late release of radioactive material to the environment. This type of event was at one time considered a generic safety issue (GSI-188) [21]. A variation of the scenario has been studied by Argonne National Laboratory in support of the resolution of that generic safety issue in which the failure mechanism of the steam generator tubes is associated with stresses in the tubes induced by differential expansion of structures in a steam generator and binding of tubes at tube support plates.

3.2 Degradation Mechanisms

3.2. 1. Steam Generator Tubes

The integrity of the SG is essential since a leak of a sufficient size or a rupture may contaminate the steam in the secondary loop and cause a release of radioactive material to the environment. To meet this challenge, special materials and heat treatments are used for the fabrication of SGs [22]. Many of the primary circuit components such as reactor pressure vessel (RPV), pressurizer, or the steam generator are produced from nickel-based alloys (referred by their trademark name as Inconels). These alloys have coefficients of thermal expansion similar to low alloy steels and are characterized by relatively high corrosion resistance in PWR primary and secondary water environments. The use of nickel-based alloys has evolved over the decades as new degradation mechanisms were identified.

In the early-to-mid 1970s, essentially all nuclear power plants (NPPs) in the United States used SGs with tubes made from mill-annealed Alloy 600 (600MA). However, the water chemistry was found to lead to excessive thinning. In response, NPPs changed their water control programs to eliminate this mechanism. In the mid-to-late 1970s, a process called denting resulting from the corrosion of the carbon steel support plates and the buildup of corrosion products in the crevices between tubes and tube support plates became a major issue. Although measures such as changes in the chemistry of secondary loop were taken to limit this problem, other mechanisms continued to cause cracking in plants with Alloy 600MA tubes. The need for extensive plugging of Alloy 600MA SGs forced the industry to begin to replace them with SG tubes made from high-temperature treated Alloy 600 (600TT). The replacement process began in the early 1980s and up until now no significant degradation issues have been observed. Nevertheless, beginning in 1989, NPPs began using SGs made from thermally treated Alloy 690, which is believed to be even more corrosion resistant due to its nearly doubled chromium content. The switch to Alloy 690 was accompanied by changes in corresponding weld metals. Alloys 152 and 52 replaced previously used weld Alloys 182 and 82.

In this study, only SCC is modelled as a SG tube degradation mechanism, specifically the initiation and growth of axial cracks in Alloy 600 MA tubes. Crack growth rate is assumed to follow the form of the Scott model:

$$\frac{da}{dt} = 2.8 \cdot 10^{-12} \cdot (K - 9)^{1.16} \tag{2}$$

where
$\frac{da}{dt}$ = crack growth rate (m/s)
K = crack tip stress intensity factor (MPa$\sqrt{m}$)

Although crack initiation time is typically modeled using a Weibull Distribution, for this study, the time for the initiation of cracks is based on a log normal distribution, as measured by Staehle [25], which indicated a constant logarithmic standard deviation of uncertainty over the range of measured data. A large number of cracks are initiated in the first few cycles of operation after which crack initiation tapers off slowly. By the end of the 40-year lifetime, a majority of the tubes have been subjected to crack initiation. In this study, crack growth was treated by dividing the steam generator tubes into 20 groups, with different inherent crack growth rates. The crack growth rate for each group was determined empirically based on data on crack lengths collected at the Ringhals plant [26] after 11 years of operation, as illustrated in Figure 2. Correction factors for the 20 groups of tubes were applied to the general form of Scott's crack growth rate formula [24] to duplicate the distribution observed at the plant. Figure 3 illustrates crack length as a function of time after crack initiation for examples from the 20 groups. In reality, it is not possible to distinguish between crack initiation timing and crack growth rate from the Ringhals data. The manner in which these processes were treated in the current study is recognized to be speculative. However, the predicted size distribution at 11 years and the timing at which PWRs have reached the economic limit on plugged SG tubes are in reasonable agreement with actual experience.

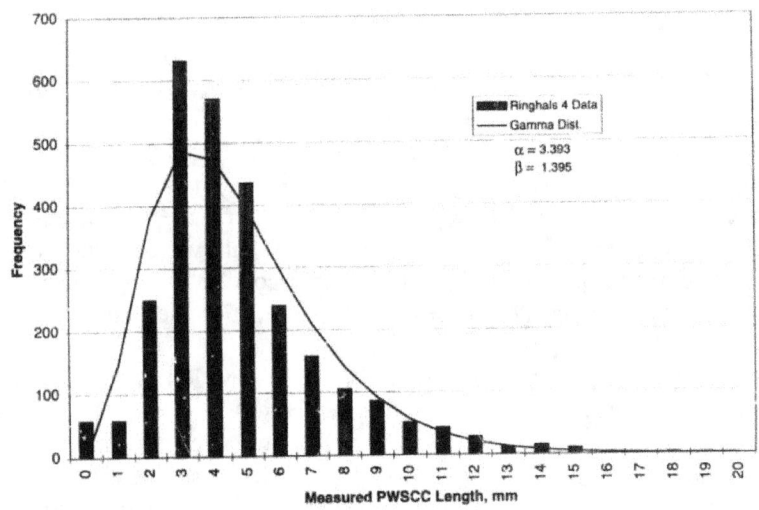

Figure 2. Measured Crack Length Distribution at Ringhals Plant [26]

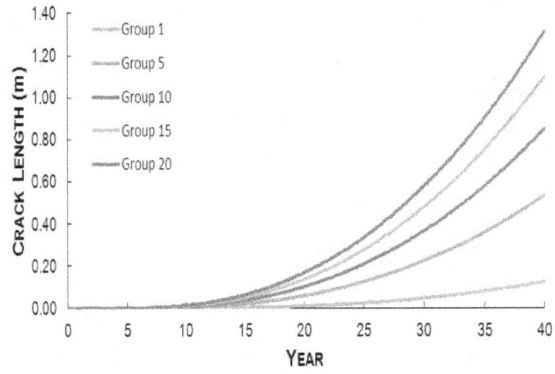

Figure 3. Crack Length as A Function of Time for Representative Groups

Data have been collected on the reliability of eddy current testing. Based on sets of repetitive tests by different teams of testers to determine the probability of failure to detect [27], it was assumed that the probability of failure to detect would be 0.02 at the first point at which a cohort of tubes reached the plugging criterion. Based on the binomial distribution, one can then determine the probability of different numbers of tubes that would be undetected from that cohort. To simplify the analysis, only the average number of undetected tubes were subsequently tracked in the dynamic analysis. At the next opportunity to detect, we have no data on which to assess whether an undiscovered tube that should have been plugged will now be detected. Thus, we used the simple approximation that at this point and subsequent tests, chances are that 50 percent of faulted tubes would be identified. Based on the binomial distribution, we can then determine the probability of one or more tubes remaining in a faulted state as a function of time:

$$P(x \geq 1|N, k) = 1 - (1 - p^k)^N \tag{3}$$

where
N = number of undetected tubes during the initial surveillance test
p = non-detection probability
x = number of undetected tubes
k = cycle of interest

3.2.2. Steam Line

Carbon steel piping of the secondary side carrying flowing water or wet steam is susceptible to FAC. This degradation process leads to wall thinning (material loss) that may result in pipe leaks or rupture. One of the most notable accidents caused by FAC occurred on December 9, 1986 at the Surry Nuclear Power Station. A rupture of an elbow in the condensate system resulted in four fatalities [28]. In PWRs, FAC has been only recorded in piping outside the containment [29]. Under normal conditions, a thin, protective layer of magnetite (Fe_3O_4) forms on the inside surface of carbon steel feedwater piping. Metal oxidation occurs at the metal-oxide interface in deoxygenated water. Ferrous species (Fe^{2+}) diffuse through the porous oxide layer and are dissolved into the water flow [28, 30]. FAC occurs in single- and two- phase flows. However, presence of water in its liquid state is necessary, and therefore it does not affect pipelines transporting dry or superheated steam [28].

The treatment of the probability of FAC leading to rupture in this paper is superficial relative to the effort that would be required in an actual risk assessment. The first step would be to examine plant drawings to determine locations on the secondary side where the conditions exist that could lead to FAC and which could not be isolated in the event of MSIV failure. The analyst would then examine the plant's FAC program to determine surveillance requirements for these locations. We also did not have data on the likelihood that an ultrasonic test of a vulnerable section of piping would not detect significant erosion of pipe wall thickness. Plant drawings were not available to us. Thus, we modeled a single characteristic location.

Industry and laboratory experience has led to the development of several semi-empirical models for predicting rates of FAC and resulting potential for pipe ruptures. The three most commonly used FAC models are contained within the WATHEC, CHECKWORKS, and BRT-Cicero packages [30]. The use of the latter two, however, is somewhat limited due to their proprietary nature. The KWU-KR model developed by Kastner and Riedle as part of the WATHEC code produced by Siemens/KWU is openly available and has been widely analyzed in published literature. For this reason, it was used for the simplified analysis performed for this paper. The KWU-KR model determines the rate of wall thinning due to flow-accelerated corrosion. This approach accounts for geometry, flow velocity, chemistry (pH and oxygen content), temperature, piping metallurgy (including chromium and molybdenum content of steel), and exposure time [30].

Failure of a component damaged by FAC can be calculated using a load-capacity formulation. In the case of feedwater carbon steel piping, the load is defined as pressure imposed on the piping during steady-state and transient conditions. The capacity is defined as the maximum pressure sustainable by a pipe subjected to wall thinning. A pipe fails when the load exceeds the capacity. NUREG/CR-5632 uses capacity expression from Wesley et al. shown in Eq. **Error! Reference source not found.** [29].

$$P_{capacity}(t) = \frac{\sigma_f \cdot h_{pipe}(t)}{[r + h_c(t)](1 + 0.25\varepsilon_f)} \tag{4}$$

where
$P_{capacity}(t)$ = pressure capcity at time t (ksi)
σ_f = failure stress (ksi)
r = nominal pipe radius (cm)

$h_{pipe}(t) =$ pipe wall thickness at time t (cm)

$h_c(t) =$ calculated thickness of pipe corroded away at time t (cm)

$\varepsilon_f =$ median hoop strain at failure

For the purposes of this study, analysis was performed for two secondary-side pipe ruptures that have actually occurred. The results obtained with the KWU correlation were consistent with the actual plant experience. In these cases, failure occurred after approximately ten years of operation. Because we did not have the information required to examine multiple secondary side locations, for this analysis we assumed a ten year delay in operating history before a steam line break could occur but used the NUREG-1150 value for steam line break frequency after that period.

3.2.3. Main Steam Isolation Valve

Considerable data on the effects of aging and service wear on MSIVs, which have been collected in the National Plant Reliability Data System, are presented in NUREG/CR-6246 [31]. One of the purposes of a MSIV is to limit the consequences of a steam line break. In PWRs, MSIVs are typically located within a valve vault external to the containment boundary. Depending on the design of the MSIV, it may only stop flow in the downstream direction, providing protection for the turbine. For those designs, it is necessary to also include a check valve in the line that limits backflow for a break upstream of the valve. MSIV failure modes can be divided into six categories: failure to open, failure to close, spurious valve closure, spurious valve open, valve stem or shaft leakage, and valve seat leakage. The probability distribution of these different failure modes depends on the type of valve: gate valve, globe valve or check valve. Failure to close of a MSIV is the mode of concern analyzed in the accident scenario. Failure of the valve can either be the result of a fault in the valve or the valve actuator. NUREG/CR-6246 [31] presents failure data for the number of times that failure to close has been experienced for MSIVs. We did not have data on the associated number of challenges. Based on an estimate of three challenges per MSIV per year of plant operation in this period, we obtained a probability of failure on demand of 1.9×10^{-2} per demand, which is an unexpectedly high value. The value used in NUREG-1150 was 1×10^{-4} per demand.

3.3 PRA Model

The chosen scenario of concern is modeled using the Zion Nuclear Power Station as a reference PWR plant. The Zion plant was a two-unit station located on the shore of Lake Michigan. Each of the two units was a four-loop Westinghouse nuclear steam supply system with a rating of 1100 MWe housed in a large, pre-stressed concrete, steel-lined dry containment [32]. The safety functions and associated probabilities of failure on demand described in NUREG/CR-4550 Vol. 7 [33] were used to develop an event tree for the primary accident of interest resulting from a steam line break. The Top Events and corresponding failure probabilities are summarized in Table 1.

Table 1. Top Events and Failure Probabilities [33]

Top Event Acronym	Top Event	Probability of failure on Demand
SLB	Steam line break outside the containment (Initiating Event)	See Section 3.2
K	Reactor trip	1.8E-4
M	MSIV failure	See Section 3.3
TR	Tube rupture given pressure gradient	See Section 3.1
RW	Refueling water storage tank	2.4E-8
SS	Safety injection system actuation signal	2.2E-5
L1	Auxiliary feedwater actuation and secondary cooling	3.4E-5
HP	High Head Injection/Feed and Bleed	2.1E-8
R2	Low pressure recirculation	4.6E-4

In addition, the event tree was developed based on several assumptions:

- NUREG/CR-4550 Vol. 7 considers scenarios with combinations of AC buses not working. The scenarios considered here assume that all AC buses work properly due to the very low frequency of the scenario, in which any of the three AC buses is not working.
- MSIV failure leads to a blowdown of affected SG (secondary side) and depressurization of the steam header. If the faulted SG is one of the two lines that provide steam to the auxiliary feedwater system, only one auxiliary feedwater pump is available. In this case, the probability of failure of auxiliary feedwater increases.
- SGTR results in a small break LOCA.
- An operator can act to depressurize the RCS to slow down the release to the environment caused by leakage of RCS inventory into the turbine building. However, if the auxiliary feedwater system fails, it is necessary to rely on the feed and bleed system at high pressure.
- Since modern seals are less likely to fail, the scenario ignores the possibility of a pump seal LOCA.

The failure of reactor trip events (Anticipated Transients without Scram) were not analyzed because of their low probability. Of the remaining end states five result in severe accident conditions:

- SLB/MSIV/SGTR/failure of low pressure recirculation (late meltdown)
- SLB/MSIV/SGTR/failure of high pressure injection and feed and bleed (early meltdown)
- SLB/MSIV/SGTR/failure of auxiliary feedwater actuation and secondary cooling (early meltdown)
- SLB/MSIV/SGTR/failure of reactor safety actuation signal
- SLB/MSIV/SGTR/failure of refueling water tank flowpath (early meltdown)

Following a successful plant shutdown, the closure of a MSIV leads to a design basis accident resulting in a successful prevention of core meltdown. In case of MSIV failure, if the integrity of SG tubes remains uncompromised, a meltdown is also avoided. However, if the SG tubes rupture, a series of safety functions act to stop core meltdown. Failure of the first line of defense (i.e. supply of water from the refueling water storage tank or high pressure injection) leads to an early core meltdown. The last line of defense is the low pressure recirculation. In case of a SLB, there is no water in the sump that would permit recirculation to work properly. In that case, even if the system functions properly from a mechanical standpoint, it does not have the necessary inventory, which will result in a late core meltdown. The two remaining emergency actions that could be undertaken would include providing additional borated water to the refueling water storage tank or the containment sump.

4. MATLAB MODEL

4.1 Modeling Assumptions

A MATLAB model was written to follow the initiation and growth of cracks in steam generator tubes. It was assumed that the cracks were initiated in each cycle according to a log normally distributed initiation time as reported by Staehle [25]. This led to the initiation of cracks in 31% of the tubes by the end of the first two-year interval. The fraction of new tubes with newly initiated cracks then decreased rapidly with subsequent cycles. The crack growth rate of tubes was assumed to be independent of initiation time. Newly introduced cracks were assumed to have a length of 0.1 mm. Furthemore, their growth was characterized by a distribution of twenty growth rates which were followed in time. Cracks were assumed to have the length-to-depth ratio of 3:1 throughout their growth. It was assumed that when a crack had grown to 40% of the tube wall thickness, it would be plugged following the next inspection unless there was a detection failure. As discussed in Section 3.1, at the first time at which a cohort of tubes (a given initiation cycle and given growth rate equation) exceeded the plugging criterion, it was assumed that 2% of the tubes in that cohort were not identified as having exceeded the criterion. In subsequent inspections, the fraction of tubes exceeding the plugging criterion that should have been detected previously was reduced by 50%.

There are three critical crack lengths of importance: the length at which the 40% through-wall criterion is reached (1.5 mm), the length at which a steam line break with failure of an MSIV could potentially induce a rupture (33 mm) and the length at which the tube could rupture spontaneously (62.5 mm). Based on some analyses of characteristic pipes that had actually failed by FAC, it is assumed that a steam line break would not occur prior to the 5th two-year cycle. At that point the steam line failure frequency is taken as 1.88×10^{-3} per yr in agreement with the NUREG-1150 study.

4.2 Results

As expected, the model predicted rapid deterioration of SG tubes. By the 4th cycle (6 to 8 years), the cohort associated with cracks initiated in the first cycle and the most rapid growth rate had grown to the point at which plugging was required. Figure 4 shows the history of plugging for each of the four steam generators. There are 3,592 tubes per SG. Thus, if the limit of plugged tubes is 20% without substantially impacting plant performance, this limit would be reached in 10 years (5 two-year cycles). Historically, SGs with Alloy 600MA have lasted longer than 10 years but substantially less than the 40-year lifetime of the plant, indicating that the degradation model somewhat underestimating the time to initiate cracks or over-estimating the crack growth rate.

Because the model used to assess the probability of failure to detect cracks larger than the plugging criterion always has a non-zero probability at some point, SG tube ruptures would also occur with some probability. Figure 4 indicates the probability of SG tube rupture per SG as a function of time. The model shows that the SGs would probably be replaced before being susceptible to a high likelihood of tube rupture. Although ruptures have occurred in steam generator tubes in U.S. PWRs, they have not necessarily ruptured as a result of SCC.

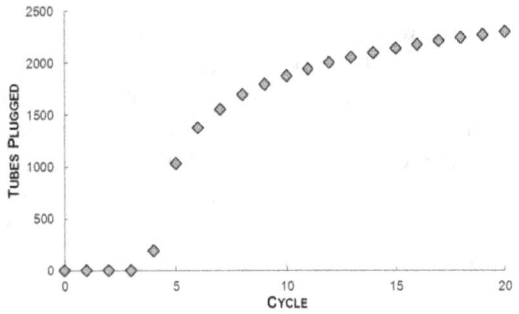

Figure 3. Number of Plugged Tubes Per SG as a Function of Time

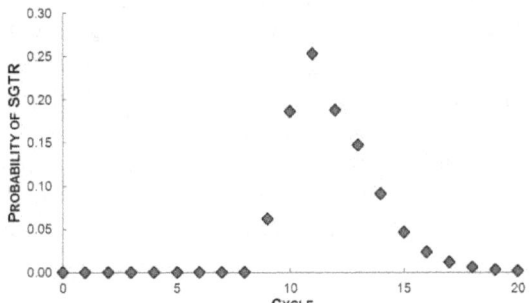

Figure 4. Probability of SG Tube Rupture vs. Cycle

Figure 5 illustrates the bottom-line objective of the analysis, the time-dependent risk of a SLB/MSIV/SGTR event. Because the critical length at which a tube becomes susceptible to failure resulting from depressurization of the SG is shorter than the critical length for a spontaneous rupture, the time at which the scenario probability becomes non-zero occurs earlier than the time at which an SGTR becomes non-zero. The time-dependent core damage frequency for this scenario peaks at 20 years. The average risk over the lifetime of the plant is 2.5×10^{-5} per yr. As indicated for the spontaneous steam generator tube rupture frequency, based on the results of this degradation model, the plant would probably have changed SGs before these conditions were reached.

5. CONCLUSIONS

The key questions regarding whether a dynamic, condition-dependent approach to PRA should be pursued are:

1. Is it technically feasible based on the state of the art of modeling degradation phenomena?

2. Is the level of effort required cost-effective?

3.Is the level of effort warranted, i.e would we believe the results sufficiently to use them in managing aging risk. Could the results help to identify unsuspected mechanisms or vulnerabilities?

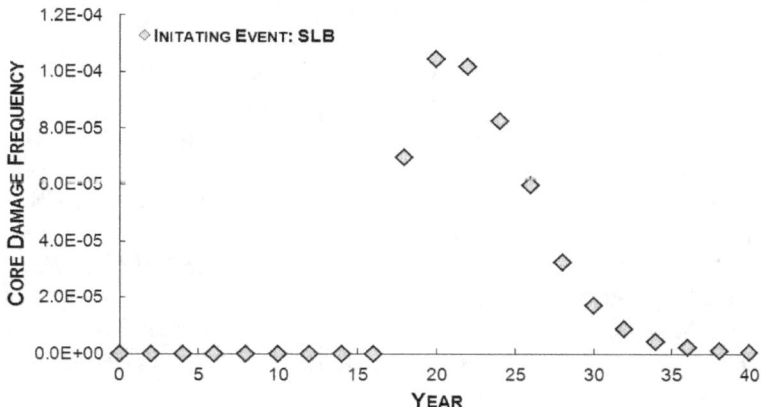

Figure 5. Core Damage Frequency for SLB/MSIV/SGTR Events

The results of the case study performed indicate that the mechanics of performing a condition-dependent risk assessment are feasible. The level of effort required would be substantially greater than for the case study because it would be necessary to identify a number of vulnerable locations in the plant and to consider alternative degradation mechanisms. There would also be a continuing effort associated with the analysis of surveillance results and modification of the parameters in the degradation model.

The case study involving Alloy 600 MA steam generator tubes was retrospective in nature. Thus, there were data available to initially tune the parameters in the degradation models and some indication of the dominant types of degradation mechanisms. In comparing the results of the degradation model for the case study with actual PWR operating experience in terms of the number of tubes plugged, the degradation model appears to have been slightly conservative. However, the model could certainly have been further tuned to be more consistent with operating experience. If we were to do a prospective analysis for the future risk associated with Alloy 690 SG tubes, we would not have a good starting model. The general form of the model might be the same as used for the case study, for example the Scott model. After each set of surveillances the model would be updated. Thus, in a boot-strap approach, by the time degradation mechanisms became important, the predictive capability of the model should be substantially improved.

An important consideration highlighted by the case study is the need for a quantitative understanding of the probability of failure of surveillance tools. This is an area in which more research is needed.

The performance of this very limited case study does not provide a definitive answer to the questions posed at the beginning of this section. Regardless whether the goals of developing a dynamic, condition-dependent risk capability for each plant are achievable today, the concept has significant potential value to risk managmeent and should continue to be pursued.

6. ACKNOWLEDGEMENTS

This research has been performed using funding received from the DOE Office of Nuclear Energy's

Nuclear Energy University Programs.

7. REFERENCES

[1]   P.L. Andersen, et al., *"Expert panel report on proactive materials degradation assessment. BNL-NUREG-77111-2006,"* NUREG/CR-6923, Brookhaven National Laboratory, Uptown, N.Y.(2007).

[2]   T. Aldemir, *"A Survey of Dynamic Methodologies for Probabilistic Safety Assessment of Nuclear Power Plants,"* Annals of Nuclear Energy, **52**(0): p. 113-124 (2013).

[3]   A. Hakobyan, et al., *"Dynamic Generation of Accident Progression Event Trees,"* Nuclear Engineering and Design, **238**, 3457-3467 (2008).

[4]   C.D. Fletcher and R.R. Schultz, *"RELAP5/MOD3 User Guidelines,"* NUREG/CR-5535, U.S. Nuclear Regulatory Commission, Washington, DC (1992).

[5]   R.O. Gauntt, *"MELCOR Computer Code Manual, Version 1.8.5, Vol. 2, Rev. 2,"* NUREG/CR-6119, Sandia National Laboratories (2006).

[6]   Chang, Y. and Mosleh, A., *"Cognitive Modeling and dynamic Probabilistic Simulation of Operating Crew Response to Complex System Accidents (AD-IDA Crew),"* Center for Technology Risk Studies (1999).

[7]   U. Catalyurek, et al., "Development of a Code-Agnostic Computational Infrastructure for the Dynamic Genration of Accident Progression Event Trees, "Reliab. Engng & System Safety,**95**, 278-304 (2010).

[8]   M. Kloos and J. Peschke, *"MCDET: A Probabilistic Dynamics Method Combining Monte Carlo Simulation with the Discrete Dynamic event Tree Approach,"* Nuclear Science and Engineering, **153**(2), 137-156,2006.

[9]   U.S. DOE, *"Light Water Reactor Sustainability Program, Integrated Program Plan,"* INL/EXT-11-23452, Rev. 0, (2012).

[10] B.W. Spencer, et al., *"Risk-Informed Assessment of Degraded Containment Vessels,"* NUREG/CR-6920, SAND2006-3772P, Sandia National Laboratories (2006).

[11] U.S. NRC, *"An Approach for Using Probabilistic Risk Assessment in Risk-Informed Decisions on Plant-Specific Changes to the Licensing Basis,"* Regulatory Guide 1.174, Washington, D.C.. (2002).

[12] U.S. NRC, *"Safety Goals,"* SECY-00-0077, Washington, D.C. (2000).

[13] K. Fleming, *"Markov Models for Evaluating Risk-informed In-service Inspection Strategies for Nuclear Power Plant Piping Systems"*, Reliability Engineering and System Safety, **83**, 27–45 (2004).

[14] U.K Health and Safety Executive, *"Probability of Detection (PoD) Curves: Derivation, Applications and Limitations,"* Research Report 454 (2006).

[15] G.A. Georgiou, "The Validity of the JCL 'Probability of Inclusion Model," Appendix C of Research Report 454, Jacobi Consulting Ltd (2000).

[16] SKI Report 2005:03, *Probability of Detection for the Ultrasonic Technique according to the UT-01 Procedure*, January 2005.

[17] Appendix G to ML-HDBK-1823, *Nondestructive Evaluation System Reliability Assessment*, April 1999.

[18] Rummel, W.D. and Matzkanin, G.A., *Nondestructive Evaluation (NDE) Capabilities Data Book*, Nondestructive Testing Information Analysis Center (NTIAC) (1997).

[19] Bruce, D., *NDT Reliability Estimation from Small Samples and In-Service Experience*, Structural Materials Center (1998).

[20] Kanzler, D., et al., *Bayesian Approach for the Evaluation of the Reliability of Non-Destructive Testing Methods: Combination of Data from Artificial and Real Defects*, Proc. 18th World Conference on Nondestructive Testing, South African Institute for Non-Destructive Testing, Durban, South Africa (2012)

[21] U.S. NRC, "Resolution of Generic Safety Issue 188: Steam Generator Tube Leaks or Ruptures with Containment Bypass from Main Steam Line or Feedwater Line Breaches," NUREG-1919, (2009).

[22]   U.S. Nuclear Regulatory Commission. *Backgrounder on steam generator tube issues.* 2009 29 March 2012 [cited 2012 December 27]; Available from: http://www.nrc.gov/reading-rm/doc-collections/fact-sheets/steam-gen.html.

[23]   U.S. Nuclear Regulatory Commission, *PMDA PIRT Report,* U.S. Nuclear Regulatory Commission, Washington, D.C. (2005).

[24]   Scott, P.M. "An analysis of primary water stress corrosion cracking in PWR steam generators", *Proc. of the Specialists Meeting on Operating Experience with Steam Generators,* Brussels, Belgium (1991).

[25]   Staehle, R.W., *Historical Review on Stress Corrosion Cracking of Ni-based Alloys – The Coriou Effect* (Presented at CEA, January 26, 2010. To be published)

[26]   Wu, G., *A probabilistic-mechanistic approach to modeling stress corrosion cracking propagation in Alloy 600 components with applications,* Mechanical Engineering, University of Maryland: College Park (2011)

[27]   Kupperman, D.S. et al., "Eddy Current Reliability Results from the Steam Generator Mock-up Analysis Round-Robin, NUREG/CR-6791, U. S. Nuclear Regulatory Commission (2009)

[28]   Dooley, R.B. and B.B. Chexal, *Flow-accelerated corrosion,* in *CORROSION 99*1999, NACE International, 12, San Antonio, TX (1999)

[29]   Smith, C., et al., *Incorporating aging effects into probabilistic risk assessment: a feasibility study utilizing reliability physics models. NUREG/CR-5632,* U.S. Nuclear Regulatory Commission (2001)

[30]   Ahmed, W.H., "Flow accelerated corrosion in nuclear power plants", *Nuclear Power- Practical Aspects,* A. Wael, Ed., InTech (2012)

[31]   Clark, R.L. and J. Jackson, *Effects of aging and service wear on main steam isolation valves and valve operators. NUREG/CR-6246,* U.S. Nuclear Regulatory Commission, Washington, D.C. (1996)

[32]   U.S. NRC, "Severe Accident Risks: An Assessment for Five U.S. Nuclear Power Plants," NUREG-1150, , U.S. Nuclear Regulatory Commission, Washington, D.C. (1990).

[33]   Sattison, M.B. and K.W. Hall, *Analysis of core damage frequency from internal events: Zion Unit 1. NUREG/CR-4550,* , U.S. Nuclear Regulatory Commission, Washington, D.C. (1989)

Risk-Informed Safety Margin Characterization Case Study: Use of Prevention Analysis in the Selection of Electrical Equipment to Be Subjected to Environmental Qualification

D. P. Blanchard[a] and R. W. Youngblood[b]

[a]Applied Reliability Engineering, Inc. (AREI), San Francisco, California USA [*]

[b]Idaho National Laboratory (INL), Idaho Falls, Idaho, USA

Abstract: Age-related degradation of electrical equipment is cited in numerous discussions of extended nuclear power plant operation as an important issue. Which SSCs matter? For which SSCs do we need ongoing assurance of performance? Replacement of all components and cables is a daunting prospect. Being able to focus on a subset of SSCs from an environmental qualification (EQ) perspective, while still maintaining plant-level safety and efficiency even if the other components and cables degrade, would be worthwhile.

This paper summarizes a case study that examines SSC aging for components within a PWR large dry containment. The case study illustrates how an understanding of SSC margin can be characterized given the overall integrated plant design, and was developed to demonstrate a method for deciding on which SSCs to focus, which SSCs are not so important from an environmental qualification margin standpoint.

The method chosen for selection of SSCs important to aging and environmental challenges is known as Top Event Prevention (TEP) or Prevention Analysis. TEP is a Boolean method for optimal selection of SSCs (that is, those combinations of SSCs both necessary and sufficient to meet a predetermined selection criterion) and allows demonstration that plant-level safety can be maintained by the collection of selected SSCs alone.

Keywords: Prevention Analysis, TEP, environmental qualification, RISMC

1. INTRODUCTION

A harsh environment is considered to be a common mode challenge to nuclear power plant components exposed to that environment, even across different component types. Considerable resources are expended on qualification of safety related equipment as well as numerous non-safety electrical equipment that may be exposed to such environments with the intent of assuring that those components are capable of performing their intended functions given the environmental challenge [1-4].

As licensees consider operating their plants well beyond the original license term, equipment aging becomes an increasingly important issue, including age-related degradation of components and cables (such that they become more susceptible to harsh environments). The DOE's Light Water Reactor Sustainability (LWRS) program includes a Risk Informed Safety Margin Characterization (RISMC) effort that considers system, structure and component (SSC) aging within the concept of "margin." This concept refers not only to the margin in individual SSCs' capability to meet the functional challenges posed to them, but also to margin in overall integrated plant design including its response to a full spectrum of transients and accidents.

---

[*] *dblanchard@ar-eng.com*

In order to examine SSC aging from an environmental qualification perspective, a case study was defined [5] that illustrates how the state of knowledge regarding SSC margin can be characterized given the overall integrated plant design. The case study demonstrates a method for deciding on which SSCs to focus, which SSCs are not so important from an environmental qualification margin standpoint, and what plant design features or operating characteristics determine the role that environmental qualification plays in establishing a safety case on which decisions regarding margin can be made. This paper summarizes the results of that case study.

The approach taken in performing this evaluation was relatively straightforward and included the following four steps:

Identify components explicitly modeled in the internal events probabilistic risk assessment (PRA) that are located inside containment

Characterize the environmental profiles to which components inside containment would be exposed for different accident sequences

Modify PRA models to include explicit failure modes associated with component exposure to a harsh environment

Quantify accident sequences and identify components important from an environmental qualification perspective.

In the case study, the latter step was performed through use of Top Event Prevention (TEP) or Prevention Analysis, a technique based on Boolean optimization. An overview of the TEP methodology is presented in Attachment 1 along with a simple example. A test of the effectiveness of the subset of SSCs selected by TEP along with a comparison with more traditional importance measures demonstrates certain important advantages of TEP for this environmental qualification related application.

2. CASE STUDY PLANT DESCRIPTION

The plant selected for the case study is a two-loop pressurized water reactor (PWR) with a large dry containment.

Case Study Plant Systems

The plant has two steam driven feedwater pumps and three auxiliary feedwater (AFW) pumps (two motor and one turbine driven). Primary system pressure relief includes three code safeties and two large power operated relief valves (PORVs), either one capable of supporting feed and bleed operation. Reactor inventory control consists of three low volume charging pumps, two high pressure safety injection (HPSI) pumps and three low pressure safety injection (LPSI) pumps. Containment heat removal consists of three fan coolers (CAC) and two containment spray (CSS) trains.

Support systems include two essential buses normally aligned to offsite power and backed up by two automatic emergency diesel generators (EDG) and one manually operated diesel that can be aligned to either bus.

The internal events PRA for this PWR has the following characteristics:

>   50 initiating events (including the following, some of which may lead to harsh environments)
>   >   Four ranges of loss of coolant accident (LOCA) break sizes
>   >   Steam line breaks (inside and outside containment)
>   >   Steam generator tube rupture (SGTR)
>   >   Interfacing system LOCA
>   >   Transients (with the potential for feed and bleed operation)
>   >   >   Turbine trips, Loss-of-feedwater (LOFW), etc.
>   >   >   Loss of support systems (service water (SW), instrument air (IA)),etc
>   >   >   Loss of ac buses (essential and non-essential)
>   >   >   Loss of instrument ac buses, dc buses

>   Consequential initiating events
>   >   Transient induced LOCA (e.g., primary coolant pump seal LOCAs, pressurizer safety relief valve (SRV) challenges, failure of letdown isolation)
>   >   Transient induced steam line breaks (e.g., stuck open steam dump valves)

>   System fault trees include extensive modeling of instrumentation and control
>   >   Auxiliary Feedwater actuation, Safety Injection Signal, Recirculation actuation
>   >   Containment spray and containment atmospheric cooler actuation,
>   >   Load shed, Emergency ac actuation and
>   >   Control room indication for credited operator actions

3. IDENTIFICATION OF COMPONENT GROUPS

The first step in the case study is to identify all of the individual components explicitly modelled in the PRA for this PWR and establish their location in the plant. To assist in identifying components located in the containment, the plant staff provided an equipment list that includes the location of each tag ID.

Of several thousand components represented in the PRA, over 200 are located in the containment. However, not all of these components are subject to failure were they to be exposed to a harsh environment. Components such as check valves, manually operated valves, tanks, and heat exchangers can be screened from the list. The remaining components are those that contain parts whose performance could be affected by harsh environmental conditions and aging phenomena.

Major active components	Major rotating equipment	Instrumentation	Miscellaneous
Motor operated valves		Transmitters	Power supplies
Air operated valves	Pump motors	Switches	Penetration seals
Solenoid valves	Fans	Temp elements	
PORVs		Signal converters	

It should be noted that there are many components and their failure modes that are not explicitly modelled in the PRA but are effectively selected for inclusion in the case study as a result of their association with the components that are modelled. Examples include power and control cables, junction boxes, and terminals. The selected basic events effectively can be considered to be modules that not only include the component in question, but supporting subcomponents needed for the component to function.

Approximately 140 basic events were selected in this manner to represent the component failures that could occur due to a harsh environment for components located inside containment.

A final grouping was undertaken for the basic events that were selected as representing the components and failure modes that could occur due to a harsh environment inside the containment. This final grouping reflects that environmental effects are common cause challenges to the components that are exposed to them. A grouping of identical components that perform the same function was performed so as to recognize that if one component in a group were to fail as a result of harsh environmental conditions, then it was highly likely that the other members of that group that perform the same function also would fail. The 140 basic events representing components inside containment and their failure modes that were assumed to occur due to environmental challenges were placed into the approximately fifty component groups shown in Table 1. Each component group represents one to eight components and their corresponding harsh-environment-related failure mode.

4. CHARACTERIZATION OF THE ACCIDENT SEQUENCE ENVIRONMENT

The next step in the analysis was to develop the general characteristics of the environment associated with the various accident sequences modelled in the PRA. For the purpose of the case study, the conditions associated with five different accident types are considered in terms of the harsh environment that each may impose on components in the containment. These five accident types each will have an environmental 'profile' (e.g., pressure, temperature, etc., versus time) that can be assumed when considering the response of selected components during these accidents.

> LOCAs - Very Small, Small, Medium/Large
> Steam line break
> Feed and bleed

Considering the approximately fifty component groups and their associated failure modes that potentially could occur when exposed to a harsh environment, along with the five environmental 'profiles' defined above, yields roughly 250 component group environmental condition combinations which must be reflected in the case study. Each of these 250 combinations is represented by a unique environmental related basic event and incorporated into the fault trees of the PRA for the case study.

5. ACCIDENT SEQUENCE QUANTIFICATION

On incorporating the harsh environment related logic into the system fault trees for the case study plant, accident sequence quantification was performed twice, once to produce the accident sequence results as a function of environmental-related events and the second time to test the effectiveness of a minimal set of component groups selected for qualification.

5.1 Initial Accident Sequence Quantification

Initial accident sequence quantification was performed in the same manner that the PRA is quantified for any application. In order to focus on components whose function could be affected by environmental conditions, however, it is useful to regenerate the cut sets as a function of the environment related basic events. This was accomplished by setting the 250 environmental qualification events to unity and regenerating the cut sets. Tens of thousands of additional cut sets were generated over those resulting from the base case PRA with up to eleventh order cut sets that included environmental-related basic events.

5.2 Selection of a subset of harsh-environment basic events and testing their effectiveness

Not all of the environment-related basic-related events that are found in the cut sets generated above need to be prevented in order to assure a reasonably low core damage frequency. A method for selecting the most important of these harsh-environment basic events is needed. A probabilistic or a deterministic approach could be taken in identifying a subset effective in managing core damage frequency.

Probabilistic Selection Of Environmental Basic Events

The cut sets produced above reflect the distribution of risk from the original PRA plus a significant additional number of cut sets that are a function of the various harsh environments that may occur throughout the accident sequences. Importance measures were developed based on the cut sets that were a function of environmental-related events. Typically, in importance measure based risk-informed applications, components having a Fussell-Vesely measure greater than 0.5% or a Risk Achievement Worth greater than 2 are candidates for being considered as important [6, 7]. (Note that as the harsh environment related events have an assigned failure probability of 1.0, Risk Achievement Worth does not play a role in determining their importance for the case study.) Harsh-environment-related basic events, representing thirteen of the 50 groups of components, are identified by importance measures as being important from a harsh-environment and possible equipment qualification perspective.

A probabilistic test of the effectiveness of the basic events in the thirteen environment groups was performed by regenerating the accident sequence cut sets after setting each of the environmental related basic events in these groups to False (effectively assuming that they were environmentally qualified) and leaving the environmental basic events in the other groups set to a failure probability of 1.0 (assuming that they would fail on exposure to a harsh environment). Table 2 shows the results of the accident sequence quantification for this test. The core damage frequency for this case is several times higher than that of the base case PRA. The majority of the increase appears to be associated with transient-initiated events that evolve into sequences in which the containment environment becomes degraded (e.g., feed and bleed) and the larger break size LOCAs. It is clear that lowering the importance measure threshold when selecting environmental related basic events (and place the components associated with those basic events in an equipment qualification program) may be necessary if the core damage frequency is to be maintained near its base case value.

Deterministic Selection Of Environmental Basic Events

An alternate method of identifying important environmental related basic events employs a deterministic criterion for selection of important events in the PRA. A method available for the selection of components in such a deterministic manner is Top Event Prevention (TEP) or Prevention Analysis [8-14]. TEP uses Boolean methods to perform a systematic examination of the accident sequence cut sets of a PRA to identify subsets of the basic events found in those cut sets whose collective prevention is effective in maintaining acceptable results (in this case, minimal degradation of CDF with respect to the baseline). A TEP analysis can be probabilistic in nature, deterministic, or a blend of both. The subsets of components (or prevention sets) identified as important to the PRA have several characteristics:

- A prevention set consists of complete paths of equipment which, if they operate successfully, will assure the accomplishment of the safety functions modeled in the PRA. TEP results are presented in terms of success paths, in this regard.

- Each prevention set emerging from TEP is minimal with respect to the prevention criterion. That is, only those components contained in a prevention set are necessary to assure an adequate level of protection from core damage or large early releases.

- Multiple prevention sets are often generated as a part of a TEP analysis. Each prevention set by itself is a complete solution. Only one prevention set need be selected to identify the success paths that are important to preventing core damage or large early releases.

As noted above, a deterministic defense-in-depth related criterion was implemented for the identification of harsh-environment related basic events that were important to the results of the PRA for the case study. The criterion employed was similar to the single failure criterion. In this regard, cut sets were considered to be adequately prevented if two or more low-probability failures were required for any given initiating event before core damage would occur. In the application of TEP to the cut sets of the PRA, events credited toward prevention of each cut set included not only random failures but harsh environment related basic events as well. For an environmental-related basic event to be considered low in probability, the components in that group would need to be subject to an environmental qualification program.

Application of TEP to the case study yielded more than 180,000 prevention sets. Each prevention set was over 400 basic events in length. Prevention sets generally contain many basic events each, because each prevention set represents a combination of success paths, and each success path consists of many individual components. Given the prevention-set criterion that each cut set should be prevented by at least two failures, the case study prevention sets each comprise at least two success paths for each initiating event.

Within each prevention set is a combination of random failures and basic events representing failure of components due to harsh environmental conditions that were added as described in the preceding sections. For purposes of illustration, a prevention set was selected having the lowest number of harsh-environment-related basic events. These environmental-related events in the selected prevention set represented 17 of the original component groups defined in Table 1. Table 1 notes which component groups are found in the selected prevention set.

A probabilistic test of the effectiveness of preventing the selected 17 groups of harsh-environment related events was performed by regenerating the accident sequence cut sets after setting each of the selected basic events to False (effectively assuming that they were environmentally qualified) and leaving the remaining environmental basic events set to a failure probability of 1.0 (assuming that their failure was guaranteed on exposure to a harsh environment). Table 2 shows the results of the accident sequence quantification for this test. It is noted that the core damage frequency is within 10% of the base case core damage frequency, suggesting that the selected components would be successful in managing core damage risk were they to be subject to an environmental qualification program that was effective in preventing them from failing if exposed to a harsh environment. This is not necessarily the most effective prevention set; it was simply chosen for illustration.

6. EXPLANATION OF THE RESULTS

Of the roughly fifty component groups located in the containment of the case study plant that potentially could be affected by harsh environmental conditions during various accident sequences considered in the internal events PRA, only seventeen of the groups appear to be important with respect to maintaining the core damage frequency at an acceptable level, assuming adoption of the overall prevention strategy implied by selection of the particular prevention set selected in the preceding section. It is these seventeen component groups for which margin with respect to qualification of the equipment to withstand the expected harsh environments may be most valuable or, alternately, for which development of an environmental fragility curve may be useful.

6.1 Component Groups For Which Qualification Margin May Be Worthwhile

The following discusses a few the seventeen selected component groups and the reasons that a characterization of the behaviour of the components within these groups under harsh conditions may be worthwhile. Note that some of the groups are non-safety related and perform functions that are considered to be beyond the design basis.

Steam generator instrumentation

Two sets of steam generator level transmitters are shown to be important with respect to environmental qualification. The first set is responsible for automatic actuation of auxiliary feedwater, whereas the second set is associated with the feedwater control system and is credited in the PRA only as backup instrumentation used by the operators to manually initiate makeup to the steam generators in the event that automatic actuation does not occur. Steam generator pressure instrumentation is used to isolate the steam generators during a steam line break. Failure to isolate the steam generators results in loss of the steam supply to the turbine driven AFW pump. (Note that this steam generator pressure instrumentation is required only immediately following the initiating event, and is not required to function for a significant period of time under harsh environmental conditions.)

Feed and Bleed

The PORVs are required to support feed and bleed operation. The accident sequences in which the PORVs would be required to operate include small LOCA, steam line breaks and feed and bleed operation itself. (Note that PORVs would be required to be functional throughout the rest of the event, once feed and bleed was initiated.)

Reactor inventory control

Both cold-leg injection and hot-leg injection are assumed to be required for LOCAs. Cold-leg injection is the primary means of makeup to the reactor from HPSI during small breaks and during recirculation for the entire break spectrum. Hot-leg injection is assumed to be required long term following a large LOCA to avoid boron precipitation and plate out on the fuel assemblies during recirculation. Pressurizer pressure is important in assuring reactor inventory control, as it is the primary means of actuating safety injection for the entire range of breaks in the LOCA spectrum. (Note that pressurizer pressure initiation of safety injection is required early in the event and is not needed once actuation has taken place.)

6.2 Component Groups Not Needing Significant Qualification Margin

Equally important in determining the need for margin is an understanding of the reasons selected component groups do not contribute significantly to the core damage frequency if it assumed that they are not qualified. In this regard there are several component groups that do not appear in the selected prevention set.

Reactor pressure control

Pressurizer sprays are not necessary for achieving a safe stable state following a transient. The accident sequences for which pressurizer spray plays its most significant role is during SGTR in support of reducing reactor pressure to near that of the affected steam generator. Again, because primary coolant loss is not into the containment for SGTR, there is little degradation of the environment that would keep pressurizer spray components from providing their safety function.

Reactor inventory control (low pressure injection)

LPSI motor operated valves (MOVs) are located inside containment and would need to open to support the low pressure injection function during a medium or large LOCA. However, best estimate analysis for the case study plant shows that HPSI in conjunction with initial injection from accumulators will provide adequate core cooling. As HPSI is necessary for the small end of the LOCA break spectrum and as it also can be aligned for recirculation, LPSI injection MOVs simply provide a redundant backup to injection from HPSI.

7. SUMMARY AND CONCLUSIONS

A methodology has been developed for the purpose of identifying the minimum set of SSCs in a nuclear power plant that need to remain functional when exposed to a harsh environment following an accident. The methodology has been demonstrated for the components located inside containment using a full scope Level 1 internal events PRA for a PWR with a large dry containment.

In performing the demonstration, equipment located inside the containment that could be affected by harsh environments or aging were binned into roughly fifty component groups where a component group was defined as identical components having the same failure mode. Each component group represented one to eight components, including not only equipment with a specific tag id but all supporting hardware or parts that are necessary for the component to perform its function (e.g., junction boxes, power and control cables, penetration assemblies, etc.).

Generation of accident sequence cut sets as a function of the component groups and their environmental challenges was performed using the PRA for the case study plant. With these cut sets as input, a minimal prevention set of component groups was then selected, whose implementation would entail formal equipment qualification: that is, demonstrating the ability of the components within the group to remain functional following exposure to a harsh environment is of significant importance. For purposes of comparing methodologies, this selection was done in two different ways: one way based on traditional importance measures, and the other way using a method known as TEP.

TEP suggested that within one candidate strategy, only seventeen of the original fifty component groups potentially exposed to harsh environmental conditions in the containment for the case study plant need to be qualified to function in these harsh environments. (TEP presents the decision-maker with different strategic options; the present discussion is based on selection of the strategy requiring EQ of the smallest number of component groups.) Verification of the effectiveness of this subset of the component groups in maintaining an acceptably low core damage frequency was performed by assuming that *all* of the components in *all* of the non-selected component groups failed when exposed to a harsh environment. Making this assumption and regenerating the accident sequence results of the PRA resulted in an increase in core damage frequency of less than 10%, demonstrating that the components within the selected seventeen component groups suffice to be successful in managing core damage risk, if they are subject to an environmental qualification program that is effective in preventing them from failing if exposed to a harsh environment. The analogous exercise performed on the importance-measure-based selection of component groups demonstrated much less successful control of EQ-related core damage frequency.

The components in the seventeen component groups not only are those for which implementation of an environmental qualification program is worthwhile, but are components for which demonstrating margin on the capability of the components to remain functional when exposed to the various harsh environments may be of value. Alternately, characterizing the fragility of the components within these groups to the environmental conditions (temperatures, pressures, humidity, etc.) to which the components may be exposed during an accident may be worthwhile. Regardless, with respect to the component groups that were *not* selected as a part of this case study, it is concluded that the rigor to which environmental qualification is applied to components within these groups appears to be of relatively low importance, nor do these components require significant margin with respect to environmental challenges and/or aging.

While the case study was limited to just those components located inside containment, the proposed approach is sufficiently straightforward that it can be applied to any component types located in a nuclear power plant that may be exposed to harsh environmental conditions during an accident or subject to aging. The methodology is sufficiently systematic that the specific accident sequences that result in the need for qualification of individual components, and hence their associated environmental conditions, can be identified. Just as important, the method supports development of the engineering rationale as to why components are or are not selected as being important from an aging perspective or during harsh environmental conditions. Using the methodology of this case study, this engineering rationale can be documented in terms of plant specific design features and operating characteristics that drive the results.

References

[1] 10 CFR 50.49, "Environmental qualification of electric equipment important to safety for nuclear power plants."

[2] Regulatory Guide 1.89, Rev. 1, "Environmental Qualification of Certain Electrical Equipment Important to Safety for Nuclear Power Plant", U.S. Nuclear Regulatory Commission, 1984.

[3] IEEE Standard 323-1983, "Standard for Qualifying Class 1E Equipment for Nuclear Power Generating Stations."

[4] Regulatory Guide 1.97, "Criteria for Accident Monitoring Instrumentation for Nuclear Power Plants," U.S. Nuclear Regulatory Commission, June 2006.

[5] INL/EXT-11-23479 Revision 1, Risk Informed Safety Margin Characterization Case Study: Selection of Electrical Equipment To Be Subjected to Environmental Qualification, D. Blanchard and R. Youngblood, April 2012.

[6] NUMARC 93-01.Rev. 2, "Industry Guideline for Monitoring the Effectiveness of Maintenance at Nuclear Power Plants", April 1996.

[7] NEI 00-04, Rev 0, "10CFR50.69 SSC Categorization Guideline", 2005.

[8] R. W. Youngblood, "Applying Risk Models To Formulation Of Safety Cases," Risk Analysis 18, No. 4, p. 433, August 1998.

[9] J. R. Schaefer, R. B. Worrell, P. Szetu, "Implementation of an Air-Operated Valve Program at Northern States Power Company", 8th International Conference on Nuclear Engineering (ICONE8), April 2000.

[10] R. A. White and D. P. Blanchard, "Development of a Risk-Informed IST Program at Palisades Using Top Event Prevention," Proceedings of ICONE10, April 2002.

[11] P. Szetu, S. Hesler and W. Reuland, "Risk-Informed Turbine Missile Analysis for a BWR 4", Proceedings of PSA '05, Sep 2005.

[12] G. B. Varnado , R. B. Worrell, and D. P. Blanchard, "Risk-Informed Physical Security"; Dynamic Allocation of Resources, Proceedings of PSA '05, Sep 2005

[13] B. A. Brogan, R. B. Worrell and D. P. Blanchard, "Focusing the Circuit Analysis Effort in Transitioning to NFPA-805 using Top Event Prevention (TEP)", Proceedings of PSA '08, Sep 2008.

[14] R. Torok and D. Blanchard, "Risk Insights Associated with Digital Upgrades", PSAM10, June 2010.

Table 1: Component Groupings

This table lists component groups and failure modes considered in this case study. The columns on the right indicate whether a given group was selected for EQ within the two methods applied (importance measures and TEP); refer to Table 2.

Component Group / Failure Mode		Importance Measures	Prevention Set
Auxiliary feedwater			
SG level transmitters AFW actuation	Fail to function	✓	✓
SG level transmitters Feedwater control (operator information)	Fail to function	✓	✓
Pressure transmitter Steam generator isolation	Fails to function		✓
Shutdown cooling			
MOV Shutdown cooling	Fails to open		
Limit switch LPSI MOV	Fails to remain closed		
Pressure transmitter LPSI suction	Fails to function		
Reactor Pressure Control			
AOV Pressurizer spray	Fails to open		
AOV Pressurizer spray	Fails to remain open		
Solenoid valve Pressurizer spray	Fails to energize		
Solenoid valve Pressurizer spray	Fails to remain energized		
Pump Primary coolant	Fails to run		
Block valve Pressurizer	Fails to open		✓
PORV Pressurizer	Fails to open		
PORV Pressurizer	Fails to remain open		✓
Pressure transmitter Pressurizer (operator information)	Fail to function		
Reactor inventory control (charging/letdown)			
AOVs Letdown flow	Fail to open		
AOVs Letdown isolation	Fail to close	✓	✓
AOVs Letdown flow	Fail to close	✓	
AOVs Charging makeup	Fail to close		
AOVs Charging makeup	Fail to remain closed		
E/P transducer Letdown flow	High output		
E/P transmitter Letdown flow	Fails to function		
Solenoid valve	Fail to deenergize	✓	✓

Component Group / Failure Mode			Importance Measures	Prevention Set
	Letdown flow			
	Solenoid valve Letdown isolation	Fail to energize	✓	
	Solenoid valve Charging makeup	Fail to energize		
	Solenoid valve Letdown flow	Fail to energize		
	Solenoid valve Charging makeup	Fail to remain energized		
	Temperature element Letdown htx	Fails to function	✓	
	Level transmitter Pressurizer	Fails to function		
	Pressure transmitter Letdown pressure	Fails to function		
	E/P transmitter Letdown control	Fail to function		
	Valve position controller Letdown control	Fail to function		
Reactor inventory control (safety injection)				
	Limit switch HPSI MOV	Fails to close		
	Limit switch HPSI MOV	Fails to remain closed		
	MOV Hot-leg injection	Fails to open		✓
	MOV Cold-leg injection	Fails to open	✓	✓
	MOV Hot-leg injection	Fails to close		✓
	MOV LPSI	Fails to open		
	Pressure transmitter Pressurizer	Fails to function	✓	✓
	MOV SIT	Fails to remain open		
Containment control				
	Fan Containment cooler	Fail to start		
	Fan Containment cooler	Fail to run	✓	✓
	AOVs SWS to containment coolers	Fail to open	✓	✓
	Solenoid Valve SWS to containment coolers	Fail to deenergize	✓	✓
	Pressure Transmitter Containment pressure		✓	✓
	Radiation monitor Containment	Fail to remain energized		
	Seal Equipment hatch	Fails to remain closed		✓
	Hatch Fuel transfer tube	Fails to remain closed		✓
	Flange ILRT penetration	Fails to remain closed		

Table 2: Accident Sequence Quantification Results

Accident Sequence Type	Base case CDF (1/yr)	Qualify components selected using importance measures[*] CDF (1/year)	Qualify components selected using TEP[*] CDF (1/year)
Transient with reactor at high pressure and failure of injection	1.7E-6	5.9E-5	1.7E-6
Transient with reactor at high pressure and failure of recirculation	8.4E-7	3.3e-6	8.4E-7
Station Blackout	2.9E-6	2.9E-6	2.9E-6
Containment Heat Removal Failure	9.5E-7	9.5E-7	9.8E-7
LOCA with reactor at high pressure and failure of injection	5.2E-6	6.3E-6	5.8E-6
LOCA with reactor at high pressure and failure of recirculation	4.1E-6	2.2E-5	5.0E-6
LOCA with reactor at low pressure and failure of injection	2.2E-7	3.4E-5	2.8E-7
LOCA with reactor at low pressure and failure of recirculation	1.7E-6	4.3E-5	1.7E-6
Anticipated Transient without SCRAM	6.9E-8	6.9E-8	6.9E-8
Steam Generator Tube Rupture	6.0E-6	6.0E-6	6.0E-6
LOCA Outside Containment	1.7E-8	1.7E-8	3.9E-8
Total	2.4E-5	1.8E-4	2.5E-5

[*]Accident sequence quantification performed with all environmental failure basic events having high importance or in the selected prevention set to False (as though they were qualified) and the remaining environmental failure basic events failed ($P_f = 1.0$).

Attachment 1 – Overview of Top Event Prevention Analysis (TEP)

Definitions and general concepts associated with Top Event Prevention analysis and the generation of prevention sets are described in this attachment. An overview of the steps in the TEP process is provided along with a simple example.

General Concepts and the Steps in TEP

Regardless of how it is obtained, the Boolean expression under consideration will be called **the top event expression**, and we will assume it takes the form of its minimal cut sets. **A prevention set** is a collection of basic events which, if they all do not occur, precludes the occurrence of the top event. Thus, a prevention set contains at least one basic event from every top event minimal cut set. A prevention set is **minimal** if it ceases to be a prevention set when any of its basic events are removed.

The idea of prevention sets can be extended to include a level of prevention. A **prevention set of level L** contains at least L basic events from each top event minimal cut set, and it is **minimal** if it ceases to be a prevention set of level L when any of its events are removed. Besides specifying a level of prevention, one can indicate which of the basic events are to count toward the prevention level. Basic events that count toward L are **credited events**; those that do not count toward L are **excluded events**. (Examples of events that the analyst may wish to exclude from the analysis include those having a high probability of failure.) Thus, prevention sets of level L contain at least L credited events from every top event minimal cut set, and minimal prevention sets of level L cease to be prevention sets of L credited events if any of their events are removed. For example, level 1 prevention sets contain at least one credited event from every minimal cut set in the top event expression; level 2 prevention sets contain at least 2 credited events from every top event minimal cut set, etc.

In general, Top Event Prevention Analysis comprises four steps:
  (1) Build and solve a model to obtain the top event expression.
  (2) Choose a prevention level L, and specify the events that are to be credited toward prevention or, conversely, those that are to be excluded.
  (3) Generate an expression for each top event minimal cut set that represents prevention of the cut set by L credited events.
  (4) Form the Boolean product of the expressions generated for each of the minimal cutsets and expand and simplify this product to obtain all minimal prevention sets of level L.

As noted above, the output of a TEP analysis takes the form of prevention sets. Prevention sets have the following characteristics.
  - A prevention set consists of complete paths of equipment which, if they operate successfully, will assure the accomplishment of the safety functions modeled in the PRA. TEP results are presented in terms of success paths, in this regard. Specifying a level of prevention (L) results in each prevention set containing multiple (L) success paths.
  - Each prevention set emerging from TEP is minimal with respect to the prevention criterion. That is, only those components contained in a prevention set are necessary to assure an adequate level of protection from the occurrence of the top event. Components not included in a prevention set are not needed to prevent the top event.
  - Multiple prevention sets are often generated as a part of a TEP analysis. Therefore, each prevention set by itself is a complete solution. The analyst needs to select only one prevention set to have identified a sufficient set of components necessary to prevent the top event.
  - Prevention sets can be tested to determine their effectiveness with respect to cut sets that likely were truncated in obtaining the top event expression developed to begin the analysis. To test the effectiveness of a prevention set on truncated minimal cut sets, the models used to obtain the top event expression are solved again crediting basic events that are in the chosen prevention set without crediting those that are not. Additional cut sets generated as a part of this test can be appended to the original cut sets to regenerate the prevention sets and produce components making up complete success paths needed to prevent the top event.

Simple TEP Example

A simple application of these steps is presented in Figure 1. The figure contains a simple line diagram of a pneumatic system typical of that found in many power plants. It includes a three-train instrument air system backed up by a single train nitrogen supply. Included in the figure are a fault tree and the cut sets and importance measures for each of the components modeled in the system.

A common use of the importance measures is to identify those components that could contribute to the risk associated with this system from two perspectives:

(1)   Those components which currently contribute most to the failure of the system (represented by the Fussell-Vesely measure of importance), and

(2)   Those components that could potentially contribute significantly if they were to degrade in reliability (represented by Risk Achievement Worth).

In practice, thresholds are often selected for each of these types of measures (e.g., Fussell-Vesely $\geq$ 0.5% and Risk Achievement Worth $\geq$ 2) above which the components are considered to be risk significant and thereby subject to focused attention to assure their reliability. Using these thresholds in the example of Figure 1, the entire train of nitrogen and the air filters would be selected as the most important components in this pneumatic supply system. But notice that none of the compressors are identified as being important. This example illustrates a practical limitation of importance measures, that is (due to combinatorial issues), the components they identify are important but those that do not meet the numerical thresholds cannot be stated to be unimportant without further analysis.

Top Event Prevention Analysis differs from the traditional use of importance measures for identifying important contributors to top event occurrence because it finds combinations of events that are necessary and sufficient to prevent the occurrence of the top event to the chosen level. If the components in a prevention set of level L receive focused attention in plant programs to assure their reliability, these components are enough to protect against the occurrence of the top event to level L.

Returning to the simple example in Figure 1, prevention sets have been generated from the cut sets developed for the pneumatic system. A level of prevention of two has been selected for this example. That is, at least two components from each cut set are required to be considered important and subject to focused maintenance or testing to assure that the top event effectively has been prevented. Using this approach, not only is the train of nitrogen identified as being potentially important but the air receivers and at least one air compressor as well. Each prevention set identifies complete paths of equipment needed for the system that is modeled to perform its function, systematically addressing the limitation to importance measures noted above. Further, the method finds every minimal prevention set of level L. Since all prevention sets are found, one can choose from among them a solution that satisfies some additional criteria such as being easier to implement or less costly than other solutions.

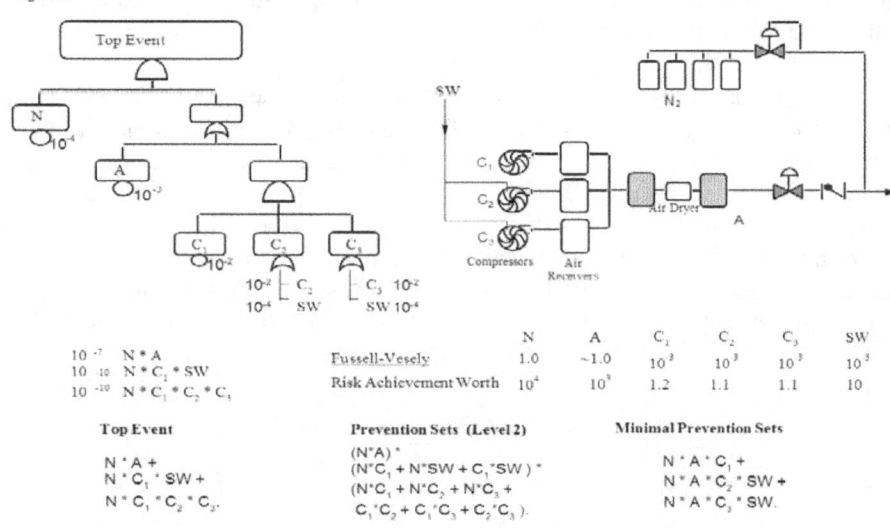

Figure 1

Risk-informed prioritization of modernization activities using ageing PSA model

Shahen Poghosyan[*a] and Armen Amirjanyan[a]
[a]Nuclear and Radiation Safety Center, Yerevan, Armenia

Abstract: Nuclear Power Plant modernization is a continuous process, which is aimed to reduce risk as low as reasonably achievable. Modernization process is especially important for old design NPPs to keep them in compliance with current safety standards. In addition, modernization process is important for plants where ageing is becoming more and more significant factor in regard with equipment reliability. Development of modernization program requires not only listing the issues to be addressed, but also to come up with common understanding of importance of proposed measures and their priority. Traditionally prioritization of modernizations is mainly done using deterministic considerations. Meanwhile parallel application of PSA models allow to come up with numerically justified and optimal solutions. Incorporation of ageing aspects in PSA model provide additional information for modernization prioritization in regard with plant's components ageing perspective. This paper describes a feasibility study aimed to use integrated risk-informed decision-making principles for prioritization of modernizations. Paper discusses proposed approach for prioritization, which implies combination of probabilistic and deterministic indicators. In addition, paper discusses comparative analysis of results obtained using base case PSA model and ageing PSA model.

Keywords: Ageing PSA, time-dependent reliability, risk-informed decision making, modernization

1. INTRODUCTION

Modernization process is integral part of Nuclear Power Plant operation, which is mainly aimed on plant safety improvement. Continuous risk reduction through modernization activities is an essential part of ALARA principle. Modernization process is especially important for old design NPPs which were designed without taking into account current safety standards. In addition, one of the problem of old design NPPs is ageing factor, which is becoming more and more significant factor in regard with equipment reliability.

Plant modernization program is usually based on results of safety assessment, operational experience analysis and best international practice. All of the mentioned aspects assist to reveal plant weaknesses and construct comprehensive list of measures to be implemented for safety enhancement. However development of modernization program requires not only listing the issues to be addressed, but also to their categorization and setting up implementation priority.

Prioritization of modernizations is done using combination of deterministic and probabilistic considerations. Incorporation of ageing aspects in PSA model provide additional information for modernization prioritization in regard with plant's components ageing perspective.

This paper describes a feasibility study aimed to use integrated risk-informed decision-making principles for prioritization of modernizations. Calculations have been performed using plant-specific PSA models developed for Armenian NPP Unit 2. Feasibility study was performed in the frame of Ageing PSA European Network organized by Institute for Energy and Transport (EC JRC, Petten).

---

[*]*e-mail: s.poghosyan@nrsc.am*

2. DESCRIPTION OF PSA MODELS AND OBTAINED RESULTS

Base case and ageing PSA models of Armenian NPP Unit 2 have been used as a base for prioritization of plant modernizations. The scope of mentioned models is following:

- Undesired event considered: Damage of the fuel located in the reactor core
- Considered regimes: 50-100% of nominal power (regimes with both turbines in operation)
- Considered initiators: Internal initiating events

Armenian NPP Unit 2 Ageing PSA model was developed using results of time-dependent reliability analysis for selected equipment [1]. Failure records have been thoroughly examined in order to identify increasing ageing trend or to assure applicability of constant failure rate model for selected equipment. Data for time-dependent reliability analysis have been gathered from several sources: plant-specific information, data from other VVER-440 reactors [2] and generic sources [3]. Time-dependent reliability analysis was performed using best-fit model criteria [4]. Results of time-dependent reliability analysis shows increasing ageing trend model for several components modeled in current PSA, it was also proved that for the rest of component constant failure rate model is applicable [5]. Identified patterns of ageing trends have been integrated in existing PSA model and ageing PSA model was created. Base case PSA model corresponds to 25 year of plant operation. 30, 35, and 40 year's prediction has been made for selected components using information about components' reliability behavior in time. Consequently, all PSA model calculations have been made for 4 lifetime points: 25, 30, 35 and 40 respectively.

Ageing PSA model quantification was implemented for all selected lifetime periods. Integrated APSA model has been recalculated for each considered case (30 years, 35 years, 40 years and 45 years) using following assumptions:
- No any modernizations or significant replacement of equipment is foreseen
- CCF models recalculated based on new reliability parameters' values (see tables 2 and 3)
- Human error probabilities are constant
- Maintenance unavailability values are correlated with increasing failure rates.
- Error factors assigned to reliability parameters are not dependent from failure rate values

Core damage frequency analysis results for base case and ageing PSA models presented in Figure 1.

Figure 1: Results of CDF calculations for base case and Ageing PSA model

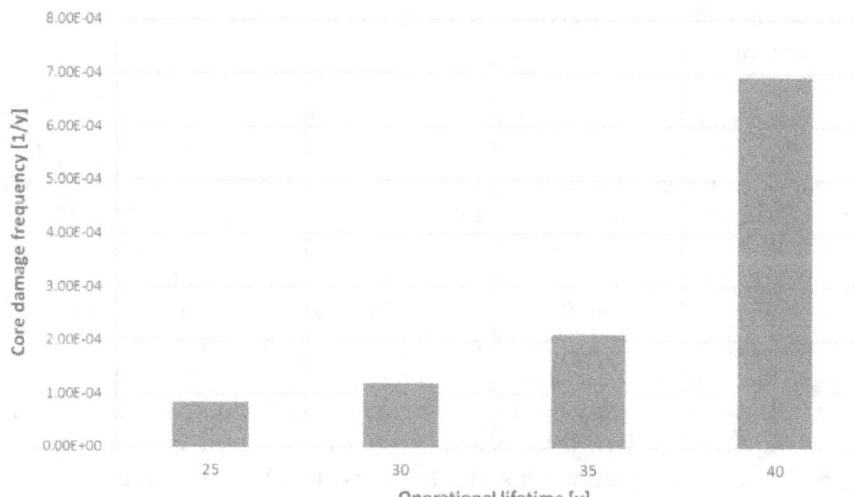

Detailed investigation of received minimal cutsets shows that risk profile is changing in time. It was noticed that though main contributors (primary and secondary side breaks) do not change significantly in time, the overall proportion of their contribution is decreased meanwhile the role of transients increased. In addition, importance analysis performed for component level reveals significant changes of risk importance parameters (e.g. Fussel-Vessely parameter) in time [6].

3. RISK-INFORMED PRIORITIZATION OF MODERNIZATION ACTIVITIES

3.1. Planned Modernization Activities

Analysis was performed for modernization activities planned for Armenian NPP Unit 2. Armenian Nuclear Power Plant (ANPP) is VVER-440/270 type reactor installation, which was designed in 70's using USSR standards. Unit 2 was commissioned in 1980 and permanently shut down for the period from 1989 to 1995. Modernizations were continuously implemented during ANPP operational lifetime (including pre start-up period from 1991 to 1995).

Currently ANPP modernization program implies implementation of several measures identified by different basis: results of safety assessment, operational experience analysis and best international practice. Major items of the program that have been selected for detailed analysis are listed below:
- Complex investigation of reactor pressure vessel resource
- Modification of the spray system Modernization against sump clogging effect (installation of effective grid system)
- Evaluation of essential service water system's pipeline breaks in boron room (flooding and spray effect analysis)
- Modernization of ECCS (reliability enhancement measures)
- Confinement safety valves reliability enhancement
- Improvement of confinement leak-tightness
- Installation of remote shutdown panel
- Verification of the capability of ventilation system to provide adequate cooling of safety system compartment after reconstruction of ECCS and spray system.
- Verification of the capability of ventilation system for cooling of reliable power supply switchgear
- Reliability enhancement of residual heat removal (RHR) system
- Complex evaluation of PTS phenomena
- Modification of emergency feedwater system (reliability enhancement measures)
- Modification of the primary circuit overpressure protection system
- Secondary side piping reliability assessment
- Investigation of external grid recovery action in case of LOSP
- Enhancement of reactor protection system reliability

3.2. Modernizations Prioritization Approach

The proposed approach for modernizations prioritization implies consideration both deterministic and probabilistic outputs for each measure listed in subchapter 3.1.

In terms of safety-related measures prioritization the main qualitative (deterministic) reference is list of safety issues specific for VVER-440 type reactors. Categorization and descriptions of VVER-440 safety issues are presented in IAEA TECDOC-640 [7]. Issues both related to design and operation are ranked according to their safety significance in four categories of increasing severity.
- Category I: Issues in Category I reflect a departure from recognized international practices. It may be appropriate to address them as part of actions to resolve higher priority issues.
- Category II: Issues in Category II are of safety concern. Defense in depth is degraded. Action is required to resolve the issue.
- Category III: Issues in Category III are of high safety concern. Defense in depth is insufficient. Immediate corrective action is necessary. Interim measures might also be necessary.

- Category IV: Issues in Category IV are of the critical concern. Defence in depth is unacceptable. Immediate action is required to overcome the issue. Compensatory measures have to be established until the safety problems are resolved.

In its turn PSA output implies information on dominant initiating events (IE) contribution and risk importance parameters for systems.

First PSA output (dominant IEs) could be interpreted as a contribution to core damage frequency from different IE categories. From this point of view following four contribution groups have been considered:
- 1st group (high importance): CDF_{IE} (contribution from particular IE category) > 30% of overall CDF
- 2nd group (medium importance): 30% of total CDF > CDF_{IE} > 20% of total CDF
- 3rd group (low importance): 20% of total CDF > CDF_{IE} > 10% of total CDF
- 4th group (negligible): CDF_{IE} < 10% of total CDF

Second PSA output (system risk importance) and its grouping was done based on Fussel-Vessely (FV) parameter which was used as an indicator for system importance. Systems have been grouped based on FV value. From this point of view following four risk importance groups are considered:
- 1st group (high importance): FV > 1E-01
- 2nd group (medium importance): 1E-01 > FV > 1E-02
- 3rd group (low importance): 1E-02 > FV > 1E-03
- 4th group (negligible): FV < 1E-03

Final priority of specific planned modernization activity was made by combining IAEA-TECDOC-640 categorization [7] with the PSA outputs in terms of IE contribution and risk importance grouping presented above. Combination of mentioned factors made using risk-informed matrix presented in Figure 2.

Figure 2: Risk-infromed prioritization matrix

		PSA output			
		1st group (high)	2nd group (medium)	3rd group (low)	4th group (negligible)
TECDOC-640	Category IV	I	II (a)	III (a)	III (b)
	Category III	II (a)	II (b)	III (b)	IV
	Categories I & II	III (a)	III (b)	IV	IV

3.3. Modernizations Prioritization Results

Modernizations prioritization was done using matrix presented in Figure 2 by applying base case PSA and ageing PSA models. Base case PSA model reflects current state of equipment reliability, whereas ageing PSA model takes into account equipment ageing factor and contain reliability parameters calculated with 15 years prediction. Comparison of results obtained by mentioned models was made with the purpose to check influence of ageing on priorities of modernization activities.

Results of modernizations prioritizations is presented in Table 1. Each modernization activity could be related both to the IE contribution (IE) and to specific system importance (RI). In such cases higher group was assigned as a PSA output information (e.g. for Reliability enhancement of RHR system the IE=3 and RI=2, the final PSA output is considered 2).

Table 1: Summary results of modernization prioritization results using base case PSA and ageing PSA models

Measure	Relevant IE	TECDOC – 640	BASE CASE PSA				AGEING PSA			
			PSA ranking			P	PSA ranking			P
			IE	RI	PSA output		IE	RI	PSA output	
Complex investigation of reactor pressure vessel resource	LOCA	1	1 (45.37 %)†	-	1	**I**	1 (38.33 %)	-	1	**I**
Modification of the spray system	LOCA	1	1 (45.37 %)	1	1	**I**	1 (38.33 %)	1	1	**I**
Modernization against sump clogging effect	LOCA	1	1 (45.37 %)	1	1	**I**	1 (38.33 %)	1	1	**I**
Evaluation of essential service water system's pipeline breaks in boron room	LOCA	1	1 (45.37 %)	1	1	**I**	1 (38.33 %)	1	1	**I**
Modernization of ECCS	LOCA	1	1 (45.37 %)	1	1	**I**	1 (38.33 %)	2	1	**I**
Confinement safety valves reliability enhancement	LOCA	2	1 (45.37 %)	-	1	**II(a)**	1 (38.33 %)	-	1	**II(a)**
Improvement of confinement leak-tightness	LOCA	2	1 (45.37 %)	-	1	**II(a)**	1 (38.33 %)	-	1	**II(a)**
Installation of remote shutdown panel	LOCA	2	1 (45.37 %)	-	1	**II(a)**	1 (38.33 %)	-	1	**II(a)**
Verification of the capability of ventilation system for cooling ECCS and spray system after reconstruction	LOCA	2	1 (45.37 %)	1	1	**II(a)**	1 (38.33 %)	1	1	**II(a)**
Verification of the capability of ventilation system for cooliing of reliable power supply switchgear	-	2	-	2	2	**II(b)**	-	2	2	**II(b)**
Reliability enhancement of RHR system	Transient	2	3 (10.78 %)	2	2	**II(b)**	3 (13.91 %)	2	2	**II(b)**
Complex PTS evaluation	SLB	2	2 (25.55 %)	2	2	**II(b)**	1 (39.75 %)	2	1	**II(a)**

† Value in brackets reflects portion of particular IE contribution in overall CDF

Modification of emergency feedwater system	Transient	1	*3 (10.78 %)*	*3*	**3**	**III(a)**	*3 (13.91 %)*	*2*	*2*	**II(a)**
Modification of the primary circuit overpressure protection system	Transient	2	*3 (10.78 %)*	*4*	**3**	**III(b)**	*3 (13.91 %)*	*4*	*3*	**III(b)**
Secondary side piping reliability assessment	SLB	3	*2 (25.55 %)*	*2*	**2**	**III(b)**	*1 (39.75 %)*	*2*	*1*	**III(a)**
Investigation of external grid recovery action in case of LOSP	LOSP	3	*4 (1.23%)*	*-*	**4**	**IV**	*4 (0.28%)*	*-*	*4*	**IV**
Enhancement of reactor protection system reliability	Reactivity accidents	3	*4 (2.26%)*	*3*	**3**	**IV**	*4 (2.29%)*	*3*	*3*	**IV**

4. CONCLUSION

Performed research is devoted to investigation of modernization activities planned at Armenian NPP Unit 2 by means of deterministic and probabilistic considerations. Risk-informed prioritization matrix was proposed in order to assure effective combination of probabilistic and deterministic considerations.

Set of criteria stated in [7] have been used as deterministic indicators. Meanwhile probabilistic considerations have been derived from plant-specific PSA models. For this purpose comprehensive ageing PSA plant-specific model has been developed for Armenian NPP Unit 2 based on the results of time-dependent reliability analysis. An attempt was done to compare results received by application of ageing PSA and base case PSA models.

From obtained results it could be concluded that the overall prioritization profile is quite similar for base case PSA and ageing PSA models. However some differences still exists, particularly measure related to "Modification of emergency feedwater system" has changed priority from III to II. Also sub-priorities of "PTS evaluation" and "Secondary side piping reliability assessment" have been changed with APSA model. It was noted that although APSA model could not significantly change the prioritization of modernizations, it could tune details related to sub-priorities between measures located at the same priority zone.

In addition, results of risk-informed prioritization of modernizations have been compared with the results of research performed for systems ranking purposes [6]. Comparison shows that advantages of Ageing PSA application are more strongly marked for system/component level application rather than for such global tasks like modernization prioritization. Summarizing mentioned research studies it is necessary to stress that application of Ageing PSA model allowed analysts to have broader view to the safety issues for considered NPP. Incorporation of ageing aspects in PSA models could reveal aspects which were hidden from analyst in base case PSA model. Having both results of current situation and prediction of risk profile analyst receive a chance to construct more precise action plan for the future.

References

[1]   Sh. Poghosyan, A. Malkhasyan and A. Rodionov. *"Component selection for ageing PSA of Armenian NPP Unit 2"*, Proceeding of ANS PSA 2008 Topical Meeting - Challenges to PSA during the nuclear renaissance - PSA'08, Knoxville, Tennessee, USA, September 7–11, 2008,

[2]   Sh. Poghosyan, A. Malkhasyan and A. Rodionov. *"A case Study on VVER-440 component age-dependent reliability data assessment"*, Proceeding of Tenth Conference on Probabilistic Safety Assessment and Management - PSAM 10, Seattle, Washington, USA, June 7-11, 2010,

[3]   Levy, I. et al. *"Prioritization of TIRGALEX Recommended Components for Further Aging Research"* (NUREG/CR-5248), US NRC, 1988,

[4]   A. Rodionov, D. Kelly and J-U. Klugel, *"Guidelines for Analysis of Data Related to Ageing of Nuclear Power Plant Components and Systems"* (EUR 23954 EN), EC JRC Institute for Energy, Petten, Netherlands, 2009,

[5]   Sh.Poghosyan and A.Amirjanyan. *"Investigation of ageing impact on safety systems' reliability"*, Proceeding of ANS PSA 2011 International Topical Meeting on Probabilistic Safety Assessment and Analysis – PSA 2011, Wilmington, NC, USA, March 13-17, 2011,

[6]   Sh. Poghosyan and A.Amirjanyan, *"Application of ageing PSA model for systems' risk-informed classification"*, ESREL2013 Conference, Amsterdam, The Netherlands, September 29 – October 02, 2013,

[7]   *"Ranking of safety issues for WWER-440 model 230 nuclear power plants"* (IAEA TECDOC-640), IAEA, Vienna, Austria, 1992.

The Reliability Effects of Transient-Induced Degradation on the Performance of Large Power Transformers

Brittany L. Guyer[a*], Carl R. Grantom[b], and Michael W. Golay[a]
[a] Massachusetts Institute of Technology, Cambridge, MA, USA
[b] CRG LLC, West Columbia, TX, USA

Abstract: Increased knowledge of the effects of severe operational transients on component reliability, in combination with currently used mechanistic component degradation models, could augment the predictive capability of reliability modeling. A new component reliability model has been developed that considers the effects of both types of degradation. An application of the new model was sought in order to provide insight into both the sources and consequences of severe component transients and how these considerations can be formulated into a new framework for component aging management supporting component reliability programs.

The large power transformer was selected for demonstration of this new reliability model. This component was selected as it is a component that has failed prematurely, has experienced strong transients during its operational lifetime, data are available about the important effects that the occurrence of strong transients have had on this component, and the transients experienced have resulted in effects that are not readily repairable (i.e., requiring component replacement). In this work, a strategy is proposed for the development of a physics-of-failure model of large power transformers that could be implemented in order to make more realistic performance predictions, supporting improved long-term plant asset management.

Keywords: transient-induced degradation, reliability, physics-of-failure, transformers

1. INTRODUCTION

Traditional component reliability models consider exclusively the effects of age-related degradation in their estimation of the component failure frequency. These reliability models could be further improved by also incorporating the effects that transient-induced (or event-induced) degradation has on the characteristic failure frequency of the component. This more realistic representation of the failure frequency that incorporates plant-specific operating experience could provide for improved asset management capabilities, as more accurate predictions could be made concerning the remaining useful life of components.

These improved reliability predictions provide strategic value specifically for those components that are characterized by a high capital cost, a long lead-time for replacement, or whose failure would result in an unplanned plant shutdown. The large power transformer is characterized by these component qualities, as it is a component that has failed prematurely, has experienced stressful transients during its operational lifetime, information is available about its important effects resulting from the occurrence of stressful transients, and the transients experienced have resulted in component effects that are not readily repairable (i.e., requiring component replacement). Therefore, it was selected as a component eligible for application of the new reliability model.

In this work, a strategy for the development of a physics-of-failure model for the large power transformer is proposed in order to be able to apply the transient-induced degradation reliability model. The application of this method demonstrates the importance of the availability of component-specific operational data pertaining to transient-induced degradation.

---

[*] Brittany L. Guyer: guyer@mit.edu

2. DESCRIPTION OF RELIABILITY MODEL

The reliability model [1] developed for application to this work is a probabilistic model that accounts for three types of failures: random failures, a random failure following a transient and a failure due to the occurrence of the transient itself. The general model provides the probability of failure for the component lifetime from the beginning of life to the time of planned shutdown (t_s) and is shown in Eqn. 1.

$$P(fail) = \int_0^{t_s} \lambda e^{-\lambda t}\, dt + \int_0^{t_s} \lambda'\, dt \int_0^t \lambda_T P(\Delta\lambda_R \mid T) e^{-\lambda_T t_T} e^{-\lambda'(t-t_T)}\, dt_T \qquad (1)$$

In Eqn. 1, the first term represents the failure probability distribution representing the occurrence of random failures and the second term represents the failure probability distribution of the failures resulting from the occurrence of the transient. (See nomenclature section for variable definitions.) The failure frequency is defined as the summation of contributions from both the random and transient-induced failures, as shown in Eqn. 2.

$$\lambda = \lambda_R + \lambda_T \cdot P(failure \mid T) \qquad (2)$$

In Eqn. 2, the total failure frequency (λ) is expressed as the sum of the random failure contribution (λ_R) and the transient-related contribution where λ_T is the frequency of the damaging transient, and *P(failure|T)* is the probability that failure occurs due to the occurrence of the transient. In this way, the total failure frequency can be dependent upon the occurrence of many different degradation-inducing transients, which are characterized by various frequencies and failure probabilities.

The occurrence of the transient(s) results in the creation of a new failure frequency defining the operation of the component, as shown in Eqn. 3.

$$\lambda' = \lambda_R + \Delta\lambda_R \qquad (3)$$

Here, λ' represents the new failure frequency characterizing the latter failure probability distribution of Eqn. 1, where the $\Delta\lambda_R$ represents the step-change increase in failure frequency due to the occurrence of the degradation-inducing transient.

3. COLLECTION OF PLANT-SPECIFIC DATA

3.1. Fault Evaluation

Because the successful application of the reliability model that we seek to apply in this work requires the use of a component-specific event history, it was necessary to identify a utility partner who would be willing to share their transformer operating history data. In choosing a partner, we looked for a utility that had experienced unanticipated transformer events at a nuclear power plant. The record of these events allows for the development of a relationship between classes of transients and resulting increases in expected failure frequencies. While general relationships between types of transients and increases in failure frequencies can be derived, the prediction of future transformer reliability is dependent upon an accurate record of its event-history, as the effects of degradation resulting from these degradation-inducing transients are cumulative.

The utility partner provided to us a record of the condition reports of the plant events that potentially posed a threat to the integrity of the transformers. There are seven large power transformers at this site, two main transformers and one auxiliary transformer for each of the two units, and one spare transformer. The record extends back to the beginning of operation for each unit, 26 years and 25 years, respectively.

Based upon the knowledge of the fault location, voltage fluctuations and physical inspection findings, the utility staff ranked the severity of the transient's effect on each of the transformers through the use of impact codes, which are defined in Table 1. In the evaluation of the fault, it was assumed that if the fault were a phase-to-ground fault, these events should contribute to relatively little through fault current in the transformers, as the damage would be limited by the neutral grounding resistors in the auxiliary transformer and main generator, and the delta windings of the auxiliary transformer and the generator step-up transformer. The faults that were considered to be more severe were those involving multiple phases, as in a phase-to-phase fault or an exciter fault. [2]

Table 1: Impact Codes

Code	Severity
0	None
1	Low
2	Low/Medium
3	Medium
4	Medium/High
5	High

The occurrence of voltage transients was also considered in the development of the impact code for each event, but their occurrence was not given as much weight as the contribution of the fault to the severity. This lesser importance derives from the likelihood that voltage transients have a more immediate effect on the transformer, rather than the through-faults, which have a cumulative effect on the transformer internal components by loosening the windings and the clamping. Also, the transformers are protected from internal damage by arrestors on their bushings. In general, since voltage transients are more severe closer to the fault, if the fault is not close to the transformer it is less of a concern. Additionally, the transient's dispersed effects are difficult to evaluate. [3]

Lastly, in the evaluation of the fault severity, if the sudden pressure relays on the transformer actuate during a fault, it is an indication that there is a cause for concern for the integrity of the transformer internals. In the management of these events, the utility performed an analysis on the transformer oil in order to see if degradation occurred based upon the test findings.

3.2. Fault Evaluation Data

During the lifetimes of the seven transformers present at the two-unit site, 17 transient events occurred that affected the transformers. The lifetime-sums of the impact codes characterizing the events affecting each transformer are shown in Table 2. Due to the different nature of each transient event, not every transient affected all transformers. Examining the lifetime sum of the individual event impact codes reveals the variability of the impact of the events over the fleet of transformers, ranging from only 2 to 26.

Table 2: Lifetime Impact Codes from Plant Data Set

Transformer Name	MT1A	MT1B	UAT1	MT2A	MT2B	UAT2	Spare
Lifetime Impact Code Sum	2	20	11	20	26	20	13

3.3 Classification of Internal and External Transformer Events

In comparing the undesirable quality of transformer-related transient events, a contrast can be drawn between events that are internal and those that are external to the transformer. Here, we define internal events as those that occur as a direct result of the malfunctioning of components internal to the transformer. External events are defined as those that could affect the future performance of the transformer by inducing degradation to the transformer, but were initiated by another component affecting the plant electrical equipment, thereby affecting the transformer. In contrasting these two classes of events, the internal transformer events are the more severe of the two event classes from the perspective of both the asset management and reliability of the transformer. Internal events are worse from this perspective because the transformer itself is the source of the problem requiring plant shutdown, versus other equipment that do not represent single point vulnerabilities for plant power generation. Furthermore, they occur presumably as a result of the existence of a degraded material state within the transformer, indicating the potential for reduced confidence in future transformer performance.

The premise behind the application of the reliability model described in Section 2 is that the occurrence of external events can influence both the frequency and the severity of the occurrence of events internal to the transformer. Therefore, as the number of events, both internal and external, increases during the lifetime of the transformer, degradation will be expected to accumulate over time. While the utility's definition of the impact codes for each transient event is not based upon a scientific physics-of-failure basis, the qualitative-engineering judgment employed is based upon the premise that the more severe the event, the more degradation induced. Also, it is plausible that a more degraded transformer will experience future events more severely than its less degraded counterpart. These two inferences from the impact code classification suggest that as the total number of events experienced by a transformer increases, the severity of the events, as indicated by the sum of the impact codes, will increase as well.

This relationship between the event severity and the number of lifetime events experienced by each of the seven transformers is depicted in Figure 1. Not only does the severity tend to increase as the number of the lifetime events increases, it also does not increase proportionally to the number of events. If a comparison is made between UAT1 at 9 events and an impact code of 11 and MT2B at 12 events and an impact code of 26, it can be seen that the severity does increase with the number of events, but as the number of events increases, the associated impact code increases more significantly. This trend suggests that the transformers are experiencing the transient events more severely as they become more degraded with the occurrence of each new event.

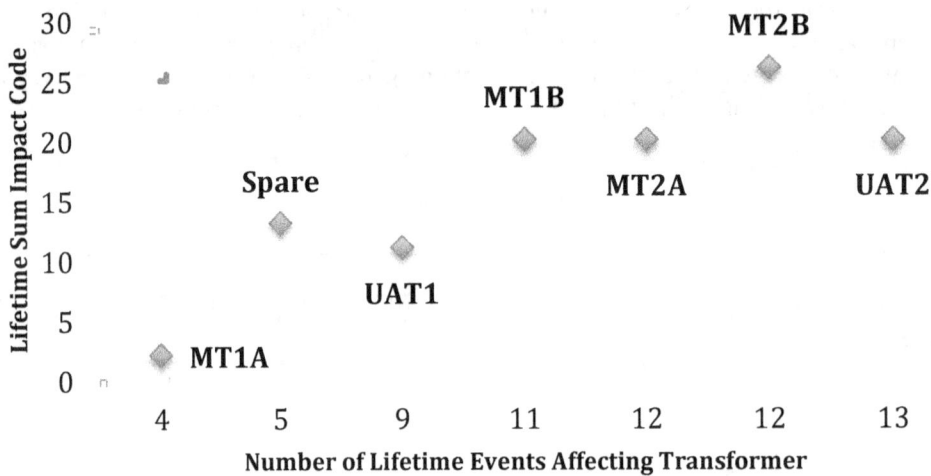

Figure 1 Comparison of the Lifetime Severity and Number of Events Experienced by Each Transformer

Examining the relationship between the number of both the external and internal events that have occurred during a transformers life also suggest that it is plausible that the number of external events influences the likelihood of the occurrence of internal events. Table 3 contains the number of internal and external events that each transformer experienced during its lifetime.

If the number of external events that a transformer experiences induces degradation on the transformer, we should expect that the number of internal events that the transformer experiences would increase with increasing occurrence of external events. In Figure 2, the relationship between the number of internal events and external events for each transformer is depicted.

Table 3: Numbers of Lifetime Events for Each Transformer

Transformer Name	MT1A	MT1B	UAT1	MT2A	MT2B	UAT2	Spare
Internal Events	0	1	1	2	1	2	0
External Events	4	10	8	10	11	11	5
Total Lifetime Events	4	11	9	12	12	13	5

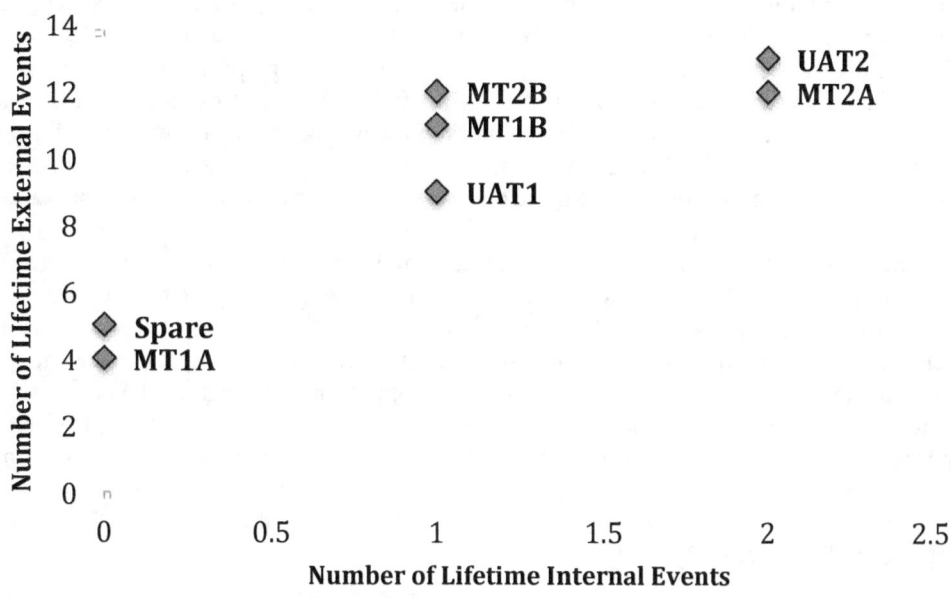

Figure 2 Comparison of Internal and External Events for Transformers Examined

Figure 2 shows that there exists a plausible physical dependency of the number of internal events upon the number of external events. Logically, the level of this dependence will depend upon the influence of the severity of the external events upon the integrity of the transformer, but even disregarding the severity of the events, the trend shown in Figure 2 is suggestive of a correlation between the number of internal and external events experienced by a transformer.

4. STRATEGY FOR DEVELOPING A PHYSICS-OF-FAILURE PREDICTIVE MODEL

While the fault impact codes provided by the utility give a good first indication of the impact of historical transient events upon future transformer reliability, the data give only a relative indication of confidence about future performance, and do not provide a means for making numerical predictions about the future failure frequencies (and reliability) of the transformers. In future work we shall develop the capability for ensuring both a more accurate characterization of the induced degradation and improved reliability predictions that are based upon the results of a physics-of-failure model. The current practice of characterizing transformer-related events with impact codes implicitly makes this assumption of induced degradation, but it does not do so in a scientific manner. The introduction of an increasingly explicit accounting of induced degradation levels will likely provide for improved reliability predictions; however, because of the possibility for many failure mechanisms, our proposed approach will not address all failure mechanisms, but will acknowledge all operational events that have occurred. We anticipate that this approach will help bridge the gap between the implicit and fully explicit approaches of degradation characterization and provide insights into improvements in plant reliability programs supporting long-term operations.

3.2. Fragility Analysis

In structuring an approach for the development of an impact code analysis with improved realism, it is most logical to consider those components and associated degradation modes that would dominate the risk of transformer failure. Ultimately, the usefulness of this new approach will be judged by one simple criterion: how it is able to predict the occurrence of transformer downtime. In considering the application of this model as an asset management tool, utilities are primarily concerned with the occurrence of an unanticipated transformer end-of-life failure event. Because transformers have a long lead-time for fabrication, this class of event has significant economic consequences for the utility. As a secondary concern, utilities are interested in avoiding unplanned shutdowns caused by transformer failures (or degradation) since these also result in lengthy and costly plant shutdowns.

Naturally, the most severe of the two scenarios is that in which the transformer experiences an ultimate failure for which the utility has not planned, as this event has the potential for the most severe economic consequences. Therefore, the goal of the development of our approach is to focus exclusively on the life-limiting failure modes of the transformer in order to enhance the predictive capability of the time at which end-of-life occurs. Here we present the steps for the development of an approach to improve the current standard of using impact codes by implementing a physics-of-failure based approach in order to develop a fragility characterization for the components most significantly contributing to transformer failures.

3.2.1. Component/Degradation Mode Identification

The first step in developing a fragility analysis for the transformer is to identify the most important life-limiting components. Industry data on transformer performance will be used to inform this selection process. After this selection has been made, the predominant modes of degradation contributing to the ultimate failure of these components will be identified.

3.2.2. Development of the Fragility Factor

The development of a fragility factor in order to characterize the level of degradation is key to improving the current methods of assessing transformer degradation. Developing this fragility factor requires a mechanistic understanding of three general factors contributing to a component's level of degradation. These three areas are the contributions of age, shocks (external events) and repairs to degradation.

In the development of this fragility factor we do not seek to develop new physical models, but instead we seek to apply those that already exist in order to provide for a more realistic assessment of transformer performance. Therefore, for each dominant mode of degradation identified for the life-limiting components, the currently existing physical models of degradation will be evaluated for application to the development of the fragility factor.

The successful application of the physical models will be determined by their ability to apply the information relevant to the occurrence of external events and repair-induced failures. Therefore, a key criterion for selecting these physical models will be their ability to be related to event-related data, such as temperatures, voltages etc. Successfully meeting this criterion will ensure that there exists continuity between the physical nature of the external events and the degradation consequently induced.

In order to predict transformer failure better, the development of the fragility factor will require the definition of the failure state. While this definition could be based upon the physical mechanism alone, the entirety of the range of potential degradation modes cannot be realistically captured in this analysis. Therefore, the definition of failure will need to be informed by the current treatment of examining transformer life, which are contained in the industry standards for transformer performance. Generally, these standards have not been developed using a physics-of-failure based approach; however, because they must be applied for transformer operation, they must be considered in this evaluation. Therefore, in the application of this fragility analysis, the definition of ultimate failure may not be different than in prior treatments. The value of its application, however, lies in its ability to make more realistic statements about the current level of degradation and the assignment of that degradation to specific components within the transformer, which can be used valuably to support better asset management.

Therefore, the complete development of the fragility analysis will require a review of the current transformer standards associated with the life-limiting components/events, such as through-faults. The limits set by these standards will be used to inform the failure definitions for each mechanistic failure mode.

The fragility factor, F, will be calculated by considering the various levels of component degradation present for all components that are considered to be life-limiting for the transformer. Eqn. 4 is the definition of the fragility factor,

$$F \equiv \frac{\sum_{i=1}^{n}[\% \; Component \; Degradation]_n}{n}, \qquad (4)$$

where, n, is defined as the number of components. The percent of component degradation, P_D, is defined by Eqn. 5,

$$P_D \equiv MAX\left[\frac{Degradation}{Degradation \; Limit}\right]_m, \qquad (5)$$

where, m, is the number of modeled degradation modes and the percentage is defined as the maximum percentage of all degradation modes considered, since that mode of degradation will likely be the first to induce a component failure.

Using these two equations as the basis for the development of the fragility factor allows for the inclusion of many degradation modes and components in the fragility analysis. By developing a physics-of-failure interpretation of the degradation associated with the life-limiting failure mechanisms, a more realistic understanding of the remaining useful life will be revealed through the results of the fragility factor analysis.

3.2.2. Reliability Predictions

The benefit of developing a scientific basis for the derivation of the fragility factor is that it can provide a means to formulate improved reliability predictions, providing improved asset management capability to the utility. The reliability prediction is made by using the knowledge of the frequencies of the occurrence of the external events in order to make predictions about the levels of degradation expected to be induced over time to specific transformer components.

In order to make the reliability prediction, a defined set of relevant external events must be categorized by both their level of induced degradation and their frequency of occurrence. This information can then be combined in order to calculate a predicted level of induced degradation per unit time. Combining this with the historical record of induced degradation as indicated by the fragility factor analysis, a more informed prediction can be made of both the transformer's characteristic failure frequency and expected time of end-of-life.

5. CONCLUSION

A review of the impact factor analysis performed on seven large power transformers at a nuclear power plant demonstrates that there exists evidence to suggest that a more scientifically based analysis of the degradation effects of external events on the reliability of transformers is warranted. A strategy for the development of this analysis was shown to include the mechanistic relationship between the occurrence of the internal event and the level of induced degradation in the transformer. The proposed analysis focuses on the life-limiting transformer components, as they are the most likely to influence the asset management capabilities of the fragility analysis. Reliability predictions can be made by implementing the physics-of-failure based model in order provide for a more realistic understanding of the timing of future transformer failures.

Nomenclature

$\Delta\lambda_R$ = increase in random failure frequency due to occurrence of transient
λ = failure frequency due to both random failures and the occurrence of the transient itself
λ' = failure frequency of the component after the occurrence of a degradation-inducing transient
λ_R = failure frequency due to random failures
λ_T = frequency of the degradation-inducing transient
σ = fatigue-induced stress level
N = number of fatigue cycles
$P(failure|T)$ = probability of failure given the occurrence of the transient
$P(\Delta\lambda_R|T)$ = probability of degradation occurring as a result of the transient occurrence
t = time
t_T = time of transient occurrence
t_s = time of planned shutdown

Acknowledgements

This research was supported by a grant from EDF R&D.

References

[1] B. L. Guyer and M.W. Golay, "The Reliability Effects of Transient Induced Degradation," *Transactions of the American Nuclear Society*, Vol 109, No. 1., November 2013, pages 2045-2048.

[2] IEEE Power and Energy Society, 2011, "IEEE Guide for Loading Mineral-Oil-Immersed Transformers and Step-Voltage Regulators," IEEE Standard C57.91-2011, New York, NY.

[3] IEEE Power and Energy Society, 2010, "IEEE Guide to Describe the Occurrence and Mitigation of Switching Transients Induced by Transformers, Switching Device and System Interaction," IEEE Standard C57.142-2010, New York, NY.

Risk-informed Simulation Optimization for engineering asset management

Jérôme Lonchampt[a], William Lair[a]
[a] EDF R&D, Chatou, France

Abstract: This paper present a general method coupling genetic algorithms and Monte-Carlo simulation to address simulation optimization issues in the field of engineering asset management. After a description of the method, parameters tuning issues are analyzed through a test-case.

Keywords: Simulation-Optimization, Genetic Algorithm, Monte-Carlo simulation, Asset Management

1. INTRODUCTION

Optimizing the maintenance schedule for a component or a system, a classic problem in Engineering Asset Management (EAM), faces two major challenges. The first one is to build a realistic model that can be used to assess the efficiency of a given maintenance strategy. The second one is to handle the important combinatory of the optimization problem since, on a year based maintenance, the solution space size is growing exponentially with the operating time remaining.

In this paper we present a general framework for risk-informed constrained maintenance scheduling optimization coupling a Genetic Algorithm (GA) and Monte-Carlo simulation algorithm. The performance and the parameters tuning of this Genetic Algorithm for Simulation Optimization (GASO) will be discussed based on a test case (finding the replacement dates minimizing the global owning cost of a single component with a Value at Risk constraint) with a special focus on the fitness function for which two alternatives have been studied.

2. SIMULATION OPTIMIZATION REVIEW

Taken separately, efficiency assessment issue and asset management strategy optimization issue have been addressed successfully and are widely described in the literature. Reliability and, more recently, industrial asset management models have been developed for decades to assess the efficiency of a maintenance strategy, on the technical point of view (reliability, availability or safety indicators) or on the financial one (discounted cash-flows, Net Present Value...). Many mathematical models and associated tools are used (Markov graphs, Piecewise Deterministic Markov Process, Petri Net...). These models are solved thanks to numerical calculation techniques or Monte-Carlo Simulation if the underlying model is more complex.

On the other hand, finding the optimal schedule for maintaining a component is a hard optimization problem, as it is often impossible to write down the goal function as an easily optimizable function (linear, convex...), whether because the asset to model or the indicators to be optimized are too complex. Exact methods are then not usable (cutting plane methods for linear problems) or not efficient enough (branch and bound). An efficient alternative is to use approximation methods such as metaheuristics, among which one of the most popular is the genetic algorithm.

Solving optimization problem for goal function assessed through simulation is a research area known as "Simulation Optimization" and it has been increasingly studied in the past fifteen years. A recent survey ([1]) identifies metaheuristics as the best methods for global integer optimization. The difficulty is then to associate Monte-Carlo simulation and Genetic Algorithm, as a matter of fact Monte-Carlo simulations need a large number of replications to narrow the results confidence interval and Genetic Algorithms need to evaluate a large number of solutions to converge toward a global optimum, a simplistic coupling could lead to calculation durations too long to be tractable. Examples of coupling such algorithms have been described in cases where simulation durations were not an issue ([2]). In other cases ([3]), the confidence interval is taken into account to penalize the goal function but the convergence of solutions is not improved throughout the process. A sequential method, named OCBA, improving the convergence of "good solutions" throughout optimization, has been described in [4] and coupled to evolutionary algorithms for manufacturing or design problems ([5]).

In the field of maintenance and asset management, there have been very few uses of these methods but it would respond to a growing concern of decision makers who want to take multi-objectives and multi-constraints decisions for complex assets. It is then impossible to avoid using simulation to assess the objectives or to check constraints (especially risk-informed ones). One of the few examples found ([6]) does not seem to have been widely used since, perhaps because the goal function (expected availability of a redundant system) could be approximated with a Markov process numerically calculable.

3. METHOD

3.1. Asset Management Simulation Model

A generic asset management model has been developed within EDF R&D to evaluate the profitability of an investments strategy for large assets of power plants. This model relies on a parallel evaluation of two strategies, the reference one and a new one that decision makers want to assess, replicated with a Monte-Carlo simulation tool. The only source of uncertainty taken into account in this model are the failure dates of the components, which is an aleatoric uncertainty as opposed to epistemic uncertainties such as maintenance costs or spare parts supply delays which aren't modeled in EDF tool but analyzed through sensitivity analysis. The generic method and tool have been discussed in [7].

For the purpose of this paper, the asset management model has been simplified to a single repairable component with a non constant failure rate. Repairs are assumed instantaneous, unavailability of the component after failure being taken into account in the sole total costs. Preventive replacement, assumed to be an As Good As New (AGAN) maintenance task, may be performed according to a date-based maintenance program. Figure 1 shows the graph for such a model, for which:

- λ is the failure rate of the component depending on its age
- δ is the dirac function
- π is the deterministic law modeling preventive replacements
- t is the time
- a is the age of the component
- CF is the cumulated cash-flows
- C_{CM} is the corrective maintenance cost (including spare part costs, forced outage costs...)
- C_{PM} is the preventive maintenance cost
- α is the discount rate

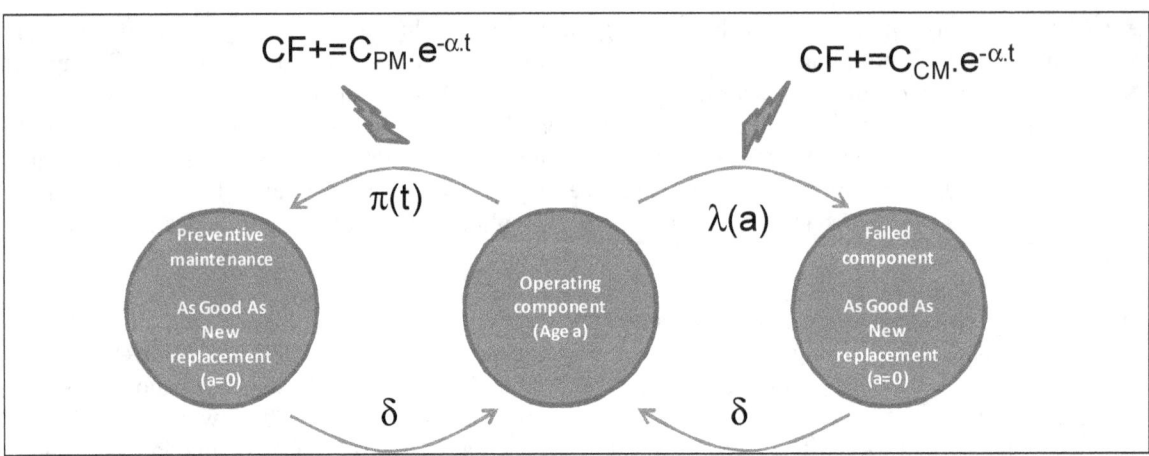

Figure 1 - Pseudo-Markov graph with transitions impacts on cash-flows for a single repairable component

The Monte-Carlo simulation used to estimate the probabilistic distribution of the cumulated cash flows is described in Table 1.

```
while r<NReplications
        Total_Cash_Flow(r)=0;
        t_failure=reliability_law(rand);
        t_preventive_replacement=[t_pr_1;t_pr_2;...];
        while min(t_failure ;t_preventive_replacement)<T_final
                t= min(t_failure ;t_preventive_replacement) ;
                if t=t_failure
                        Total_Cash_Flow(r)+=corrective_cost*exp(-alpha.t);
                        t_failure=t+ reliability_law(rand);
                else
                        Total_Cash_Flow(r)+=preventive_cost*exp(-alpha.t);
                        t_failure=t+reliability_law(rand);
                        delete(t,t_preventive_replacement);
                end;
        end;
        r+=1;
end;
```

Table 1 - Pseudo-code for the single repairable component Monte-Carlo simulation

3.2. Genetic Algorithm

The Genetic Algorithm (GA) was introduced by Holland in [9] and popularized by Goldberg [10]; it is a evolutionist meta-heuristic widely used for combinatorial optimization. It is based on an analogy to Darwin's evolution theory on natural selection stating that, within a population, organisms with a high fitness are more likely to reproduce and to create offsprings with higher fitness.

Genetic algorithms include different operators mixing global search (selection of best solutions, cross-over to build new ones) and local search (mutation) in order to find a good approximation to an optimization problem. Figure 2 presents a generic scheme of such an algorithm.

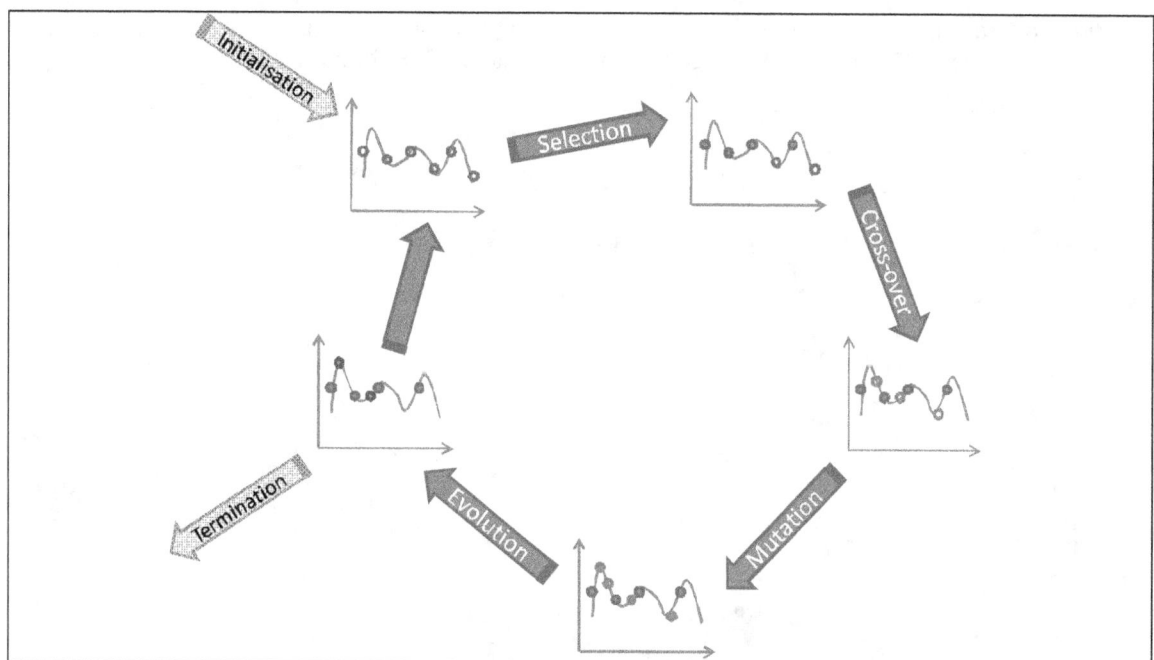

Figure 2 – Generic Genetic Algorithm

This method is one of the most popular meta-heuristic used in EAM optimization problems [8]. It has been implemented in IPOP software to solve investments planning problems with budget constraint for EDF (for a complete description of the specific algorithm and a discussion on parameters tuning, one should read [11]). The goal function used in IPOP is a mean indicator whose evaluation is fast enough to be efficiently computed with regular Genetic Algorithms.

3.3. Genetic Algorithm for Simulation Optimization (GASO)

When the goal function is expensive to compute, which is often the case for simulation evaluation, it is very difficult to use GA as described previously. As a matter of fact, for complex models, industrial assets simulation of one given strategy may take up to several minutes or hours to compute converged estimators. On the other hand, GA may need the evaluations of several thousands of different strategies, leading to calculations that would last days or months. Even if it does not seem impractical, and that calculations time could be shortened using supercomputers, it is not a workable method as this kind of calculations needs to be assessed daily by system engineers or business planners, often with very short delays.

The main idea of the method described in this paper is to improve the convergence of the simulations throughout the optimization process. At the initialization step, all solutions, chosen randomly, are evaluated with a small number of replications N then offspring are created using usual cross-over and mutation operators except that, when it comes to evaluating the offsprings, the simulation is ran in two different ways:

1. If the new solution already exists: M replications (with M<N) are added to the existing solution and the number of offsprings is not incremented
2. If the new solution is not present in the population: it is initialized with N replications and considered as an offspring.

When the number of offsprings reaches the limit required to update the population, a $(\lambda+\mu)$ evolution strategy is applied keeping the best solutions of both the original population and the offsprings, and N replications are added to all solutions.

To the regular termination criteria of a Genetic Algorithm (maximum number of generations reached or no improvement of the best solution after X generations) are added criteria on the simulation convergence, such as a maximal width of confidence intervals, non overlapping intervals for the best solutions or reaching a maximal total number of replications (resource-limited computing). **Figure 3** gives a simple view of such a Genetic Algorithm for Simulation Optimization (GASO).

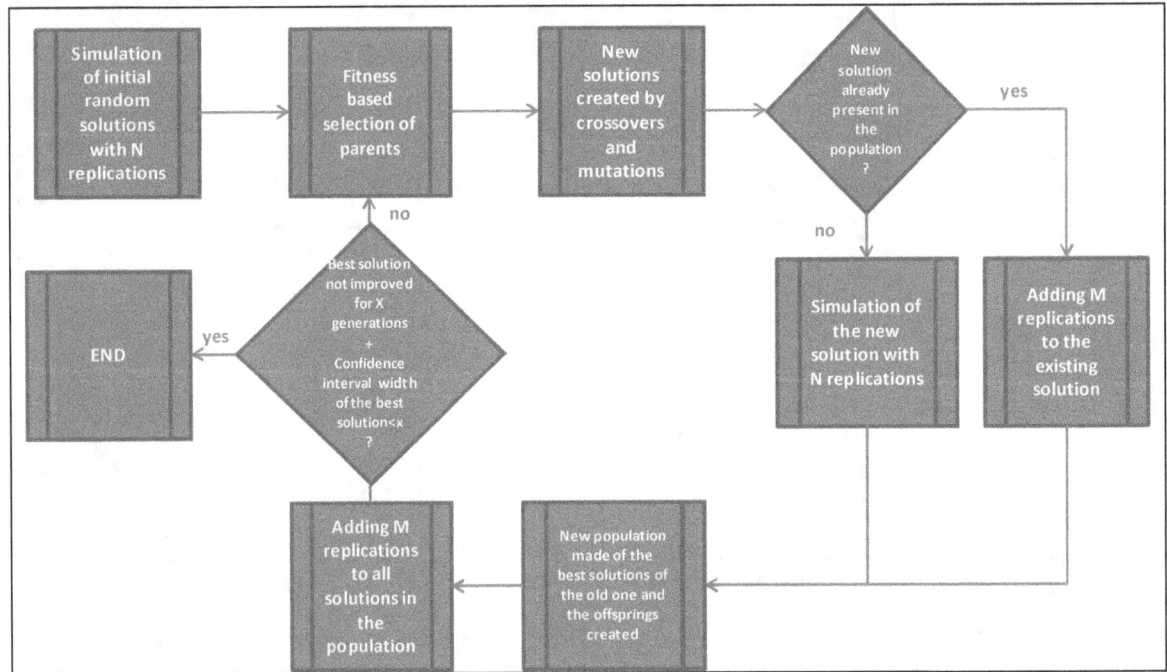

Figure 3 – Genetic Algorithm for Simulation Optimization (GASO)

Three specificities of this method, compared with regular GA, should be addressed.

- Parameters tuning: it is a well-known limitation of Genetic Algorithms that their efficiency is highly dependent on the different parameters like the size of the population or the mutation rate. No simple rule actually exists to evaluate these parameters according to the problem characteristics, so the right tuning of parameters often depends on the skills of the analyst. Adding two more parameters corresponding to the numbers of replications M and N makes the tuning more difficult.
- Goal function: in this method the fitness of a solution will change throughout the optimization process, as it is an estimator depending on the growing sample of replications. A solution then needs to be associated with more than one single piece of data per objective, the number of values needed will depend on the type of statistic. For a statistic like a probabilistic moment two values are sufficient as the couple (empirical value, number of replications) contains all the information needed to update the estimator when replications are added. On the opposite, using a quantile for objective is more complicated and it seems necessary to store the results of all replications, leading to memory issues.
- Confidence interval: as the value of a solution is estimated, solutions may be compared taking into account the accuracy of the estimators. Depending on the type of statistic the confidence interval may be more or less easily calculated for a given solution and updated when new replications are added. If the goal function is the mean value for instance, the solution in the population will need to be associated with the empirical standard-deviation so that an approximation of the confidence interval may be computed according to the central limit theorem. Once a confidence interval is available and easily updatable for the solutions it may be used at two different steps of the GA:
    1. Selection of solutions candidates to cross-over: whatever the selection method is, as long as it is based on a ranking of the population (that is to say all usual selection methods except the random one), a mathematical order should be defined. The question then comes whether solutions must be ranked according to the estimator of the goal function, regardless of the number of replications, or according to the

confidence interval bounds. These two different possibilities will be discussed on a test-case in §4.

2. <u>Termination criterion</u>: for regular GA the termination criterion is usually the fact that the best solution has not been improved for several generations. This kind of criterion proved to be not sufficient enough in the case of GASO as a solution with a very large confidence interval may dominate the population of solutions for many generations. This is the reason why the termination criterion for GASO is a double one with the best solution not improved for X generations **and** having a confidence interval width lower then Y%. Another possibility would have been that the top Z solutions have non-overlapping confidence intervals, but, as it will be discussed in §4.2., this criterion is often impractical.

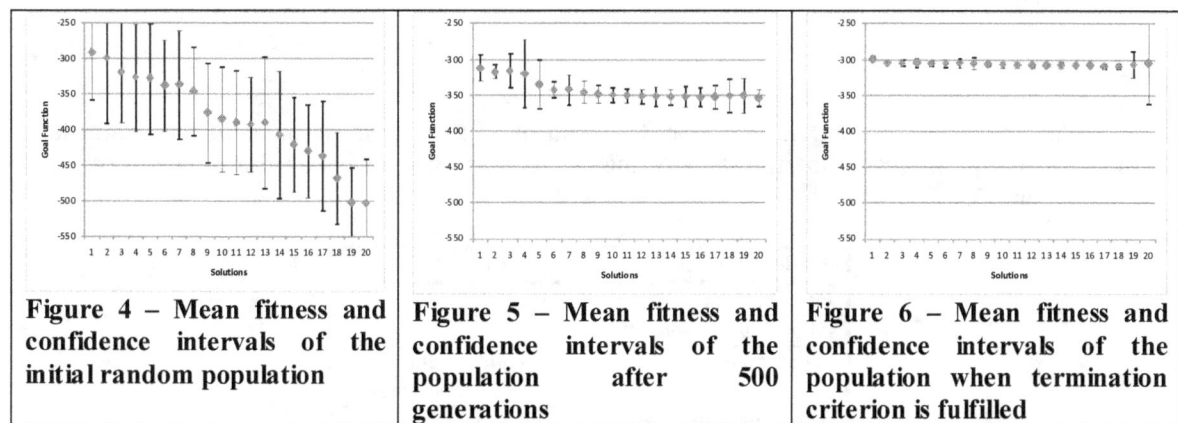

Figure 4 – Mean fitness and confidence intervals of the initial random population	Figure 5 – Mean fitness and confidence intervals of the population after 500 generations	Figure 6 – Mean fitness and confidence intervals of the population when termination criterion is fulfilled

Figure 4 to **Figure 6** give an example of the evolution of a GA for a maximization problem. After 500 generations (**Figure 5**) the best mean fitness is actually worse than the best mean fitness of the initial population (**Figure 4**), but the average confidence interval width is much smaller. The best solution, at this point, does not have the narrowest confidence interval, showing that it is a "young" solution present in the population for few generations. When the algorithm terminates the top solutions all have narrow confidence intervals. This example also highlights the fact that the best solution may be present in the population before being considered optimal, as the red mark, representing the optimal solution, is identified as a good one but ranked number four after 500 generations (**Figure 5**).

4. TEST-CASE

4.1. Test-case description

The test case consists in finding the optimal preventive replacement planning for a repairable component. Both preventive and corrective actions are perfect maintenance. The component reliability is modeled by a Weibull distribution. All costs are discounted using a discount rate. The parameters of the test case are given in Table 2.

Parameter	Value
Scale parameter λ	0.05
Shape parameter β	2.3
Time horizon	40 years
Corrective cost	1000
Preventive cost	100
Discount Rate	7.5%/year

Table 2 - Asset parameters

As for the optimization problem, it is a cost minimization one with a risk constraint:

$$\begin{cases} \min Cost(x) \\ subject\,to\,Prob(Cost(x) \geq 800) \leq 0.06 \end{cases} \qquad (1)$$

The replacements schedules being annual ones, the size of the search space is $2^{40} \approx 1,1.10^{12}$.

The test-case aim is to study the impact of both the numbers of replications at each step of the GASO and the selection criterion based on the mean values of the different objectives or the upper bound of the confidence intervals.

4.2. GASO

As explained in §3.3, GASO is a regular GA with an iterative enhancement of the simulations convergence. For the regular part of the GA, the methods and parameters used for this test case are given in Table 3.

Method/Parameter	Value
Population size	20
Offspring per generation	1
Selection method	Tournament with three candidates
Selection elitism rate	0.9
Crossover method	Uniform
Crossover rate	0.7
Mutation method	Neighbors swap
Mutation rate	0.1
Evolution strategy	λ+μ

Table 3 – GA features

As for the GASO specific features, the aim of the test-case is to study the impact of both the numbers of replications and the selection criterion. The impact of the termination criterion was not studied in this test case. A limit on the total number of replications is applied to control the computing cost of the optimization.

Method/Parameter	Value
Initial number of replications (N)	100/1000/10000
Enhancement number of replications (M)	10/100/1000
Selection order	• Mean values • Upper bound of the 95% confidence interval
Termination criterion	**Dispersion (width of the confidence interval over the mean estimator) of the best solution lower 1%** **AND** **Best solution ranked first for at least 20 generations** **OR** **Total number of replications higher than 50.10^6**

Table 4 – GA features

The design of experiment is to run 10 trials for each of the 18 sets of parameters (initial number of replications, enhancement number of replications and selection order measures).

4.3. Proof of optimality and efficiency measure

It is a well-known limitation of GA that the convergence towards a global optimal solution is often difficult to prove. For GASO the difficulty is even higher as the fitness is based on an estimator and not the real value; if a GASO is run twice, two close solutions may actually be ranked differently if the convergence of the simulation is not sufficient. The number of replications needed to achieve, with a high confidence, an optimization problem with a flat optimal neighborhood may happen to be too important to be feasible. A practical method is to run the algorithm several times and to study the frequency of the different best solutions and the dispersion of their estimators. If no solution happens to appear as the best one, additional replications should be added to narrow the width of the confidence intervals.

Such an analysis have been performed for the test-case. **Figure 7** shows the different best solutions of all instances of the design of experiment, with the mathematical order used for the selection mechanism being the mean value. It represents 90 trials of the algorithm. Not all solutions found are shown, but only the ones with a confidence interval width lower than 1% of the mean estimator (some of the trials did not converge, the algorithm ending because it reached the maximum number of replications awarded to the calculation) and appearing more than once. This leads to eight different solutions. If their confidence intervals show some intersections, making impossible to demonstrate the dominance of one of the solution, the fact that the first one appears 12 times as the optimal one, the second best appearing only 3 times is a good indication that it is a good candidate to optimality. This indication is confirmed by the fact that the best (i.e. minimal) estimator value is found for this solution.

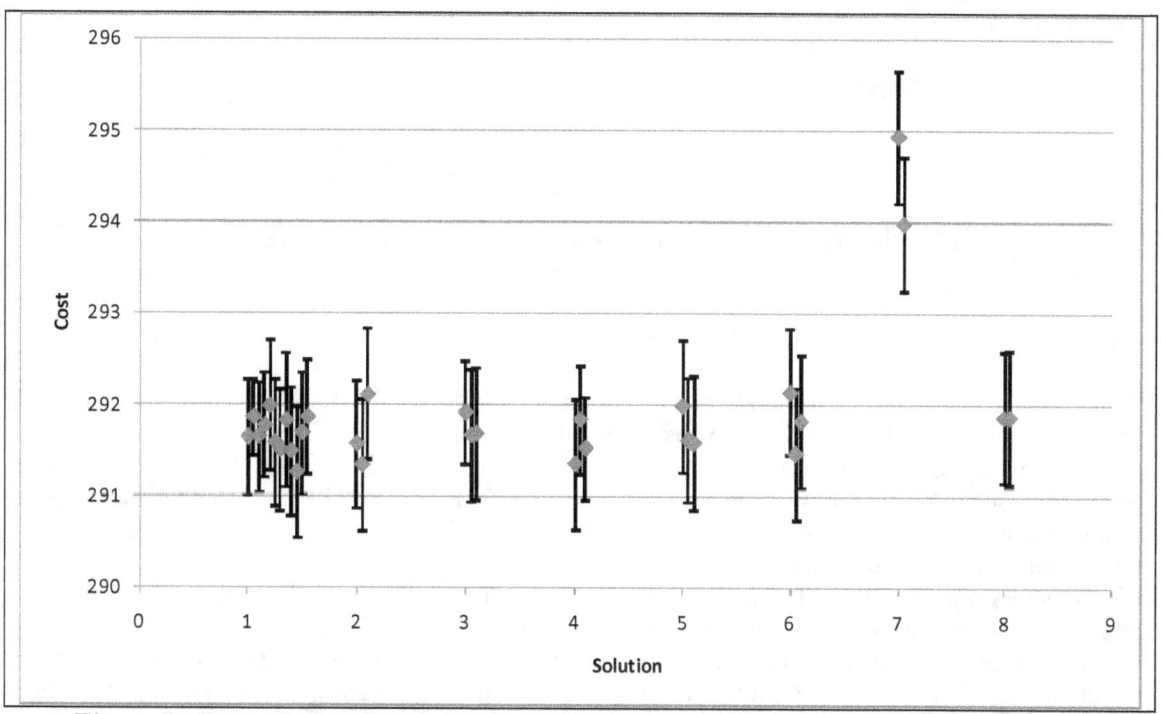

Figure 7 – Best solution dispersion for the test-case with their 95% confidence intervals

This solution corresponds to replacements scheduled at years (5, 12, 19, 26, 33). It is then used as the reference one (and called optimal solution) to evaluate the efficiency of each set of parameters of the design of experiment. This efficiency will be measured by counting the number of trials for which the algorithm finds this solution, and the number of trials for which this solution is present in the top ten solutions of the population.

The optimal replacement schedule without risk constraint is (8, 16, 24, 32). It was also obtained using a GASO. **Table 5** presents the valuation for both solutions with 10^6 replications. The estimated cumulative distribution function is shown in Figure 8: it clearly shows that the optimal constrained solution has a higher minimal cost, corresponding to the preventive replacements without any failure (five replacements instead of four), leading to a better control of the failure risk and the respect of the constraint.

	Replacement schedule (5, 12, 19, 26, 33)	Replacement schedule (8, 16, 24, 32)
Mean cost	291.9±0.5	284±0.5
Prob(C>800)	0.0567±0.0001	0.0719±0.0002

Table 5 – Evaluation of optimal schedules with and without constraint

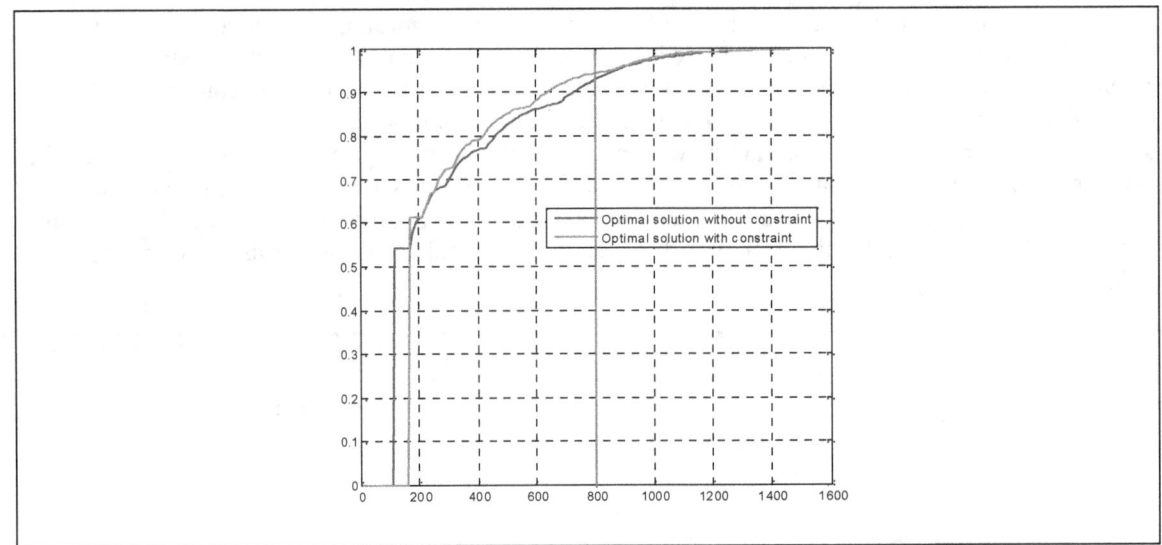

Figure 8 - Cumulative Distribution Function of the cost for both optimal strategies

4.4. Results

Table 6 and Table 7 present the number of successes of the algorithm to converge towards the optimal solution, out of ten trials. The first conclusion that can be made is that using the upper bound of the confidence interval is not efficient at all, and that it is better not to take into account the convergence of the simulations to compare two candidates in the selection step of the GASO. Using the mean value of the estimator presents a maximal frequency of success of 40% with N=100 and M=1000.

Low numbers of replications are not efficient because low convergence tends to create singular values of estimators very far from the real values. Such solutions will be at first considered as good candidates, dominating the population and expelling other good solutions, until its estimators start to converge towards its real value and is, in turn, expelled by a new solution. A detailed analysis of the evolution of such a case actually showed a cycle of the solutions, with specific solutions appearing and disappearing several times during the optimization.

As for high numbers of replications, they are not efficient because the maximal total number of replications awarded (50.10^6) is reached before the optimal termination criterion is met.

		Initial number of replications (N)		
		100	1000	10000
Enhancement number of replications (M)	10	0	3	0
	100	4	3	2
	1000	1	1	3

Table 6 – Efficiency of the algorithm to converge towards the optimal solution with the mean value as the mathematical order for selection mechanism

		Initial number of replications (N)		
		100	1000	10000
Enhancement number of replications (M)	10	0	0	0
	100	0	0	1
	1000	0	0	0

Table 7 – Efficiency of the algorithm to converge towards the optimal solution with the upper bound of the confidence interval as mathematical order for selection mechanism

Even if using the mean value, no matter the number of replications, as a selection indicator is better, it still gives mixed results. This can be explained by the fact that the dispersion of 1% considered for the termination criterion is not small enough to rank the solutions with no doubt because of overlapping confidence intervals, as illustrated in **Figure 7**. Narrowing the width of confidence intervals for the termination criterion would lead to computing durations making the calculation impractical. Another way to avoid the non-convergence of the GASO is to terminate the algorithm with the criterion defined in §4.2 and then to add replications to all solutions present at the last generation, without applying the GA mechanisms (static population), until the confidence intervals lengths drop down to 0.1%. The efficiency of the algorithm after this post-treatment is given in Table 8 and Table 9.

		Initial number of replications (N)		
		100	1000	10000
Enhancement number of replications (M)	10	1	6	7
	100	9	10	9
	1000	1	3	4

Table 8 – Efficiency of the algorithm to converge towards the optimal solution after post treatment with the mean value as the mathematical order for selection mechanism

		Initial number of replications (N)		
		100	1000	10000
Enhancement number of replications (M)	10	0	1	9
	100	0	1	7
	1000	0	0	0

Table 9 – Efficiency of the algorithm to converge towards the optimal solution after post treatment with the upper bound of the confidence interval as the mathematical order for selection mechanism

These last results confirm that using the upper bound of the confidence interval as an order for the selection mechanism is not as effective as using the mean estimator. If M=100 appears to be an optimal value, it is not as clear for N with an efficiency between 90% and 100%.

One of the criteria that could be used to decide between the different values of N would be the total number of replications during the GASO as computing duration is often at stake in this kind of problem (cumulative number of replications for all solutions simulated throughout all generations of the algorithm). Table 10 shows that choosing N=100 is one third less expensive with very close success frequency.

		Initial number of replications (N)		
		100	1000	10000
Enhancement number of replications (M)	10	5756012	21035744	50005685
	100	19190330	30544850	50006930
	1000	23068760	29321300	31433300

Table 10 – Average total number of replications (using the mean value as the order measure)

This work is just a first step, applying the GASO methodology on a very simple test-case, and it seems difficult to edict generic rules based on this unique example, although the following assumptions may be stated:

1. The fitness function, used to compare and select solutions in the GA, should take into account mean estimators without consideration of their convergence.
2. Number of replications should be chosen carefully as large values of M and N will lead to a very local exploration of the search space because of the replications budget limit applied to ensure the tractability of the method. On the opposite, small values of M and N will lead to a global search based on inaccurate estimators.
3. The termination criterion does not have to be too strict on simulations convergence as it can be improved after the GA has terminated.

It would be interesting to apply the GASO to a set of various problems in order to try to identify a relation between the parameters of the algorithm and some specific data, such as the order of magnitude of a random solution, its dispersion, the size of the space-search...

5. CONCLUSION

This paper presented a methodology to couple GA with Monte-Carlo simulation function when addressing Simulation Optimization issues in the field of engineering asset management. The test-case presented here proved the GASO to be an effective answer to realistic issues a system engineer or a business manager could face. If the method may be easily implemented in any EAM tool the success of a study will depend on the user skill to configure the calculations. As a matter of fact, even if the studied example showed some interesting results, such as the existence of an optimal number of replications for each simulation, or the efficiency of a post treatment of the best solutions to narrow the confidence intervals and then to finalize the optimization, it is not sufficient to edict generic rules, even empiric ones, to tune the parameters of a GASO.

The next step would be to apply this algorithm to different and more complex cases to confirm its efficiency and try to build a rule linking the parameters to the characteristics of the problem.

References

[1]   A comprehensive literature classification of simulation optimisation methods, Wak Hachicha et al., Munich Personal RePEc Archive (2010)

[2]   A hybrid simulation optimization method for production planning of dedicated remanufacturing , Jianzhi Li _, Miguel Gonzalez, Yun Zhu, International Journal of Production Economic, 117 (2009) 286–301

[3]   Multi-response simulation optimization using genetic algorithm within desirability function framework, Seyed Hamid Reza Pasandideh, Seyed Taghi Akhavan Niaki, Applied Mathematics and Computation 175 (2006) 366–382

[4]   Optimal computing budget allocation for Monte-Carlo simulation with application to product design. Chen, C. H., Yücesan, E., Lin, J., & Donohue, K., Simulation Modeling Practice and Theory 11 (2009), 57–74.

[5]   Evolutionary algorithm assisted by surrogate model in the framework of ordinal optimization and optimal computing budget allocation, Shih-Cheng Horng, Shin-Yeu Lin, Information Sciences, Volume 233, (2013), Pages 214-229

[6]   Condition-based maintenance optimization by means of genetic algorithms and Monte Carlo simulation, Marzio Marseguerra, Enrico Zio, Luca Podofillini, Reliability Engineering and System Safety 77 (2002) 151–166

[7]   Stock effects on exceptional maintenance tasks taking into account risks, J. LONCHAMPT et al., 2006 ASME Pressure Vessel and Piping Division Conference (2006).

[8]   Genetic algorithms in optimizing surveillance and maintenance of components, Munoz A, Martorell S, Serradell V., Reliability Engineering Systems Safety 57 (1997) 107–20.

[9]   Adaptation in Natural and Artificial Systems, J. H. Holland, University of Michigan Press, Ann Arbor, MI (1975)

[10]   Genetic Algorithms in Search, Optimization, and Machine Learning, David Goldberg, Addison-Wesley Professional, (1989)

[11]   On the use of genetic algorithm to optimize industrial assets lifecycle management under safety and budget constraints, J. Lonchampt and K. Fessart, Mathematics and Computation (2013)

A Usage-Informed Preventive Maintenance Policy to Optimize the Maintenance Free Operating Period for Multi-Component Systems

Romain Lesobre[*,a,b]**, Keomany Bouvard**[a]**, Christophe Bérenguer**[b]**, Anne Barros**[c]**, Vincent Cocquempot**[d]

[a] Volvo Group Trucks Technology, Advanced Technology and Research, 1 avenue Henri Germain, 69806 Saint Priest Cedex, France
[b] Laboratoire Grenoble Image Parole Signal Automatique, Gipsa-Lab, Grenoble INP, UMR 5216 CNRS, Domaine Universitaire, BP46, 38402 Saint Martin d'Hères, France
[c] Laboratoire de Modélisation et Sûreté des Systèmes, UTT, Institut Charles Delaunay, UMR 6279 CNRS, 12 rue Marie Curie, BP2060, 10010 Troyes Cedex, France
[d] Laboratoire d'Automatique, Génie Informatique et Signal, Université Lille1, UMR 8219 CNRS, 59655 Villeneuve d'Ascq Cedex, France

Abstract: This paper deals with the concept of Maintenance Free Operating Period (MFOP). This MFOP is defined as a period of operation during which the system should be able with a given level of confidence to carry out all its assigned missions without system fault or performance limitation. Based on this concept, a dynamic maintenance policy for a multi-component system is implemented. The main objective of this paper is to propose a method to integrate the usage information of the system components in order to optimize the implemented policy. The method is evaluated considering the Total Maintenance Cost (TMC) value.

Keywords: Maintenance, Reliability, MFOP, Usage

1. INTRODUCTION

Nowadays even if the vehicle configuration is important for any customer, the development of an efficient maintenance management system appears as another key of success. Aware of this opportunity, the commercial heavy vehicle industry propose, to its customers, service contracts in order to manage the vehicle maintenance.

These contracts are built from information on the vehicle configuration and on the estimation of vehicle operating conditions provided by the customer. Based on this information, a maintenance planning is created to inform the customer on the planned service operations during the maintenance contract period.

Currently the maintenance planning is static. It means that the maintenance intervals defined at the vehicle purchase date aren't updated during all the vehicle life. Moreover this planning is based on a component perspective in which the interactions at the system level are not taken into account. As a consequence, the total maintenance cost is impacted by unplanned maintenances generating high immobilization costs.

In this framework, a dynamic maintenance policy for a multi-component system integrating the possibilities offered by new information and communication technology solutions can be investigated. To increase the operational reliability of the system and decrease downtime and maintenance costs, a reliability based maintenance policy can be used. Note that most of the time, the optimization of these policies aims to define the best moments to perform maintenance tasks or inspections in order to find the best balance between preventive and corrective maintenance. The problem and the constraints are different in the heavy vehicle industry. Indeed the maintenance can be performed in a preventive way

---

[*] *romain.lesobre@volvo.com*

exclusively when the vehicle returns in the workshop. Out of these occasions, in period of operations, the maintenance is almost impossible or generates high immobilization costs. To overcome this problem, the developed maintenance policy should be able to ensure failure free operation on given period with a high confidence level and select maintenance operations to be performed during stops at the workshop.

Aware of this issue, the Royal Air Force proposed in 1996 the concept of the Maintenance Free Operating Period (MFOP) [1,2] with the objective to obtain better operational planning capability, improved operational availability and reduce running costs.

The main contribution of this paper is to propose a dynamic maintenance policy for a multi-component system. This policy, based on MFOP concept, integrates the usage information of the system components to optimize the total maintenance costs.

The remainder of the article is organized as follows. Section 2 defines the MFOP concept and the implemented maintenance policy. Section 3 illustrates the impact of usage information on the maintenance decision. Section 4 develops the use of mixture models to support the usage-informed maintenance policy. Section 5 deals with the total maintenance cost definition and the maintenance strategy optimization. The last section illustrates the method on a numerical example.

2. MAINTENANCE POLICY BASED ON MFOP CONCEPT

2.1. MFOP Concept Definition

The MFOP is defined as a period of operation during which the equipment must be able to carry out all its assigned missions without any maintenance action and without the operator being restricted in any way due to system faults or limitations [3]. The MFOP measure assumes that success is attainable and that failures can be accurately forecast [4].

According to its definition, the main objective is to avoid unplanned maintenance operations in moving all upcoming corrective maintenances to a schedule period of time of preventive maintenance. Based on this objective, the concept appears as a method to group maintenance operations at the end of MFOP (or cycle of MFOP) during stop at the workshop.

In [5], Tinga and al. argue that in this form of grouping, called time-driven clustering, the moment of maintenance is not driven by the failure of one of the components but must be planned carefully. Thereby, contrary to other forms like block replacement policy or opportunity-based maintenance, this clustering method could be very interesting for systems with high immobilization costs and where the number of maintenance opportunities is quite limited such as transport systems.

To ensure this MFOP, maintenance policies based on this concept have been introduced [6]. Nevertheless these policies are developed for single component system and do not integrate the possibility to take into account the available information on component usage in the maintenance decision process.

2.2. Dynamic Maintenance Policy

The dynamic maintenance policy implemented (see Fig. 1) consists in estimating at each end of MFOP or when a failure occurs, the probability that the multi-component system survives for the duration of MFOP given the available information [7].

If the reliability requirement is a MFOP of t_{MFOP} life units for the ith cycle of MFOP, this probability called Maintenance Free Operating Period Survivability ($MFOPS$) is given by:

$$MFOPS(t_{MFOP}) = \frac{R_{syst}(i * t_{MFOP})}{R_{syst}((i-1) * t_{MFOP})} \qquad (1)$$

where $R_{syst}(t_{MFOP})$ is the system reliability after t_{MFOP} life units.

Thereby if the $MFOPS$ at the end of a MFOP is higher than a specified confidence level, no maintenance operation is necessary and the system can be deployed for the next period. In the opposite case, where the $MFOPS$ is lower than a specified confidence level, maintenance occasion is needed to reach again the confidence level. Consequently, the $MFOPS$ allows to define if a maintenance occasion is needed or not.

Figure 1: Maintenance Policy Based on MFOP Concept

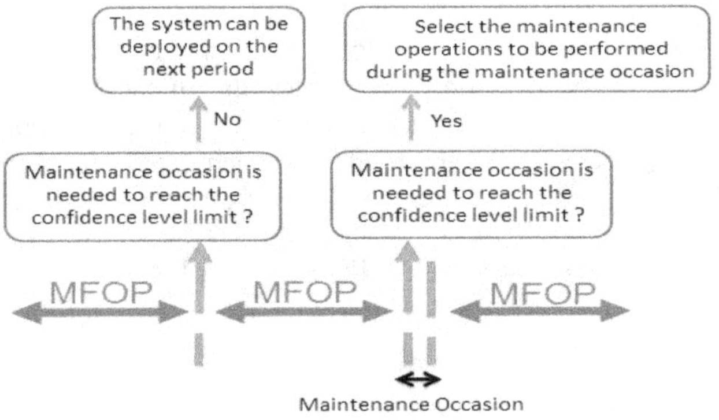

When a maintenance occasion is needed, a maintenance decision rule to select the maintenance operations to be performed during this occasion should be defined. In this paper a maintenance decision rule based on the cost minimization on the MFOP horizon is introduced [8]. In this case the problem can be mathematically formulated as follows:

$$\min_{\{x_i\}} \sum_{i=1}^{n} x_i * C_i \qquad (2)$$

$$s.t \; MFOPS > CL$$

where n is the number of system components, C_i is the operation cost including labor and spare part cost of component i, x_i is a binary variable which indicates the selection of a maintenance operation on the component i and CL is the specified confidence level. Further the following assumptions are made to solve this optimization problem. Assumption 1: After each maintenance operation where one or several components are replaced, their reliability performances are considered "as good as new". Assumption 2: The reliability performance of the other components is considered unchanged or "as bad as old".

The interesting feature of the $MFOPS$ is its update with the reliability of the components at the end of each period. Based on this feature, the uncertainty of the $MFOPS$ strongly depends on the available monitoring information. In a previous paper [8], the impact of different information levels on the components state has been illustrated. The impact of usage information will be investigated in the next section.

3. IMPACT OF USAGE INFORMATION ON THE MAINTENANCE DECISION

3.1. Lifetime Models and Usage Information

According to the operating condition in which a component is used, its degradation mechanism will be different, more or less variable. Note that disregarding this usage information, especially for systems which operate in variable conditions, generates a large uncertainty in the lifetime models and an efficiency decrease of the maintenance policy [9].

Figure 2: Usage Uncertainty and the Consequences on the Lifetime Models

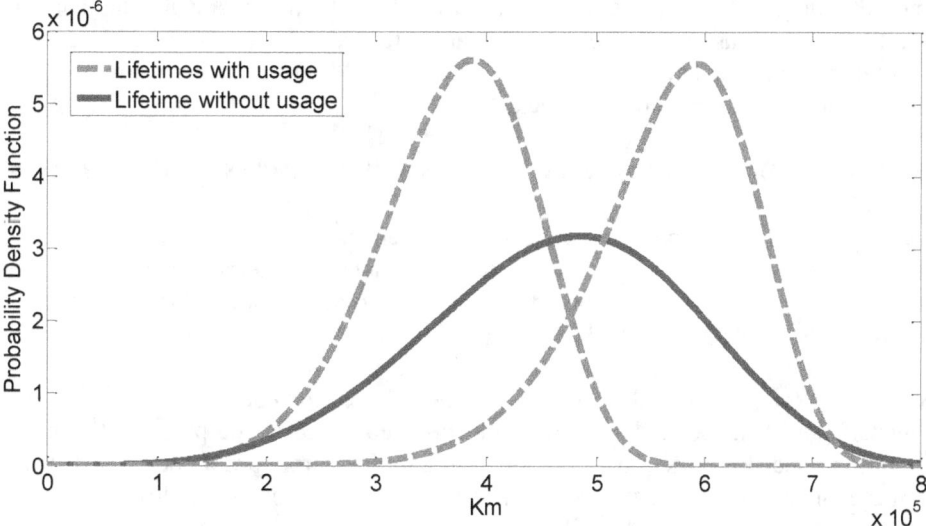

The lifetime models are obtained from Volvo databases where maintenance and repair events have been recorded on a per-vehicle basis. These models are built per component for a given vehicle range and purchase year. In these models a significant statistical variance appears that makes them economically inefficient and unprofitable on larger scales.

The recorded failures come from components used in variable usage conditions. Currently in the databases, no direct link is available to correlate the usage information and the failure date. In this framework, considering a unique lifetime model for the component datasets could be inappropriate and can explain the significant variance. To avoid this kind of problems, Tinga [9] mentions that removing the uncertainty in usage reduces the width of distributions and increases the reliability models accuracy (see Fig. 2). The proposed method to define the various lifetime models correlated with the component usage will be presented in the next section.

3.2. Information Levels Definition

In order to demonstrate the impact of the usage information on the maintenance policy, three information levels are considered per component (see Tab. 1). For the first information level, no usage information is available. For the components under this information level, a unimodal lifetime model will be considered.

Table 1: Information Levels Definition

Information Level	Usage Information	Component Lifetime Models
Level 1	No available information	Unimodal lifetime model
Level 2	Past usage information	Lifetime model selected according to the usage information
Level 3	Past usage information + Future usage estimation on the next MFOP	Lifetime model selected according to the usage information

For the second information level, it is assumed that different lifetime models are defined per component according to the operating conditions and that the usage information is known in a continuous way. Thereby the selection of component lifetime model may be updated at the end of each MFOP cycle thanks to the exact knowledge on the past operating conditions.

For the last level, the previous assumptions made for the second level are considered and additionally it is assumed that the predicted usage information for the next MFOP is available. This prevision of future operating conditions can be obtained thanks to close relationship with customers. The component lifetime model can be updated at the end of each MFOP cycle according to the knowledge on the past operating conditions and on the next MFOP.

4. MIXTURE MODELS TO CONNECT USAGES AND COMPONENT LIFETIMES

As mentioned previously, the main objective is to implement a usage-informed maintenance policy which selects, at the end of each MFOP cycle, the component lifetime models according to the usage information and uses it in the maintenance decision rule. To achieve this objective, it is necessary to be able to connect the usage profiles and the component life consumption [5].

In this section, an experience-based method is proposed in order to define this connection. Note that the component failure behavior is based on failure data collected in the past. Under the assumption that the operating conditions affect the component reliability, a mixture models identification can be applied on these failure data to identify the lifetime models conditional to the different usages.

4.1. Mixture Model Definition

Considering a mixture of lifetime distributions consists of assuming that failure data come from several sub-populations. Each sub-population can be modeled, in a separate manner, by a unique lifetime distribution. The total dataset is thus a mixture of these sub-populations. Each sub-population is assumed to represent a type of usage for the specified component.

Zaman and al. [10] explained that a mixture model of distributions is a weighted average of probability distributions with positive weights that sum to one. The density function of mixture distribution is given by:

$$g(x) = \sum_{j=1}^{k} w_j f_{j(x)} \tag{3}$$

where k represents the assumed number of sub-population in the mixture under study, w_j is the proportion of the jth sub-population in the mixture and $f_{j(x)}$ the density function of the jth sub-population.

4.2. Parameter Estimation

According to Razali and al. [11], a number of methods for estimating the parameters for mixture distribution represented by Eq. 3 including Maximum likelihood (MLE), moment method, Bayesian method and least square method can be investigated. Currently, MLE became more popular for parameter estimation in mixture model. Therefore in this paper, the MLE method will be considered.

The MLE parameter estimators can be obtained by finding the log likelihood function of Eq. 3 as follows:

$$LL = \sum_{l=1}^{m} \log(\sum_{j=1}^{k} w_j f_{j(x_l)}) \tag{4}$$

where m represents the number of data observed in the total population. The maximum of Eq. 4 can be obtained by taking the first derivative of LL with respect to all parameters and set it to be zero [12].

To solve this mathematical problem, the most popular method is the Expectation-Maximization algorithm (EM algorithm). In practice, this method needs a parameters initialization to start the iterative process. In this paper, the K-means method will be considered to increase the EM-algorithm efficiency [13]. This method aims to divide the initial population into K clusters in which each observation belongs to the cluster with the nearest mean.

The following process allows the parameters estimation for mixture distribution assuming a given number of sub-populations. Nevertheless, this sub-populations number is most of the time unknown. Only expert statements can be used to define the maximum possible number of sub-populations. Thereby a criterion based on the coefficient of determination R^2 will be used to determine the best number of sub-populations for a specified mixture distribution. R^2 is given by:

$$R^2 = 1 - \frac{\sum_{l=1}^{m}(P_l - \hat{P}_l)^2}{\sum_{l=1}^{m}(P_l - \bar{P})^2} \tag{5}$$

where m represents the number of data observed in the total population, P_l are the values of relative frequencies of observed data, $\hat{P}_l$ are the forecast values using the mixture distribution function and $\bar{P}$ is the mean of relative frequencies of observed data. Usually the coefficient of determination R^2, which is a measure of goodness fit, increases with the number of sub-populations. In order to avoid selecting always the maximum number of sub-population, a threshold equals to 0.99 is introduced on the R^2 value from which the number of sub-populations is validated.

4.3. Allocation Method

According to the methods presented in the two previous sub-sections, the number of sub-populations and the lifetime models for each sub-population can be determined from the initial mixed dataset. Then from these results, an allocation method should be implemented in order to classify the initial failure time values in each sub-population.

The natural idea is to allocate the failure data in the sub-population from which it is most likely to be seen from the observed value and characteristics of sub-populations. The probability that the observed failure a belongs to the sub-population k given the value of x_a is given by:

$$\tau_{ak} = \frac{w_k f_k(x_a)}{\sum_{j=1}^{k} w_j f_j(x_a)} \tag{6}$$

Based on this probability, a maximum a posteriori classification method can be used. This allocation method imposes to compute for each observed failure a the probabilities τ_{ak} relative to each sub-population and to allocate the failure in the sub-population with the maximum probability τ_{ak}.

As mentioned previously, no direct link is available in the current databases to correlate the usage information and the failure date. In this framework, this allocation method is able to cluster the initial failures in various sub-populations facilitating the highlighting of covariates explaining the emergence of these sub-populations. These covariates could be determined thanks to vehicle signals analysis for the different sub-populations. Thereby with the monitoring of these covariates, the current operating conditions can be correlated with the identified sub-populations and the lifetime model can be updated. Note that in this paper the covariates will be assumed to be known.

5. MAINTENANCE STRATEGY OPTIMIZATON BASED ON TOTAL MAINTENANCE COST

In order to evaluate the alternative maintenance strategies and to optimize the usage-informed maintenance policy based on MFOP concept, the Total Maintenance Cost (TMC) could be evaluated over five years which represents the nominal contract duration.

The TMC is expressed as:

$$TMC = C_{repl} + C_{cor} + C_{diag} \qquad (7)$$

where C_{repl} is the replacement cost, C_{cor} is the corrective cost and C_{diag} is the diagnosis cost. The C_{repl} can be defined as:

$$C_{repl} = \sum_{i=1}^{n} C_i * (N_{i,prev} + N_{i,cor}) + C_{setup} * N_{MS} \qquad (8)$$

where n is the number of system components, C_i is the operation cost including labor and spare part cost of component i, $N_{i,prev}$ is the number of replacements of component i during a system preventive stop, $N_{i,cor}$ is the number of replacements of component i during a system corrective stop, C_{setup} is the setup cost and N_{MS} is the total number of maintenance stops.

Then the C_{cor} is given by:

$$C_{cor} = \sum_{i=1}^{n} D_i * N_{i,cor} * \tau_{immo} + \left((D_{setup} + D_{tow}) * \tau_{immo} * N_{sfailure} \right) + (C_{tow} * N_{sfailure}) \qquad (9)$$

where D_i is the replacement duration of component i in hour, τ_{immo} is the hourly rate for a system immobilization, D_{setup} is the setup activities duration, D_{tow} is the tow duration, $N_{sfailure}$ is the number of system failures and C_{tow} is the tow cost. Thereby a failure at the system level is considered to impact the customer by the tow cost but also by the total stop duration which leads to a loss of production.

Finally the C_{diag} is expressed as:

$$C_{diag} = (C_{udiag} * n * N_{sfailure}) + (D_{udiag} * n * N_{sfailure} * \tau_{immo}) \qquad (10)$$

where C_{udiag} is the unitary diagnosis cost and D_{udiag} is the unitary diagnosis duration. Indeed when the system failed, a diagnosis for each system components is considered as mandatory to repair the system.

By Monte Carlo simulation, various maintenance strategies can be examined. The optimal solution is the strategy corresponding to the lowest value of TMC.

6. NUMERICAL EXAMPLE

6.1. Initial Database Implementation and Mixture Models Application

In the real databases of the company, no direct link is currently available to correlate the usage information and the failure date. To overcome this problem, a simulated database is built to be able to highlight the covariates effects responsible of the possible sub-populations.

In this sub-section, the aim is to define a method to build an initial failure database per component. The mixture models method will then be applied on these failure databases in order to connect lifetime models and assumed usages. Note that the way to build the initial failure databases is totally independent of the mixture models method.

Consider a deteriorating component subject to a failure mechanism due to an excessive deterioration level L. A Gamma process is considered to describe its evolution. Assume that the component operates in variable conditions and consider that the operating environment influences the speed and the variance of the degradation process [14].

In order to build the failure database relative to each component, assume that the component operates under a two-stages environment: "normal" and "stressed", and the deterioration follows a homogeneous Gamma process for each of the environment states (see Fig. 3). Let $(\alpha_n.t, \beta_n)$ and $(\alpha_s.t, \beta_s)$ denote the couples of parameters respectively for the "normal" and "stressed" environments [15]. Note that these couples of parameters will be different for each considered component.

Figure 3: Degradation Process in a Dynamic Environment

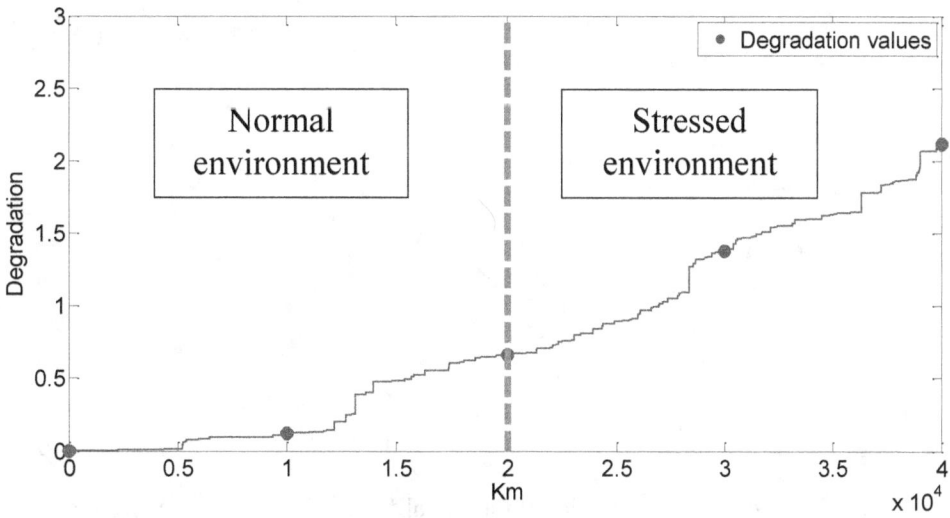

Consider the component life as a succession of running period during which its environment evolves between the "normal" and "stressed" state. A Normal distribution $N(10000, 1000)$ will be defined to simulate each running period length in kilometer. A probability of being in a "stressed" state between 0% and 60%, considered as the minimum and the maximum threshold, is affected at each component history. This probability will be used at each running period to define the environment state.

Figure 4: Histogram of Initial Failures Database

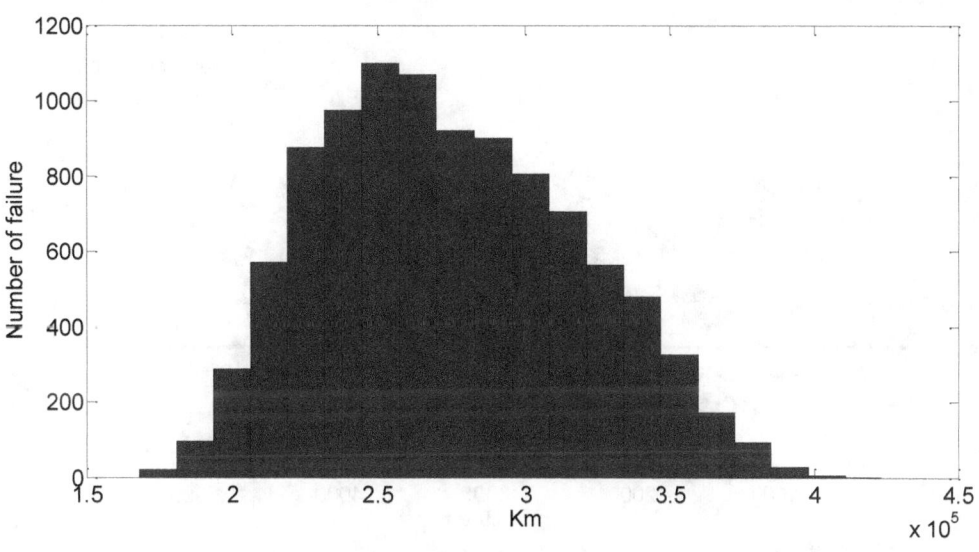

Based on this process, 10000 histories are simulated for each system component and for each history the failure date in kilometer and the kilometer ratio in the "stressed" environment are computed. Note that the kilometer ratio in the "stressed" environment is defined as the covariate. Further the Gamma processes used to build the initial failure database per component will be thereafter assumed unknown.

To illustrate the mixture models application assume that a component follows a Gamma process $Ga(7e^{-4} * t, 20)$ in a "normal" environment and $Ga(1.4e^{-3} * t, 20)$ in a "stressed" environment and that the degradation threshold is fixed at $L = 12$. The histogram given in Fig. 4 represents the obtained failure database for a component according to the implemented methodology.

Figure 5: Weibull Lifetime Models for $k = 1$ and $k = 2$ Sub-Populations

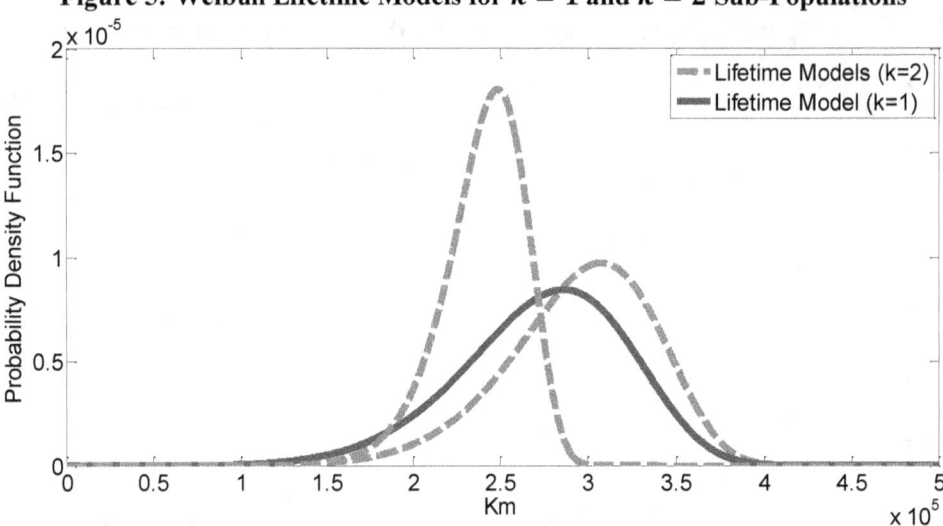

The main objective consists then to determine if the initial failure database comes from one or several sub-populations. The first step is to consider only the case with $k = 1$ sub-population. A Weibull distribution is used to model the component lifetime. Note that the Weibull distribution is the most widely used distribution for modeling failure datasets. For this first step, the Weibull lifetime model $W(2.9e5, 6.7)$ is obtained and the measure of goodness gives $R^2_{k=1} = 0.984$. In general this value is sufficient to validate the unique model nevertheless the variance obtained in this case is very width (see Fig. 5) and the use of mixture models seems to be appropriate.

Figure 6: Initial Failure Allocation and Limit Definition on the Covariate

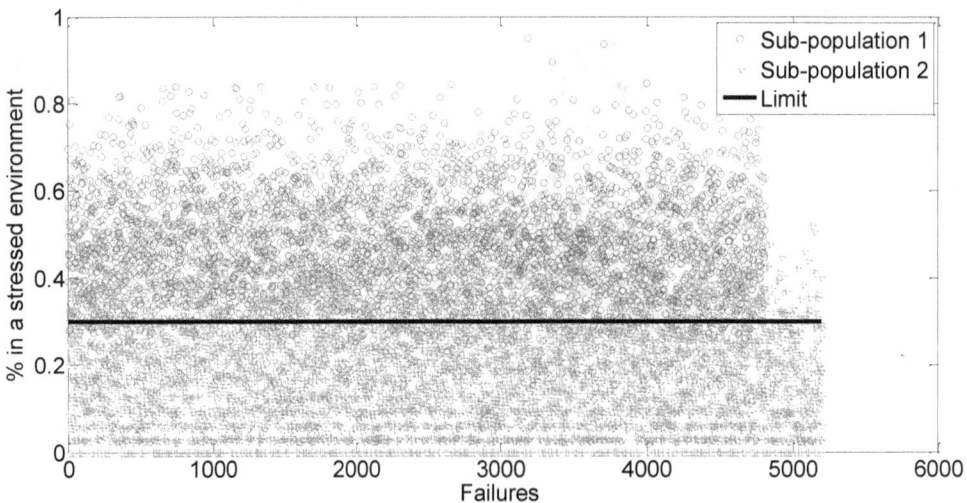

The second step is to consider the case with $k = 2$ sub-populations. In applying the defined process, the Weibull lifetime models $W(2.5e5, 12.2)$ and $W(3.1e5, 8.2)$ are obtained and the measure of goodness of fit gives $R^2_{k=2} = 0.999$. According to the rule previously mentioned, $R^2_{k=2} > 0.99$ thus the case with $k = 2$ sub-populations is validated.

Once the parameters and the number of sub-populations are defined for the mixture model, the next step is to allocate the initial failure data in each sub-population thanks to the maximum a posteriori classification method. Assume that the first sub-population represents the model $W(2.5e5, 12.2)$ and that the second sub-population represents the model $W(3.1e5, 8.2)$.

In the Fig. 6, each classified failure is associated with the kilometer ratio in the "stressed" environment. This measure defined as the covariate is assumed to be able to explain the emergence of these two sub-populations. To select, at the end of each MFOP cycle, the lifetime model according to its operating conditions, a limit to distinguish the two sub-populations is fixed on this covariate. To determine this limit, the aim is to minimize the allocation errors on the total population. Based on this requirement, the limit of 30% is fixed. Thereby at the end of each MFOP cycle, if the kilometer ratio in the "stressed" environment is inferior at 30% for the specified component, the lifetime model $W(3.1e5, 8.2)$ is selected and in the other case $W(2.5e5, 12.2)$.

6.2. System Definition

Figure 7: System Structure Definition

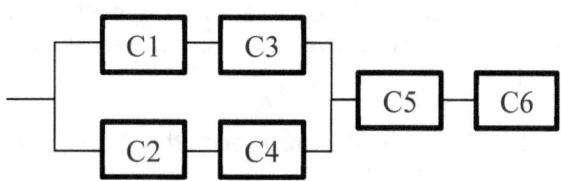

In order to illustrate the usage-informed preventive maintenance policy based on MFOP concept, the following multi-component system is defined (see Fig. 7).

For this system, the hourly rate for a system immobilization is fixed at $\tau_{immo} = 100€$, the unitary diagnosis cost and duration are respectively fixed at $C_{udiag} = 20€$ and $D_{udiag} = 5min$, the tow cost and duration are respectively fixed at $C_{tow} = 1\ 500€$ and $D_{tow} = 5h$ and finally the setup cost and duration are respectively fixed at $C_{setup} = 100€$ and $D_{setup} = 30min$.

Table 2: System Parameters

	C1	C2	C3	C4	C5	C6
Gamma processes	G(7e-4,20) G(1.4e-3,20) L=12	G(7e-4,20) G(1.4e-3,20) L=12	G(8e-4,20) G(1.6e-3,20) L=12	G(8e-4,20) G(1.6e-3,20) L=12	G(6e-4,20) G(1e-3,20) L=11	G(6.4e-4,20) G(1.4e-3,20) L=11
Lifetime models and limit $k = 2$	W(2.5e5,12.2) W(3.1e5,8.2) Limit = 30%	W(2.5e5,12.2) W(3.1e5,8.2) Limit = 30%	W(2.2e5,12.1) W(2.7e5,8.2) Limit = 31%	W(2.2e5,12.1) W(2.7e5,8.2) Limit = 31%	W(2.9e5,15.1) W(3.5e5,9.9) Limit = 32%	W(2.4e5,10.8) W(3.1e5,7.4) Limit = 31%
Lifetime model $k = 1$	W(2.9e5,6.7)	W(2.9e5,6.7)	W(2.6e5,6.7)	W(2.6e5,6.7)	W(3.3e5,8.6)	W(2.9e5,5.9)
C_i	458	458	407	407	842	1268
D_i	1	1	1.5	1.5	3.7	1.8

Tab. 2 describes the Gamma processes used to build the initial failure database per component, the lifetime models obtained for $k = 1$ and $k = 2$ sub-populations as well as the defined limit on the covariate and the specific maintenance cost and duration per component. Note that for each system component, the mixture model for $k = 2$ sub-populations has been validated.

6.3. Cost-Optimized Maintenance Policy Based on MFOP and Usage Information

A maintenance model is developed in order to calculate the TMC index over five years based on Monte Carlo simulation. Assume that, first at the end of each MFOP cycle, only the information at the system level is available and the components state inside the system is unknown. No implementation cost, including for example sensors costs or technology solutions costs, are considered.

In order to demonstrate the impact of the usage information on the TMC, the computation of TMC is realized for the same MFOP and confidence level values in integrating the first, the second and the third information level on usage for each system component.

Firstly, the Fig. 8 represents the TMC index for different MFOP between 20000 km and 60000 km by step of 10000 km and confidence levels between 75% and 95% by step of 5% when no usage information is available for each system component. In this case, the TMC is minimal when MFOP and the confidence level are respectively equal to 60 000 km and 90%. This cost-optimized solution provides the best balance between corrective and preventive maintenance operations.

Note that for some configurations, the TMC increases with the confidence level. This behavior is explained by the fact that the additional preventive maintenance cost can be higher than the gain saved by the immobilization costs reduction.

Figure 8: The TMC (Euro) with Information Level 1 on Usage

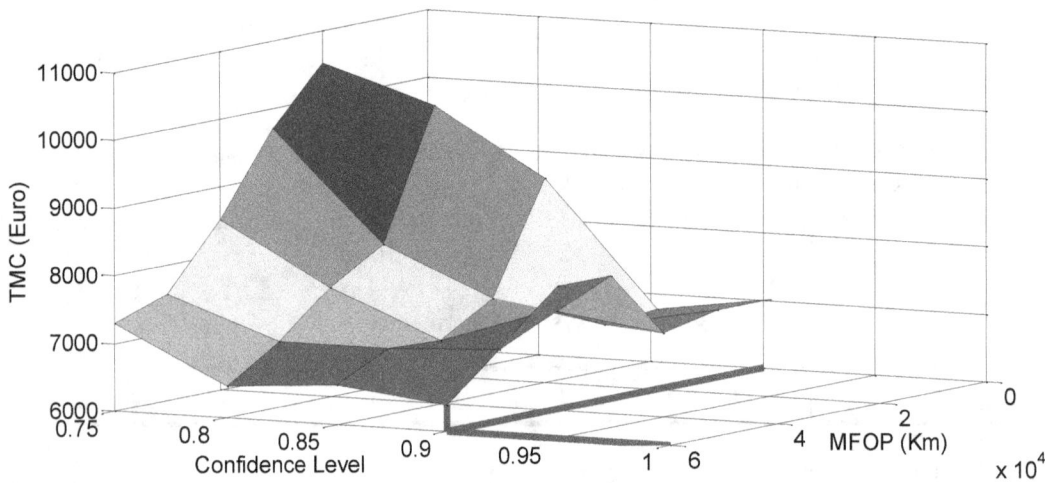

Secondly a comparison between the three information levels is performed in Tab. 3. These results justify the positive impact created by the increase of information level on usage. If the second information level is implemented for each system component, the saved cost in considering each TMC value is in average of 15.6% comparatively with the first level. For the third information level the saved cost is in average of 16.2% comparatively with the first level.

Table 3: TMC with the Three Information Levels on Usage

Information Level	Optimal TMC (Euro)	Saved costs on the optimal TMC comparatively with Level 1 (%)	Mean Saved Costs comparatively with Level 1 (%)
Level 1	6387€	-	-
Level 2	5808€	9.1%	15.6%
Level 3	5724€	10.4%	16.2%

Note that the cost-optimized solution decreases of 9.1% with the second information level and of 10.4% with the third information level. Thereby this example illustrates how the usage information can be used to optimize the dynamic maintenance policy based on MFOP concept.

7. CONCLUSION

In this article, a usage-informed preventive maintenance policy based on MFOP concept has been proposed. This dynamic maintenance policy is able to take into account, at the end of each MFOP, the usage information of each system component to update the maintenance decision process. The connection between the usage information and the component life consumption is performed thanks to an experience-based method named mixture models. The alternative maintenance strategies in considering various information levels on usage are evaluated based on *TMC* value. The results presented on a specified system allow illustrating the positive impact of usage information to define the cost-optimized maintenance strategy.

References

[1]   U. D. Kumar, J. Knezevic and J. Crocker. *"Maintenance Free Operating Period – an alternative measure to MTBF and failure rate for specifying reliability"*, Reliability Engineering & System Safety, 64, pp. 127-131, (1999).
[2]   M. Brown and C. J. Hockley. *"Cost of specifying maintenance/failure free operating periods for the Royal Air Force aircraft"*, IEEE Reliability and Maintainability Symposium, pp. 425-432, (2001).
[3]   C. J. Hockley. *"Design for success"*, Proceedings of the Institution of Mechanical Engineers Part G: Journal of Aerospace Engineering, 212, pp. 371-378, (1998).
[4]   A. A. Shaalane and P. J. Vlok. *"Application of the aviation derived maintenance free operating period concept in the South African mining industry"*, South African Journal of Industrial Engineering, 24, pp. 150-165, (2013).
[5]   T. Tinga and R. H. P. Janssen. *"The interplay between deployment and optimal maintenance intervals for complex multi-component systems"*, Proceedings of the Institution of Mechanical Engineers Part O: Journal of Risk and Reliability, pp. 1-14, (2013).
[6]   J. Long, R. A. Shenoi and W. Jiang. *"A reliability-centred maintenance strategy based on maintenance free operating period philosophy and total lifetime operating cost analysis"*, Proceedings of the Institution of Mechanical Engineers Part G: Journal of Aerospace Engineering, 223, pp. 711-719, (2009).
[7]   U. D. Kumar. *"New trends in aircraft reliability and maintenance measures"*, Journal of Quality in Maintenance Engineering, 5, pp. 287-295, (1999).
[8]   R. Lesobre, K. Bouvard, C. Bérenguer, A. Barros and V. Cocquempot. *"A maintenance free operating period for a multi-component system with different information levels on the components state"*, Chemical Engineering Transactions, 33, pp. 1051-1056, (2013).
[9]   T. Tinga. *"Application of physical failure models to enable usage and load based maintenance"*, Reliability Engineering and System Safety, 95, pp. 1061-1075, (2010).
[10] M. R. Zaman, M. K. Roy and N. Akhter. *"Chi square mixture of gamma distribution"*, Journal of Applied Sciences, 5, pp. 1632-1635, (2005).
[11] A. M. Razali and A. A. Al Wakeel. *"Mixture Weibull distributions for fitting failure times data"*, Applied Mathematics and Computation, 219, pp. 11358-11364, (2013).
[12] W. L. Hung, Y. C. Chang and S. C. Chuang. *"Fuzzy classification maximum likelihood algorithms for mixed-Weibull distributions"*, Soft Computing, 12, pp. 1013-1018, (2008).
[13] E. Lebarbier and T. Mary-Huard. *"Polycopié de classification non supervisée"*, AgroParisTech, 2008.
[14] K. Huynh. *"Quantification de l'apport d'information de surveillance dans la prise de décision de maintenance"*, Thèse de doctorat, Université de technologie de Troyes, France, 2001.
[15] E. Khoury, E. Deloux, A. Grall and C. Bérenguer. *"On the Use of Time-Limited Information for Maintenance Decision Support: A Predictive Approach under Maintenance Constraint"*, Mathematical Problems in Engineering, 2013, pp. 1-11, (2013).

Model of improvement of maintenance policies for electrical substations

Cristiano Cavalcante*[*a], Marcelo Alencar[a], Adiel Almeida[a], Ana Paula Costa[a], Rodrigo Ferreira[a], Maxwell Luna[b], Rogério Sá[b], Alison Ferreira[b] and Adilson Vieira[b]

[a] UFPE - Universidade Federal de Pernambuco, Recife, Brazil
[b] CELPE - Companhia Energética de Pernambuco, Recife, Brazil

Abstract: We observe that in electrical substations, issues often arise that directly influence the requirements for maintenance actions to be adequate. Maintenance policies are sometimes inappropriate because the aging of assets has been incorrectly evaluated, because technological upgrades are not properly reflected in maintenance plans, or because the operational regime is not taken into account. Thus, once the need for adjustments because of the presence of one or more of the issues mentioned above has been identified, it is essential that a different systematic be implemented to achieve the expected performance of the affected substation. Accordingly, this article proposes a model for establishing adequate maintenance policies to produce more effective results, taking into account not only the possible consequences of failure to which the system under study is subject but also the various specific concerns associated with the performance indices of the electricity system. A real electrical substation is used as a pilot system.

Keywords: Maintenance effectiveness, multi-criteria decision-making, additive veto model, maintenance of electrical substations

1. INTRODUCTION

According to [1], electrical systems suffer from wear and tear, causing them to deteriorate over time. This often leads to failures that can interrupt the power supply. As a result, the absence of appropriate maintenance planning inevitably leads to economic losses and unnecessary downtime.

High costs are incurred by such failures of planning in two different aspects: First, there may be failures caused by the insufficient implementation of preventive actions, which may have serious consequences in the form of prolonged interruptions or damage of another nature as a result of equipment failure. On the other hand, it is possible to perform excessive maintenance, which generates high maintenance costs and may lead to damage or malfunction caused by errors made at the time of maintenance. Thus, in either extreme, serious consequences may result from inadequate maintenance planning.

Therefore, effective maintenance planning should be conducted to ensure that the maintenance actions to be performed are truly the most effective and that the available resources are used rationally. For this purpose, the development of a maintenance plan should be guided by reliable methods of identifying effective and ineffective maintenance actions to achieve continuous improvement through the identification of ineffective maintenance actions and the replacement of such actions with new, effective ones [2].

---

* cristianogesm@gmail.com

2. THE MULTI-CRITERIA MODEL FOR THE SET OF ACTIONS TO BE IMPROVED

Multiple-Criteria Decision Making (MCDM) is a set of methods and techniques that has been developed to support individuals and organizations in resolving decision-making problems [3]. According to [4], multi-criteria decision-making analysis structures and analyzes complex decisions in which several criteria must be considered, some of which conflict with each other. In this context, [5] argues that MCDM offers a broad variety of tools that support the decision maker in solving problems while considering different points of view, often ones that are contradictory and heterogeneous.

Multi-criteria decision models have found diverse applications in fields such as water resources ([6]), multi-criteria risk analysis ([7], [8]) and maintenance. As an illustration a brief description of various maintenance problems from a multi-criteria perspective is presented. [9], [10], and [11] use multi-criteria models to address repair and outsourcing contracts. [12] presents a preventive-maintenance decision model for addressing conflicting criteria that takes into account the decision maker's preferences. [13] proposes a multi-criteria decision model concerning inspection intervals for condition monitoring and the decision maker's preferences regarding downtime and cost.

In this context, this paper proposes a multi-criteria decision model to support maintenance planning by allowing actions that are not contributing to adequate equipment performance to be identified by a prioritization model based on a multi-criteria decision method.

In fact, it is plausible to incorporate the desires of the maintenance manager into the organizational culture to create an ongoing process of reviewing proposed maintenance plans to correct actions that are not effectively contributing to good system performance, thus ensuring the best possible use of maintenance resources.

An ineffective action is one that is problematic in at least one of the following aspects: (1) it may be not feasible, (2) it may be too difficult to perform, (3) it may be likely to cause damage or malfunction, (4) it may not be procedurally appropriate, (5) the time investment required for the action may not be commensurate with the degradation of the equipment, and so on.

It is evident that an ineffective action could be identified as ineffective for any of a number of different reasons, as elucidated above, so it is possible for such a set of actions to have very different characteristics. Therefore, to identify a critical set of actions, it is necessary to define an appropriate approach to address this diversity of potential problems.

It is not a trivial task to distinguish effective maintenance actions from those that are not effective, and such an evaluation involves analysis from multiple perspectives. Thus, a multi-criteria decision model is developed here to support the selection of the critical set of actions that should be revised for the next iteration of a maintenance plan.

The construction of a multi-criteria model relies on certain steps related to the specific multi-criteria method that was chosen. In this article, we propose a model based on the compensatory method with veto, which was proposed in [14]. According to [14], the multi-attribute or multi-criteria decision methods that are most commonly found in the literature are additive methods. The reason for this preference is related to the very intuitive approach that is taken in the aggregation step of these methods. As was suitably noted by this author, despite the popularity of these methods, there are some situations in which the DM is not willing to select a particular alternative to compensate for a criterion whose performance is below a certain level. In such a case, a veto function should be invoked to avoid the selection of such alternatives.

Thus, the construction of the model proposed here consists of 3 distinct phases:

The first step consists of the definition of the set of alternatives. The alternatives are actions associated with a piece of equipment that fulfill at least one of the following conditions: (1) the action is specific to a piece of very critical equipment, (2) factors can be clearly identified that indicate the inefficiency of the action, or (3) there were changes in the system such that adjustments of the maintenance plans associated with this action are required.

The second step consists of the definition of the criteria set. Because this problem is intrinsically related to the measurement of the effectiveness of maintenance actions, the criteria set defined by the decision-maker should reflect his own philosophy regarding the efficiency of maintenance actions. As noted in several different papers (see [15]), the evaluation of the efficiency of maintenance action is not a trivial task. Thus, some criteria will inevitably be subjective. In this case, we invoke the concept of constructed attributes, as defined in [16]. According to [16], unlike natural attributes, which are simultaneously appropriate to a variety of contexts, a constructed attribute is developed for a given decision context. We discuss attributes in the next section, where all details regarding the aspects involved in the decision-making process are described.

In the third and final phase, the preferences of the decision-maker, the actions and the attributes are organized to produce a decision through the aggregation process. Here, as mentioned above, the third phase consists of following all steps that constitute the additive veto method [14].
In the additive model, we consider that the overall evaluation of alternatives $v(a)$ is the result of the additive aggregation of each criterion $v_i(a)$. Thus, the global value of alternative a is expressed by equation (1). The additive sum is a common approach to aggregating various aspects in the process of multi-criteria evaluation [17]. However, the inconvenience of unlimited compensation sometimes must be addressed, as stated by [14]. Equation 3 provides an alternative to handling this type of situation.

$$v(a) = \sum_{i=1}^{n} v_i(a)k_i \tag{1}$$

According to [14], for the ranking problem, the decision-maker is not interested in rejecting any alternative outright, but he is willing to reject the positions of certain alternatives in the ranking process. For this purpose, the method presented in [14] uses the following expression for the weighted veto function:

$$r_i(a) = z_i(a)k_i \tag{2}$$

where

$z_i(a)$ corresponds to the veto function for criterion i
and the k_i is the weight of criterion i.
The veto function for each criterion i ($z_i(a)$) is as follows:

$$z_i(a) = \begin{cases} 0, & \text{if } v_i(a) \leq l_i \\ 1, & \text{if } v_i(a) \geq u_i \\ \frac{v_i(a) - l_i}{u_i - l_i} & \text{if } l_i < v_i(a) < u_i \end{cases} \tag{3}$$

Finally, for the veto be taken into account in a general manner, the function $r_i(a)$ should be summed for all criteria. Thus, let us consider a specific alternative a; the overall role of the veto (the veto index of the alternative a) in a ranking problem is represented by the following expression:

$$r(a) = \sum_{i=1}^{n} r_i(a) \tag{4}$$

This veto index is introduced into equation 1 in such way that the analysis will account for the effect of the veto.

$$v(a) = r(a) \sum_{i=1}^{n} v_i(a)k_i \tag{5}$$

In the next section, we define all parameters of the model. We also describe the application of the model.

3. A CASE STUDY IN A REAL ELETRICAL SUBSTATION

Substations may be associated with transmission or distribution depending on the level of their operating voltages. Transmission substations operate at voltage levels of 230 kV and above. Substations with operating voltages below 230 kV are designated as distribution substations, which is the case for the particular substation under study.

Substations may also be classified based on the installation of their equipment in relation to the environment. Thus, there are the following categories: (1) external or outdoor substations and (2) internal or sheltered substations.

External or outdoor substations are those whose equipment is installed without any protection against the weather and is subject to unfavorable atmospheric conditions of temperature, rain, pollution, wind, etc. These conditions affect the wear of component materials and reduce the effectiveness of insulation, and the equipment at such facilities therefore requires more frequent maintenance. The substation under study falls into this category.

3.1. Defining the set of actions

Alencar et al. [18] have argued that the decision-making process frequently addresses the necessity of making choices among alternatives. In this case study, it is important to emphasize that each alternative is a combination of a piece of equipment and an action. An alternative may be invoked by three different procedures. It is worth noting that the number of possible combinations could be very large, as each piece of equipment may be associated with an extensive action list. Therefore, the set of alternatives is a set of ordered pairs (ac_i, q_j), which must be ranked.

The ranking could be performed according to the overall value of the alternative in such a way that the best alternative is one that has the largest value of $v( )$. In this case, a larger value of $v( )$ indicates a lesser necessity for improvement associated with the alternative. Thus, the alternatives that should be selected for consideration should be those at the bottom of the list, with the smallest values of $v( )$.

For the electrical substation under study, the set of alternatives is summarized in table 1.

Table 1: Set of actions

Alternatives	Actions	Equipment
A1	Collection of Insulating Oil	Main tank and of the transformer
A2	visual inspection	transformer
A3	thermographic inspection	transformer
A4	Measurement of contact resistance	Breaker Mechanism and contacts
A5	Disturbance	Breaker Mechanism and contacts
A6	visual inspection	switch
A7	Measurement of Resistance and Insulation FP	Insulation-System Breaker
A8	Thermographic inspection	Buses, switching and connections
A9	Thorough visual inspection (c / maneuver)	Buses, switching and connections
A10	Measurement of insulation resistance	Insulation-System Recloser
A11	Measurement of contact resistance	Recloser Mechanism and contacts
A12	overhaul	recloser
A13	visual inspection	recloser
A14	thermographic inspection	recloser
A15	visual inspection	Capacitor bank
A16	thermographic inspection	Capacitor bank

A17	Monitoring of neutral current	Capacitor bank
A18	overhaul	Capacitor bank
A19	Collection of Insulating Oil	main tank and voltage-regulator switch
A20	overhaul	Voltage regulator
A21	Measurement of resistance	Ground grid
A22	Measurement of Potential	Ground grid
A23	general revision	Keyswitch under load - OLTC
A24	Overhaul	Key oil capacitor bank

3.2. The objectives and attributes of the decision-making problem

The maintenance manager's objective when preparing the maintenance plan is to ensure that the planned maintenance actions are effective. In other words, the manager desires the resources that were allocated to the maintenance department to be put to their best possible use.

To achieve this objective, there are several aspects to be considered. For example, it is important to evaluate whether the time required for an activity is commensurate with the degradation of the equipment. In fact, it is very important to know whether the time elapsed since maintenance was last performed is indeed a good indicator of degradation. It could be argued that instead of time, the amount of use since the last maintenance activity may be a more precise indicator of degradation. Therefore, an evaluation of this issue should be performed to assess the effectiveness of maintenance.

A constructed attribute related to this issue is proposed for the decision maker. According to [16], a constructed attribute is typically meant to measure more than one facet of a complex problem, and the descriptions of the levels are very important to the correct understanding of the attribute.

Table 2 below offers descriptions of the different levels for the attribute related to age appropriateness.

Table 2: The constructed attribute related to age appropriateness

Value	Description of the attribute level
1	The variable used to measure the time elapsed since the last maintenance action has no correlation with the time to degradation and must not be used to describe the age of the equipment.
2	The variable used to measure the time elapsed since the last maintenance action has a very weak correlation with the time to degradation and should not be used to describe the age of the equipment.
3	The variable used to measure the time elapsed since the last maintenance action has a weak correlation with the time to degradation, and it is preferable that it not be used to describe the age of the equipment.
4	The variable used to measure the time elapsed since the last maintenance action has a non-negligible correlation with the time to degradation and could be used to describe the age of the equipment.
5	The variable used to measure the time elapsed since the last maintenance action has a strong correlation with the time to degradation, and it is preferable that it be used to describe the age of the equipment.
6	The variable used to measure the time elapsed since the last maintenance action has a very strong correlation with the time to degradation and should be used, in addition to other indicators, to describe the age of the equipment.
7	The variable used to measure the time elapsed since the last maintenance action is directly associated with the time to degradation and is the best variable to describe the age of the equipment.

Another aspect that should be considered is whether performing the procedures necessary to execute a maintenance action could cause damage or malfunction. Thus, in some sense, the possibility of causing damage or malfunction should be considered in the effectiveness analysis of a maintenance action.

Table 3 below offers descriptions of the different levels for the constructed attribute related to the likelihood of causing damage or malfunction.

Table 3: Constructed attribute: Possibility of causing damage or malfunction

Value	Description of the attribute level
1	Damage or malfunction is very unlikely to be caused when performing the specific maintenance action
2	Damage or malfunction is unlikely to be caused when performing the specific maintenance action
3	Damage or malfunction is somewhat likely to be caused when performing the specific maintenance action
4	Damage or malfunction is likely to be caused when performing the specific maintenance action
5	Damage or malfunction is more than likely to be caused when performing the specific maintenance action
6	Damage or malfunction is fairly certain to be caused when performing the specific maintenance action
7	Damage or malfunction is almost guaranteed to be caused when performing the specific maintenance action

Another aspect that might be important in the assessment of the effectiveness of maintenance actions is the probability of a false negative. Because a substation is a complex system, a maintenance plan for this type of system typically includes a large number of inspections. When an inspection is performed, the team may overlook some failure or defect present in the equipment.

Table 4 below offers descriptions of the different levels for the constructed attribute related to the likelihood of a false negative at inspection.

Table 4: The constructed attribute: The possibility of a false negative at inspection

Value	Description of the attribute level
1	It is very unlikely that a false negative will be encountered when performing the specific maintenance action
2	It is unlikely that a false negative will be encountered when performing the specific maintenance action
3	It is somewhat likely that a false negative will be encountered when performing the specific maintenance action
4	It is likely that a false negative will be encountered when performing the specific maintenance action
5	It is more than likely that a false negative will be encountered when performing the specific maintenance action
6	It is fairly certain that a false negative will be encountered when performing the specific maintenance action
7	It is almost guaranteed that a false negative will be encountered when performing the specific maintenance action

Once each of the attributes has been described, it is important to consult with the decision-maker to ensure that there is no doubt or ambiguity regarding the understanding of each level of every criterion. Additionally, the decision-maker must evaluate each alternative with respect to each attribute.
Table 6 presents the decision matrix, which contains the evaluation of each alternative with respect to each criterion.

Table 5: Decision Matrix

Alternatives	age appropriateness	likelihood of causing damage or malfunction	likelihood of a false negative at inspection
A1	6	6	5

A2	4	6	3
A3	6	6	3
A4	6	5	1
A5	5	6	1
A6	4	6	3
A7	6	5	1
A8	6	7	5
A9	5	5	5
A10	6	4	1
A11	6	5	1
A12	6	2	3
A13	4	3	3
A14	3	5	3
A15	4	6	3
A16	6	5	3
A17	6	7	3
A18	6	3	3
A19	6	6	1
A20	6	5	1
A21	5	2	4
A22	5	5	5
A23	6	3	5
A24	6	2	5

3.3. Defining the preference functions

Once the decision-maker has confirmed that he has a good understanding of the attributes, the next step involves the quantification of his or her preferences. This process consists of the assessment of the utility function for each attribute. For constructed indices, these functions are assessed directly for the defined points.

Let us define x as a value chosen by the decision-maker to represent the performance of an alternative with respect to a specific attribute based on the consideration of all descriptions associated with all levels.

In our specific case, x may be ordered as follows: x^0, x^1, .., x^5, x^*, where x^0 is the least preferred value and x^* is the most preferred value. Thus, for the assessment of the utility values $u(x^j)$, $j=1, ..., 5$. The decision-maker must identify for each x^j the correspondent probability p^j such that the decision-maker's preference for this situation is equivalent to his preference for a lottery that yields either x^* with probability p_j or x^0 with probability $(1-p_j)$. Then, by equating the utilities, we find

$$u(x^j) = p_j u(x^*) + (1 - p_j)u(x^0) = p_j, j=1,...,5 \qquad (6)$$

Here, it is important to emphasize that this direct assessment could yield multiple different functions for the association of x^j with $u(x^j)$ because the different levels related to a specific attribute might not be equally spaced.

In our case, for this first application, the decision-maker responded to the lotteries in such a way that the results were nearly linear functions of $u(x^j)$. Thus, for all criteria, for the sake of simplification, linear utility functions were used. Because the number of levels was the same for all attributes, the values of the utility function that were used to maximize the attributes were the same for each attribute; these values are presented in table 6.

Table 6: Values of the utility function for the maximization of attributes

j	x^j	$u(x^j)$

	1 (x^0)	0.00
1	2	0.17
2	3	0.33
3	4	0.50
4	5	0.67
5	6	0.83
	7 (x^*)	1.00

Similar to the case of the attributes to be maximized, the relation between the level of the attribute and the utility function for each attribute to be minimized was also a linear relation. The values of the utility function used for this purpose are presented in table 7.

Table 7: Values of the utility function for the minimization of attributes

J	x^j	$u(x^j)$
	1 (x^*)	1
1	2	0.83
2	3	0.67
3	4	0.5
4	5	0.33
5	6	0.17
	7 (x^0)	0

Once the utility functions have been defined, the overall evaluation of the effectiveness of the maintenance actions should be performed; however, before this can be done, it is necessary to define the scale constant for each decision axis. The values obtained by following the process for the assessment of the scale constant for each attribute presented in [19] are presented in table 8.

Table 8: Scale constants

Attribute	Value of the scale constant
age appropriateness (k_1)	0.2
likelihood of causing damage or malfunction (k_2)	0.4
likelihood of a false negative at inspection (k_3)	0.4

Once the scale constants have been defined, it is possible to calculate the global values using equation (1).. The interesting aspect is that the decision maker said to be annoyed with low level for some attributes. For each attribute, he defines the values of l_i and u_i that correspond to the lower and upper thresholds for the veto function, as summarized in table 9. To reiterate, the upper threshold corresponds to the minimum value of performance $v_i(a)$ with respect to criterion i that is acceptable to the decision maker, whereas l_i corresponds to the maximum value of performance $v_i(a)$ with respect to criterion i that the decision maker is certain to reject.

For the criteria to be minimized, these interpretations of the veto thresholds are reversed. The upper threshold becomes the maximum value of performance $v_i(a)$ with respect to criterion i that is acceptable to the decision maker, whereas l_i becomes the minimum value of performance $v_i(a)$ with respect to criterion i that the decision maker is certain to reject.

Table 9: Lower and upper thresholds for the veto function for each criterion

Attribute	l_i	u_i
age appropriateness	*1*	*2*
likelihood of causing damage or malfunction	*7*	*5*
likelihood of a false negative at inspection	*7*	*6*

3.4. Overall evaluation

Once all parameters have been defined, it is possible to determine the final global value for each alternative. It is worth emphasizing that the final global value is a measure of the effectiveness of the maintenance actions, so the worse the action is placed in the ranking, the greater is the necessity for improvement that is associated with it. Thus, our ultimate objective is to address the worst alternatives. This ultimate objective is out of the scope of this work, but the model presented here defines the very first step toward implementing the desired improvement. Therefore, we propose that this multi-attribute model should be run every year to ensure sufficient time for new maintenance actions, far superior to those identified by the model, to be incorporated into the next annual maintenance plan.

Table 9 below summarizes the overall evaluation and the final ranking.

Table 9: Overall values

Alternatives	age appropriateness	u(x)	likelihood of causing damage or malfunction	u(x)	likelihood of a false negative at inspection	u(x)	v(a)	r(a)	v´(x)	Ranking
A1	6	0.83	1	1.00	5	0.33	0.70	1.00	0.70	5
A2	4	0.50	2	0.83	5	0.33	0.57	1.00	0.57	15
A3	6	0.83	2	0.83	5	0.33	0.63	1.00	0.63	10
A4	6	0.83	6	0.17	5	0.33	0.37	0.80	0.29	20
A5	5	0.67	6	0.17	5	0.33	0.33	0.80	0.27	22
A6	4	0.50	1	1.00	5	0.33	0.63	1.00	0.63	10
A7	6	0.83	6	0.17	5	0.33	0.37	0.80	0.29	20
A8	6	0.83	2	0.83	3	0.67	0.77	1.00	0.77	2
A9	5	0.67	2	0.83	3	0.67	0.73	1.00	0.73	4
A10	6	0.83	2	0.83	4	0.50	0.70	1.00	0.70	5
A11	6	0.83	2	0.83	4	0.50	0.70	1.00	0.70	5
A12	6	0.83	2	0.83	4	0.50	0.70	1.00	0.70	5
A13	4	0.50	2	0.83	4	0.50	0.63	1.00	0.63	10
A14	3	0.33	2	0.83	4	0.50	0.60	1.00	0.60	13
A15	4	0.50	2	0.83	3	0.67	0.70	1.00	0.70	5

A16	6	0.83	2	0.83	3	0.67	0.77	1.00	0.77	2
A17	6	0.83	1	1.00	3	0.67	0.83	1.00	0.83	1
A18	6	0.83	7	0.00	6	0.17	0.23	0.60	0.14	23
A19	6	0.83	7	0.00	6	0.17	0.23	0.60	0.14	23
A20	6	0.83	5	0.33	6	0.17	0.37	1.00	0.37	18
A21	5	0.67	1	1.00	6	0.17	0.60	1.00	0.60	14
A22	5	0.67	2	0.83	6	0.17	0.53	1.00	0.53	16
A23	6	0.83	5	0.33	6	0.17	0.37	1.00	0.37	18
A24	6	0.83	5	0.33	4	0.50	0.50	1.00	0.50	17

3.5. Some discussion of the results

It is interesting to note that among the 5 worst alternatives (A18, A19, A5, A4, A7), at least 2 actions were recognized by the decision-maker as actions with serious problems to be corrected, and the decision-maker also stated that these actions required deeper study before improvement would be possible. In the case of alternative A18, the decision-maker realized while reviewing the results that this alternative was no longer under consideration for inclusion in the maintenance plan because serious potential problems with this alternative had been identified. By contrast, the decision-maker confirmed that no problems at all had been identified regarding the highest-ranked alternatives. In fact, the best alternative (A17) was designed to replace the worst alternative (A18); however, when the case study was performed, the decision-maker accidentally neglected to delete the alternative A18 from the set of alternatives.

4. CONCLUSION

An electrical substation is a complex system that demands effective maintenance procedures because of the serious consequences associated with its failure. The effectiveness of such maintenance actions depends on various factors; therefore, the discrimination of effective maintenance actions from ineffective ones is non-trivial.

For such a complex system, mistakes that are introduced in standard maintenance procedures are very dangerous because these procedures may be propagated to other electrical substations, potentially leading to countless problems caused by the spread of an incorrect procedure.

Here, the compensatory veto model was applied to the assessment of the maintenance plan of an electrical substation. It is worth emphasizing that in fact, the decision-maker in this case was not comfortable with the idea of unlimited compensation. Thus, the veto function offered an excellent method of ensuring that the results of the analysis satisfied practical expectations.

Finally, based on the results of the model, the decision-maker can select the alternatives to be improved for the next year. Therefore, the proposed model should permit the initiation of a continuous process of improvement that should benefit any segment of any industry that is concerned about improving its results.

Acknowledgments

This paper was produced as a result of research studies funded by the Brazilian Research Council (CNPq). We also extend our thanks to the ANEEL and CELPE.

References

[1]   T. Xia, T.; L. Xi; X. Zhou,; S. Du. *Modeling and optimizing maintenance schedule for energy systems subject to degradation.* Computers & Industrial Engineering. 63, 607–614. (2012.)

[2]   S.O. Duffuaa and K.S. Al-Sultan, *Mathematical Programming Approaches for the Management of Maintenance Planning and Scheduling.* Journal of Quality in Maintenance Engineering. 3, 163-176, (1997).

[3] Vincke, Philippe (1992) *Multicriteria Decision – aid.* Bruxelles:John Wiley & Sons.

[4] V. B. S. Silva ; D. C. Morais ; A. T. Almeida. *A Multicriteria Group Decision Model to Support Watershed Committees in Brazil.* Water Resources Management, v. 24, p. 4075-4091. (2010).

[5] C.M.M. Mota; A.T. de Almeida; L.H. Alencar. *A multiple criteria decision model for assigning priorities to activities in project management.* International Journal of Project Management 27, 175-181. (2009).

[6] D.C. Moraes; A.T. de Almeida. *Group decision making on water resources based on analyses of individual rankings.* Omega 40, 42-52. (2012).

[7] T.V. Garcez; A.T. de Almeida. *A risk measurement tool for an underground electricity distribution system considering the consequences and uncertainties of manhole events.* Reliability Engineering & Systems Safety v. 124, p. 68-80. (2014).

[8] Alencar, M. H., & Almeida, A. T. *Assigning priorities to actions in a pipeline transporting hydrogen based on a multicriteria decision model.* International Journal of Hydrogen Energy, 35, pp. 3610-3619, (2010).

[9] A. T. de Almeida. *Multicriteria decision making on maintenance: spares and contracts planning.* European Journal of Operational Research 129 235 – 241. (2001).

[10] A. T. de Almeida. *Multicriteria modeling of repair contract based on utility and Electre I method with dependability and service quality criteria.* Annals of Operations Research 138 113 – 126, (2005)

[11] A. T. de Almeida. *Multicriteria decision model for outsourcing contracts selection based on utility function and ELECTRE method.* Computers & Operations Research 34 3569 – 3574, (2007).

[12] C.A.V. Cavalcante; R.J.P. Ferreira; A.T. de Almeida. *A preventive maintenance decision model based on multicriteria method PROMETHEE II integrated with Bayesian approach.* IMA Journal of Management Mathematics. 21, 333–348, (2010).

[13] Ferreira R.J.P. Ferreira; A.T. de Almeida; C.A.V. Cavalcante. *A multi-criteria decision model to determine inspection intervals of condition monitoring based on delay time analysis.* Reliability Engineering & System Safety, 94:905-912, (2009).

[14] A. T. de Almeida, *Additive-veto Models for Choice and Ranking multicriteria Decision Problems.* Asia-Pacific Journal of Operational Research, 30, 1350026-1- 1350026-20 (2013)

[15]   P. A Scarf. *On the application of mathematical models in maintenance.* European Journal of operational research, 99, p. 493-506 (1997).

[16]   R. L Keeney, and Ralph L. Keeney. *Value-focused thinking: A path to creative decisionmaking.* Harvard University Press, 2009, Havard.

[17] W., Edwards, and F. H. Barron. *SMARTS and SMARTER: Improved simple methods for multiattribute utility measurement.* Organizational Behavior and Human Decision Processes 60. 306-325, (1994).

[18] L. H. Alencar; C.M.M Mota.; M. H. Alencar *The problem of disposing of plaster waste from building sites: Problem structuring based on value focus thinking methodology.* Waste Management 31 2512–2521. (2011).

[19] Keeney, Ralph L., and Howard Raiffa. *Decisions with multiple objectives: preferences and value trade-offs.* Cambridge university press, 1993, Cambridge.

A stochastic production planning optimization for multi parallel machine under leasing Contract

Medhioub Fatma, Hajej Zied, and Rezg Nidhal
LGIPM-University of Lorraine, Metz, France

Abstract: In this paper, we aim at optimizing the production planning. The problem consists on a several identical machines mounted in parallel and which are leased depending on a fluctuating demand over a finite time horizon under given service level. The objective of the production plan is to determine the best combination of leased machines numbers, production time (or level) and inventory levels, by developing a mathematical model, that minimize the average total costs over the finite time horizon. The contribution and newness of this work is that it treats this approach under new constraints related especially to leasing techniques and consequently we assume that the number of workstations varies from a production period to another. This characteristic is due to the leasing principle as well as to the fluctuating demand that we have to take into account. A numerical example confirms the analytical study.

Keywords: Optimization, Production planning, Leasing constraint, Random demand, Service level, Multi parallel machine.

1. INTRODUCTION

Ameliorating the industry situation requires certainly good making decisions and good management. These decisions can be achieved in order to improve the product quality, to reduce the costs and maximize the customer service level. In this context, we can consider the leasing as a very important solution for many manufacturing to minimize costs and guard against important competitiveness. In order to reduce the manufacturing cost, more industries have preferred leasing, newer and better equipment appears on the market and that these owning costs is increasing, rather than owning it. In this context, we can cite the work of [6] where based on the causes that newer equipment appears on the market and that the cost of owning the equipment is increasing, more businesses have started leasing equipment rather than owning it [1]. [1] considered a business model in which a company leases new equipment and sells the remanufactured one at the same time. They developed a dynamic program formulation to determine the optimal price of the remanufactured equipment and the optimal payment structure for the leased equipment to maximize the profit. [4] deal with the choice between purchase and lease in the context of radiotherapy equipment.

In the other hand, the optimization of production which minimizes the total production and inventory cost is one of the principal activities of a hierarchical decision manufacture process. In the context of production/inventory optimization problem, [2] proposed a model which defines an inter-temporal quadratic cost minimization program and that by approximating the cost functions for hiring labour and lay-off, overtime, inventory and shortage by suitable quadratic functions. As a result, and considering some constraints, this model provides an optimal smoothing solution for aggregate inventory, production and workforce. The HMMS approach has been extensively used and a source of inspiration in literature [7]. It is usually applied as a benchmarking tool in order to compare different production planning approaches and to provide managers and decision makers with perspectives and ideas about how to manage the firm's material resources. But, some other works such as [3] proved that this quadratic approach is useful to evaluate the production process. So, for example, the quadratic inventory cost describes and takes into account both possible status of inventory: negative (rupture and backorders) and positive (overstocking). [9] established a model in three phases. The first one consists in considering products requiring the same resources to find an aggregate production plan. In this phase, the objective is to minimize the costs of set up, production, inventory, workforce and maintenance. The second step is to determine a director production plan minimizing the disparity between a normal production plan and the aggregate plan. The last step consists in simulating the workstations breakdowns.

Recently, more works, concerning the production optimization, have integrated new constraints. The model of [5] is based on the control aspect of the production of a system, which is composed of m identical machines able to produce n types of products and subjected to recuperation periods after a troubleshooting. The decision variables in this research are the production rate and the repairing date. The main objective of the work is to minimize the expression of the total cost under these two variables.

Inspired from the HMMS model, we had the idea to make emphasis on the machines instead of workers, production rate and inventory levels in order to make the best and optimal production planning. Also, in our work, we make some changes on the model keeping its linear quadratic form. Furthermore, we take into account some constraints on the decision variables to make our approach more realistic and to ensure its applicability in real industrial cases.

The objective of this paper is to determine the economical production planning taken into account leasing machines number and random demand to minimize the sum of production, inventory and leasing costs.

This remainder of this paper is organized as follows: Section 2 proposes a general stochastic production, inventory model. Section 3 presents the problem formulation with the total cost expression as the diverse Costs of production system. Section 4 presents and develops the policy and analytical expression of production (works) and inventory. A simple numerical example is presented in section 5. Finally, the conclusion is given in Section 6.

2. PROPOSED MODEL

2.1. Notations

We used the following notations in this paper:

Δt : length of a production period

H : number of production periods k ($k=0, 1,\ldots, H$).

$H.\Delta t$: finite time horizon.

$\hat{d}_k$: average demand during period k ($k=0, 1,\ldots, H$).

V_{d_k} : variance of demand during period k ($k=0, 1,\ldots, H$).

M_k : number of leasing workstations during period k.

U_k : production rate during period k ($k=0, 1,\ldots, H$).

u_{ik} : quantity produced by machine i in period k.

S : inventory level at the end of period k ($k=0, 1,\ldots, H$).

M : maximum number of leasing machines per period.

m : minimum number of leasing machines number per period.

u : production rate of a machine per time unit (hour); (the same for all machines).

X_{ik}^N : number of time units (hours) that machine i works in period k (Normal time).

X_{ik}^S : number of time units (hours) that machine i works in period k (Over time).

$X_{\max}^N$: maximum of time units that a machine could perform during a period and in normal time.

$X_{\max}^N$: maximum of time units that a machine could perform during a period and as overtime.

c_1 : Cost of a machine leasing;

c_2 : fixed costs of leasing;

c_3 : Cost of variation in leasing machines number;

c_4 : Coefficient describing asymmetry between increasing / decreasing machines number;

c_5 : Cost of overtime work (per time unit);

c_6 : Cost of a time unit not used in normal time;

c_8 : Cost of storage/rupture (per unit);

2.2. Problem description

In this work, we consider a problem of production optimization planning and that under new approach considering a leasing constraint. The manufacturing system is composed of a several identical leasing machines, mounted in parallel, which produces one type of product in order to meet a random demand and minimizes the total production/inventory/leasing cost. We can consider two categories of the work time of equipment's. First category: where each machine works normally during maximum time units X_{ik}^N per period. Second category: in order to satisfy the random demand that exceeds the production, it's preferable to make machines work in overtime which not exceeds a maximum time units X_{ik}^S. The number of leasing machines varies from period to another depending to the production quantity and the random demand and consequently it represents decision variables to be determined under some constraints. We assume that the production horizon H is divided into equal period's Δt. The customer demand which is random and given by a normal distribution. See figure1

Figure 1: Problem Description

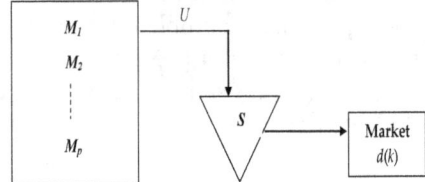

Our problem consists in determining the best combination of leased machines numbers, production time or production rates and inventory levels under service level and leasing constraints. That minimizes the average total costs over the finite time horizon in order to satisfy the fluctuating demand.

3. PROBLEM FORMULATION

3.1. Total production cost

The idea is to minimize the expected costs of leasing machines numbers, production rates and inventory levels take into account some constraints on main variables. This kind of problem can be formulated as a stochastic quadratic optimization problem under a stock threshold level constraint, with numbers of leasing machines, normal time unit of machine work and overtime unit of machine work $(M_k^*, X_{ik}^N, X_{ik}^S)$ corresponding to each period as the decision variables and which is equivalent also to (M_k^*, U_k^*).

The stochastic problem as follows:

$$\underset{(M_k, U)}{Min} \left(\sum_{k=1}^{H} E \left\{ \left[\begin{matrix} \left(c_1 \cdot M_k + c_2\right) + \left(c_3 \cdot \left(M_k - M_{k-1} - c_4\right)^2\right) + \left(c_8 \cdot \left(S_k - \left(a_1 + a_2 \cdot d_k\right)\right)^2\right) \\ + c_5 \cdot \theta_t \cdot \Delta t \cdot \sum_{i=1}^{M_k} X_{ik}^S + \left(1 - \theta_k\right) \cdot c_6 \cdot \Delta t \cdot \sum_{i=1}^{M_k} \max\left[0, \left(X_{\max}^N - X_{ik}^N\right)\right] \end{matrix} \right] \right\} \right)$$

Subject to:

$$U_k = u \times \Delta t \times \sum_{i=1}^{M_K} \left(X_{ik}^N + X_{ik}^S \right) \qquad (1)$$

$$S_k = S_{k-1} + U_k - d_k \qquad (2)$$

$$\text{Prob}\left(S_k \geq 0 \right) \geq \beta \qquad (3)$$

$$m \leq M_k \leq M \qquad (4)$$

$$0 \leq X_{ik}^N \leq X_{\max}^N \qquad (5)$$

$$0 \leq X_{ik}^S \leq X_{\max}^S \qquad (6)$$

The first constraint determine the production rate of each period which calculated according to the time of each machines works (normal and overtime) and the production rate of each machines per time unit (hour) u. According to first constraint, determining the decision variable U_k is equivalent to determine X_{ik}^N and X_{ik}^S. Constraint (2) describe the inventory level of the store S at each production period, is formulated in the form of flow balance constraints where the inventory level of S at the period k equals to the inventory level at period k-1 plus the production rate during period k, minus the demand during period k. The service level requirement constraint is determined by the probability constraint on the stock level at each period k expressed by the constraint (3). The probability variable β which can be interpreted as the degree between the high and low customer service level. Constraint (4) defines the maximum and minimum number of leasing machines of each period. Similarly for constraints (5) and (6) that gives the assumption on normal and over time work variables.

3.2. Costs of works time (production), Inventories and leasing machines

Using the model carried out by [2], the purpose of this subsection is to develop the expected works time (production), inventories and leasing machines costs over the finite time horizon H.

- Cost of leasing machines:

$$c_1 \cdot M_k + c_2 \qquad (7)$$

Where c_1 is the average cost for leasing a machine and c_2 represented the fixed cost term in the leased contract.

- Cost of variation in machines number:

$$c_3 \cdot \left(M_k - M_{k-1} - c_4 \right)^2 \qquad (8)$$

This equation defines the costs due to variation in the number of the machines required for period k. Constant term c_4 presents the asymmetry in costs of adding and eliminating machines.

- Costs of overtime work and unexploited capacities:

These costs expressions can be defined as follows:

- Cost of overtime work:

$$c_5 \cdot \theta_k \cdot \Delta t \cdot \sum_{i=1}^{M_k} X_{ik}^S \qquad (9)$$

These costs depend to binary variable θ_k. We can noted that θ_k is equal to 1 if the random demand of each period k exceeds the production quantity when all used machines M_k works during their maximum normal time and equal to 0 otherwise.

- Cost of unexploited capacities:

Other hand, if the demand is not very important and all or some machines could work less than their maximum normal work time ($\theta_k=0$). In this case, we can consider as if those machines were over paid. This cost is given as follows

$$(1-\theta_k) \cdot c_6 \cdot \Delta t \cdot \sum_{i=1}^{M_k} \max\left[0, \left(X_{\max}^N - X_{ik}^N\right)\right] \tag{10}$$

- Inventory Costs:
  Using HMMS model, we can define the inventory cost

$$c_8 \cdot \left(S_k - \left(a_1 + a_2 \cdot d_k\right)^2\right) \tag{11}$$

with an optimal level of net inventories is equal to:

$$a_1 + a_2 \cdot d_k$$

4. ANALYTICAL STUDY

4.1. Deterministic problem

Due to the stochastic nature of our problem, the constraints and the dimensionality, to try to obtain an optimal solution can become a hard task. An approach that transforms the stochastic problem into a deterministic equivalent is necessary. This deterministic problem maintains the main properties of the original problem. Also, such a technique has the advantage of giving the possibility of exploiting different mathematical programming methods in order to solve the equivalent obtained model.

The principle step of an equivalent deterministic problem for the above stochastic one may be obtained by setting the variables equal to their means values. In this context, we can cite the works of [8] which uses the linearity of the model and assumes that the demand random variation can be described as a Gaussian process.

Before proceeding, the following notation is introduced:

Mean variables:

$E\{S_{i,k}\} = \hat{S}_{i,k}$

$E\{U_k\} = U_k$

$E\{M_k\} = M_k$

$E\{X_{ik}^N\} = X_{ik}^N$

$E\{X_{ik}^S\} = X_{ik}^S$

Variance variables: $V_{U_k} = 0$, $V_{M_k} = 0$, $V_{X_{ik}^N} = 0$, $V_{X_{ik}^S} = 0$. (Note that this reflects the fact that the control

variables U_k, M_k, X_{ik}^N and X_{ik}^S are deterministic).

Using the above notations, the first constraint describing the stock level can be written under a deterministic form as follows:

$$\hat{S}_k = \hat{S}_{k-1} + U_k - \hat{d}_k \tag{12}$$

- Service level constraint

Generally in stochastic cases, it is complicated just to guarantee feasibility, though one possibility of overcoming such difficulty is to consider probabilistic constraints. Another important transformation changes the service level constraint into equivalent, but deterministic inequalities by specifying, through the following lemma, a minimum cumulative production quantity depending on the service level requirements.

Lemma 1:
We recall that β defines the targeted service level as expressed by constraint (3), repeated below:

$$\text{Prob}\lfloor S_k \geq 0 \rfloor \geq \beta$$

Then, for $k=1,..,H$ we have:

$$\text{Pr}ob(S_k \geq 0) \geq \beta \;\Rightarrow\; \left(U_k \geq \left(V_{d(k)}\right) \times \varphi^{-1}(\beta) - S_{k-1} + \hat{d}_k\right) \qquad k=1,....,H \tag{13}$$

φ : Cumulative Gaussian distribution function with mean $\hat{d}_k$ and finite variance $V_{d(k)}$

φ^{-1} : Inverse distribution function

Using the above notations, the formulation of our planning problem becomes as follows:

$$\underset{(M_k,U,N)}{Min} \left(\sum_{k=1}^{H} \left\{ \begin{array}{l} \left[(c_1 \cdot M_k + c_2) + \left(c_3 \cdot (M_k - M_{k-1} - c_4)^2 \right) + c_5 \cdot \theta_t \cdot \Delta t \cdot \sum_{i=1}^{M_k} X_{ik}^{S} \right. \\ \left. + (1 - \theta_k) \cdot c_6 \cdot \Delta t \cdot \sum_{i=1}^{M_k} \max\left[0, \left(X_{\max}^{N} - X_{ik}^{N} \right) \right] + c_8 \cdot \left(\hat{S}_k - \left(a_1 + a_2 \cdot \hat{d}_k \right) \right)^2 \right] \\ + (1 + a_2)^2 \cdot V_d \cdot \frac{H}{2} \cdot (H+1) + a_2^2 \cdot V_d \cdot \frac{H}{2} \cdot (H-1) \end{array} \right\} \right)$$

Subject to:

$$\hat{S}_k = \hat{S}_{k-1} + U_k - \hat{d}_k$$
$$\left(U_k \geq \left(V_{d(k)} \right) \times \varphi^{-1}(\beta) - S_{k-1} + \hat{d}_k \right)$$
$$m \leq M_k \leq M$$
$$0 \leq X_{ik}^{N} \leq X_{\max}^{N}$$
$$0 \leq X_{ik}^{S} \leq X_{\max}^{S}$$

4.1. Numerical Resolution Method

Regarding the complexity of our problem, the analytical resolution remains difficult and finding an exact optimal solution remains difficult. That is why; we proposed an algorithm or a numerical procedure and a method in order to determine an approximate possible solution. Among the difficulties are the random demand and the dependence between the decision variables. The objective of this algorithm is to determine the number of machines M_k for each period and then determine their work time's vectors together with the production rate of each period k.

The principle of this algorithm is based on Branch and Bound algorithm. It is used to generate randomly the vector of number of machines during the finite horizon and then generate randomly also the vectors of work times (normal and extra time). This generation necessitate the used of smaller intervals by reducing the first interval $[m,M]$ and dividing it into two major smaller intervals.

The following chart illustrates the syntax of the principal algorithm that we implement on Matlab in order to find approximate solutions:

Figure 2: Problem Description

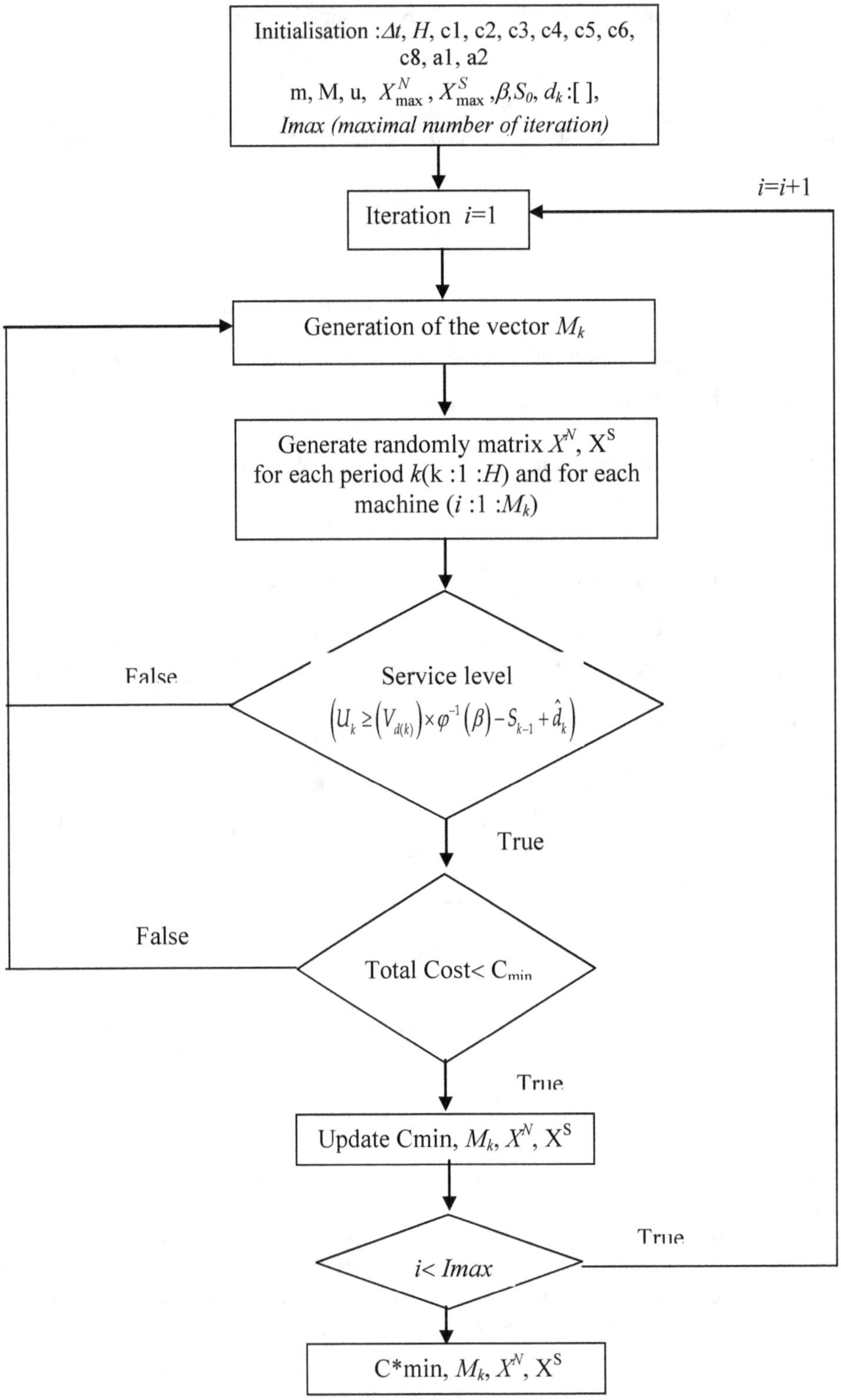

Initialisation :Δt, H, c1, c2, c3, c4, c5, c6, c8, a1, a2
m, M, u, X_{max}^N, X_{max}^S, β, S_0, d_k :[ ],
Imax (maximal number of iteration)

Iteration $i=1$

$i=i+1$

Generation of the vector M_k

Generate randomly matrix X^N, X^S for each period k(k :1 :H) and for each machine (i :1 :M_k)

Service level
$$\left(U_k \geq \left(V_{d(k)}\right) \times \varphi^{-1}\left(\beta\right) - S_{k-1} + \hat{d}_k\right)$$

False

True

Total Cost< C_{min}

False

True

Update Cmin, M_k, X^N, X^S

$i< Imax$

True

C^*min, M_k, X^N, X^S

5. NUMERICAL EXAMPLE

Let us consider a system that produces one type of products to meet the random demands below. Using the models described in previous sections, we will determine the optimal production plan minimizing the total cost over a finite planning horizon: $H=5$ periods each of $\Delta t=25$ days duration. We supposed that the standard deviation of demand of product is the same for all periods $\sigma_d=1.1$ and the initial inventory level we assume that $S_0=0$.

The data required to run this model are given in sequence.
For the cost coefficients c_i and the constants a_i :
$c_1=500$; $c_2=0$; $c_3=3.88$; $c_4=2.42$; $c_5=7$; $c_6=5$; $c_8=10$; $a_1=250$; $a_2=0$;
The number of machines as well as the work time hours should obey to the following bounders:
$m=6$ machines; $M=12$ machines; $u=15$; $X_{max}^N=6$; $X_{max}^S=2$;

The customer satisfaction degree, associated with the stock constraint, is equal to 90% ($\beta=0.9$).
The average demand is presented in tables 1 below:

Table 1: Average Demands

$\hat{d}_1$	$\hat{d}_2$	$\hat{d}_3$	$\hat{d}_4$	$\hat{d}_5$
2286	22853	11880	15123	20653

We used the numerical procedure method with Matlab, in order to realize this optimization. The economically production plan and leasing machines number, the normal work time planning and an optimal overtime work scheduling are presented respectively in table 2, 3,4 with a minimal total cost equal to 1, 27108 (Monetary Unit).

Table 2: Economical leasing machines, inventory and production plans

Periods k	$k=1$	$k=2$	$k=3$	$k=4$	$k=5$
M_k	7	8	7	7	8
S_k	8990	4991	5129	4378	3888
U_k	11250	18250	12750	12750	19500

Table 3: Normal work time planning of each leasing machine

Periods k	$k=1$	$k=2$	$k=3$	$k=4$	$k=5$
Machine 1	4	6	4	4	6
Machine 2	4	4	5	4	5
Machine 3	4	6	4	6	6
Machine 4	4	4	5	5	4
Machine 5	4	4	4	4	6
Machine 6	5	5	4	6	5
Machine 7	5	6	4	6	5
Machine 8	0	6	0	0	6
Machine 9	0	0	0	0	0
Machine 10	0	0	0	0	0
Machine 11	0	0	0	0	0
Machine 12	0	0	0	0	0

Table 4: Optimal overtime work planning of each leasing machine

Periods k	$k=1$	$k=2$	$k=3$	$k=4$	$k=5$
Machine 1	0	0	0	0	2
Machine 2	0	0	0	0	0
Machine 3	0	2	0	0	2
Machine 4	0	0	0	0	0
Machine 5	0	0	0	0	2
Machine 6	0	2	0	2	0
Machine 7	0	2	0	2	0
Machine 8	0	0	0	0	0
Machine 9	0	0	0	0	0
Machine 10	0	0	0	0	0
Machine 11	0	0	0	0	0
Machine 12	0	0	0	0	0

According the tables, we remark that on the normal and over time planning that the machines not working in the normal time for all periods; we cannot be used in the overtime. Other, some machines work less than their maximal normal work time that generates other cost but can be an advantage to optimize maintenance cost for these machines. Other hand, we remark that the stock levels relatively important. That is related to the service rate constraint which is important.

4. CONCLUSION

This paper dealt new approach consisting in optimizing stochastic production planning problem for a several leased machines considering a random demand and a service level. The contribution and originality of this study is that it treats this approach under new constraints related especially to leasing techniques by considering the variation of machines number from production period to another in order to satisfy the random demand and minimize the total production, inventory and divers cost. Given a service level, we have formulated and solved the related stochastic production problem. An optimization has been performed obtaining an optimal production plan as well as the best combination of the leased machines number and the time work of each one over the different production periods.

In order to go ahead with this work and ameliorate it, many extensions may be recommended. In fact, we can assume a maintenance strategy of the different leased machines taking into account the influence of the production rates on the degradation degree of each machines and consequently on the preventive maintenance plan.

References

[1]   N. Aras, R. Güllü, and S. Yürülmez. *"Optimal inventory and pricing policies for remanufacturable leased products"*. International Journal of Production Economics, 133, 1, pp. 262-271, (2011).
[2]   C.C. Holt, F. Modigliani, J.F. Muth, and H.A.Simon, *"Planning Production, Inventory and Work Force"*, Prentice-Hall, (1960), NJ.
[3]   A. Hax and D. Candea *"Production and Inventory Management"*. Prentice Hall, (1984), Englewood Cliffs.
[4]   A. Nisbet and A.Ward *"Radiotherapy equipment – Purchase or lease?"*, The British Journal of Radiology, 74, pp.735–744, (2001).
[5]   J. P. kenne and L. J. Nkeungoue *"Simultaneous control of production, preventive and corrective maintenance rates of failure prone manufacturing system"*. Applied Numeriacl Mathematics, 58, pp180–194,(2008).

[6]   J. Pongpech, and D.N.P. Murthy *"Optimal periodic preventive maintenance policy for leased equipment"*, Reliability Engineering & System Safety, 91, pp. 772-777, (2006).

[7]   J. Singhal and K. Singhal *"Alternate approach to solving the holt and al. model and to performing sensitivity analysis"*. European Journal of Operationnal Research, 91, pp. 89–98,(1996).

[8]   O. S. Silva Filho and S. D. Ventura *"Optimal feedback control scheme helping managers to adjust aggregate industrial resources"*. Control Engineering Practice, 7, pp. 555–563. (1999).

[9]   L. Weinstein and C. Chung *"Integrating maintenance and production decisions in a hierarchical production planning environment"*. Computers and operations research, 26, pp. 1059–1074. (1999).

Review of the Preventive Maintenance Requirements for the Safety Systems of the Mochovce NPP

Zoltan Kovacs[a*], Robert Spenlinger[a]
[a] RELKO Ltd., Bratislava, Slovak Republic

Abstract: A requirement to optimize the Preventative Maintenance (PM) tasks assigned to specified safety systems has been identified at Mochovce Nuclear Power Plant (NPP). RELKO Ltd has been tasked with optimising the PM tasks via application of the Reliability-Centered Maintenance (RCM) and PSA methodology. This paper details the results of the RCM analysis performed on the Core Cooling Systems. It is concluded that the PM tasks assigned to the Core Cooling Systems were, in the main, based upon the original equipment manufacturers' (OEM) recommendations. Following the accumulation of about ten years of operating and maintenance experience it was concluded that many of the current task types and task frequencies required major revision in order to maintain the optimum levels of both reliability and availability of the Core Cooling Systems. It is also concluded that in several cases, specific components within the Core Cooling Systems will benefit from a shift in maintenance strategy from fixed interval invasive routines to a predictive maintenance (PdM) based strategy. Such a strategy will ensure close monitoring of system and component performance without compromising nuclear safety or availability. It is recommended that the Mochovce NPP replaces the current maintenance catalogue assigned to the Core Cooling Systems with new PM tasks detailed in the paper. In addition, the paper presents the impact of changes on CDF and LERF after implementation of the new PM tasks.

Keywords: Preventive maintenance, reliability-centered maintenance, PSA, core damage frequency

1. INTRODUCTION

The construction, operation and maintenance of the nuclear power plants has been a well-subsidised business for a long time. However, the financial sources are considered as a very important aspect in this matter during the last decades. The budgets are limited and taken into account very carefully. They are needs to make reliable and cost effective choices with respect to the improvements that have to be made in case of modernization or maintenance projects.

Probabilistic Safety Assessment (PSA) is a tool which is used to evaluate the effects of improvements that are being implemented in the plant. Importance measures are defined which can show us the importance of the components from the safety point of view. These importance measures are applied mainly in the following areas: 1) optimization of the test and maintenance activities, 2) optimization of the plant design by adding, removing and modification of systems or components and 3) configuration control with the effect of taking a component out of service.

The paper is focused on optimization of maintenance activities in the Mochovce plant using the RCM methodology. The level 1 full power, low power and shutdown PSA model of the plant is used to identify the most important systems and components and to provide their importance ranking.

A requirement to optimize the PM tasks assigned to specified safety systems has been identified at Mochovce NPP. RELKO Ltd has been tasked with optimising the PM tasks via application of the RCM and PSA methodology. This paper details the results of the RCM analysis performed on the Core Cooling Systems. The focus of the study is entirely upon major electro-mechanical and mechanical components, i.e. pumps, valves, tanks, etc. It should be noted that the pipework is subject

---

* kovacs@relko.sk

to a discrete programme of specialist inspections, e.g. pipe wall thickness checks, etc., the pipework associated with the Core Cooling Systems is excluded from this review.

After introduction the usage of PSA importance measures is described in section 2 to identify the dominant list of components. Then, the overview of RCM methodology is provided in section 3. Section 4 describes the functional breakdown of the Core Cooling System. Section 5 provides the discussion of the RCM results. The results of risk assessment for the state of the plant after implementation of the proposed changes in the maintenance activities are presented in section 6. The conclusions are described in section 7.

2. USAGE OF IMPORTANCE MEASURES

The nuclear power plants are designed using the defence-in depth principle. Therefore, a single failure of a component or other basic event will probably not result in a large accident. Such accidents will be the result of combinations of multiply basic events. The PSA determines all important minimal cutsets that could lead to a large accident. The final results of a PSA study are then represented in the form of core damage frequency, early release frequency, etc. The risk importance measures give an indication of the contribution of a certain component (item) to the total risk.

The following importance measures are defined and more frequently used [2]:

1. Risk reduction worth $\quad$ $RRW = R(base) \,/\, R(x_i = 0)$
2. Risk achievement worth $\quad$ $RAW = R(x_i = 1) \,/\, R(base)$
3. Fussel-Vesely importance $\quad$ $FV = [R(base) - R(x_i = 0)] \,/\, R(base)$

The following definitions are used in the formulas:

- $R(x_i = 1)$ - the increased risk level with basic event x_i assumed failed
- $R(x_i = 0)$ - the decreased risk level with basic event x_i assumed to be perfectly reliable
- $R(base)$ - the present risk level with baseline unavailability of component i.

The RRW represents the maximum decrease in risk for an improvement to the component associated with basic event. The RAW presents a measure of the worth of the basic event in achieving the present level of risk. In addition, it also indicates the importance of maintaining the current level of reliability for the basic event. FV importance is a normalised risk reduction importance and is comparable to RRW.

Nowadays the most important area where important measures are applied is in test and maintenance programmes of the plants. The influence of test and maintenance is completely connected to the change of unavailability and not to a change in the defence in depth against a failure of the component. The components are considered important from the risk point of view during maintenance and testing if : RRW > 1.005 or RAW > 2 or FV > 0.9 [3].

3. METHOGOLOGY OF RCM

After identification of the safety important components using the importance measures the application of RCM methodology is performed. In this part the overview of the methodology is presented [1].

3.1 The RCM Overview

RCM logic is primarily concerned with the preservation of the ability of a plant item to perform its prescribed function. Note that the term 'item' may apply to a system, sub-system or a single component.

The process begins with a functional breakdown of the 'item' to ensure that all functions performed by the item undergo analysis. Once this is achieved, the second stage of the process, significant item selection, is completed. A 'significant item' is defined as any item the functional failure of which would result in an adverse effect on safety, an adverse effect on operational capability or high repair or recovery costs.

Following the compilation of the list of significant items, a comprehensive Failure Modes and Effects Analysis (FMEA) is performed for each item. The result of the FMEA for each function of each item is then subjected to the main RCM algorithm with a view to defining the optimum PM task (if one exists) to protect the function. If no suitable task exists, the RCM logic may, under some circumstances, direct that redesign is necessary, i.e. modification for in-service plant.

Assuming a suitable task is defined, the next stage is to derive the task frequency. The method of calculating task frequencies will vary dependent upon the type of PM task, however, for in-service plant, all rely heavily on historical data and operator experience.

The final stage of the analysis is to package the PM tasks into the maintenance plan for the item analysed. In the case of each pilot study it is expected that the list of tasks will form the basis for the revision of the existing maintenance schedule.

3.2 Functional Breakdown

In order to ensure that all functions of the system under scrutiny are taken into account during the RCM analysis, it is of paramount importance that the analysis is performed at the correct level of complexity. This level is known as the 'indenture level'. The normally preferred indenture level is that of 'system', however, due to the complexity of some systems it may be necessary to break down the system into sub-systems, assemblies or components.

3.3 RCM Logic

The RCM logic is embedded in RCM analysis. It provides answers to specific basic questions presented by the RCM logic. The first series of questions define the failure consequences:

- Evident Safety (evident failure with safety consequences)
- Economic/Operational (evident failure without safety consequences)
- Hidden Safety (hidden failure with safety consequences)
- Hidden Non-safety (hidden failure with economic/operational consequences only)

The failure modes of components are classified using the logic tree. The priorities are determined in this step. All failure modes are evaluated using the following issues: A) safety related failure mode, B) failure mode leading to plant shutdown or C) failure modes with minor economical problems. There are also hidden failures (D) or failures evident for the operator. After termination of analysis each failure mode is classified as A, B, C, D/A, D/B or D/C. There are the following priorities of evident and hidden failure modes:

1. A or D/A
2. B or D/B
3. C or D/C

The logic tree to identify priorities is presented on Figure 1.

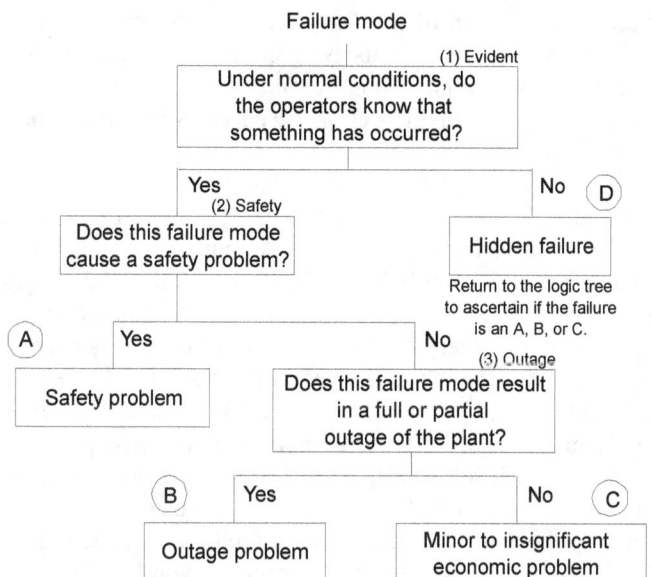

Figure 1: RCM Logic Tree

3.4 Types of Preventive Maintenance Tasks and Task Frequency Calculation

PM tasks are grouped into four basic types:

1. On-Condition Task - calls for corrective maintenance on the condition that an item does not meet a specified performance standard.
2. Hard-Time Task - an item is discarded or restored before a specified life limit
3. Combination Task - a combination of Task 1 and 2
4. Failure-Finding Task - to find the failure of hidden functions.

On-condition task frequencies are derived from factoring the calculated interval between the Potential Failure condition and the Functional Failure condition, i.e. the total time between the item no longer performing to its specified performance parameters and complete functional failure. This interval, known as the P-F interval, is then factored depending upon the failure consequences. If the functional failure of the item has safety consequences, the P-F interval is divided by a factor of 3 to produce the task interval. Should the failure consequences be economic or operational, a factor of 2 is sufficient. Therefore, should the P-F interval of an item be calculated using historical data as 2016 hours, and the failure would have safety consequences, the task frequency would be calculated as:

$$\frac{2016 \div 24 \div 7}{3} = 4 \, \text{weeks}$$

'Operator Monitoring' is a recognized On-Condition Task which involves no specific maintenance effort. It may be assigned to an item where a reasonably high degree of routine monitoring already exists. It must not be assigned to place an additional task on the operator.

Hard-Time Task frequencies are derived from calculating the 'Safe Life' of an item and then factoring the life by 2 or 3 depending upon the failure consequences. Often the Safe Life of an item, e.g. a structural or pipework support, is supplied by the manufacturer. Therefore, an item with a calculated Safe Life of 8 years, the failure of which would have only economic consequences, should be replaced or overhauled every 4 years.

A combination task is an amalgamation of an On-Condition Task and a Hard-Time Task, hence two task frequency calculations are required. This type of task is usually assigned to an item which required scheduled replacement or overhaul, but, possibly due to a harsh operating environment, it is judged prudent to also apply a series of inspection based tasks to ensure that degradation is not so rapid as to shorten the perceived life of the item.

Failure-Finding Task frequencies for safety-related hidden functions are derived primarily by calculating the acceptable degree of unavailability of the redundant or protective plant item. This is calculated by dividing the acceptable cumulative probability of failure, i.e. a Safety Case claim, by the failure rate of the primary or protected plant item. However, in order to perform this calculation precise historical data is required. Where no such data is available it is acceptable to derive the task frequency as a percentage of the Mean Time Between Failures (MTBF) dependent upon the required confidence level. For items the failure of which may carry nuclear safety consequences, a ninety-five percent confidence level is appropriate, for those without a safety consequence or plant/operator safety only, a ninety per cent confidence level is sufficient. It should be noted that for plant items that have not suffered any failures, the MTBF is taken to be the total usage to date. Some task frequencies within an RCM study may be the result of a necessary default due to constraints imposed by the Safety Case, i.e. should a revised task frequency cause a disruption to a probabilistic claim in the Safety Case, the RCM derived frequency should be ignored in favour of that stated in the Safety Case. It should also be noted that changes in the system operating regime may impact upon some task frequencies.

Should the analyst be unsure of the answer to any question posed by the main decision algorithm, he may elect to resort to the 'default logic'. For example, if it were unclear whether or not a functional failure would have an adverse effect on operating safety, the default logic would instruct the analyst to default to 'yes'. Likewise, should the analyst be aware that a functional failure would be evident to the operator whilst performing normal duties, but the failure consequences would be rapid and severe, he may default to 'hidden' thereby allowing the possibility of generating a Failure-Finding Task not available for evident functional failures.

4. FUNCTIONAL BREAKDOWN OF THE CORE COOLING SYSTEMS

The Core Cooling Systems are broken down into four major sub-systems:

- High Pressure Safety Injection System
- Low Pressure Safety Injection System
- Core Flooding System
- Containment Spray System

4.1 High Pressure Safety Injection System

The High Pressure Safety Injection System is further broken down into the following components:

- Boric acid tank
- Injection pump
- Motor operated valves
- Check valves
- Manual valves

4.2 Low Pressure Safety Injection System

The Low Pressure Safety Injection System is further broken down into the following components:

- Boric acid tank
- Injection pump

- Motor operated valves
- Check valves
- Manual valves

4.3 Core Flooding System

The Core Flooding System is further broken down into the following components:

- Hydro-accumulators, including pressure relief valves
- Check valves
- Motor operated valves

4.4 Containment spray system

The Containment Spray System is further broken down into the following components:

- Boric acid tank
- Heat Exchanger
- Injection pump
- Motor operated valves
- Check valves
- Manual valves

5. DISCUSSION OF THE RCM RESULTS

In this part the main results of the RCM analysis are discussed for the different systems and components.

5.1 High Pressure Safety Injection System

The HPSI tanks are subject to daily operator monitoring during which any signs of external corrosion or fatigue cracking would become evident and prompt timely repair action. The tank contents are also dipped monthly to test the boric acid concentration. Any internal corrosion mechanism would be revealed during water analysis as the presence of foreign material would be evident. Accordingly, no discrete PM tasks are assigned to the tanks.

The ongoing performance and reliability of the HPSI injection pump is a cause for concern. The pump is currently of poor reliability due to design issues, especially with the hydro disc. It is noted that many other WWER440 NPPs have overcome similar historical problems via successful modification action. Following a beneficial liaison visit of these plants, Mochovce NPP engineers have emulated part of their modification programme. However, some pump problems continue to occur. Accordingly, it is recommended that the Mochovce NPP HPSI pumps undergo a programme of modification in the same manner as that undertaken at other WWER440 NPPs. In the meantime, it is recognised that the current HPSI pump overhaul strategy is unnecessary and may even be counter-productive to successful pump operations by introducing maintenance-induced failure. As with the majority of pumps associated with Core Cooling Systems, the HPSI pumps are currently subject to both invasive overhaul and a condition-based PM strategy. Modern techniques associated with rotating plant tend towards PdM-based strategies (on the proviso that the plant in question runs frequently enough to render the PdM data as effective in predicting impending failure). In the case of the HPSI pumps this is clearly the case and as a result, the existing invasive overhaul routines are recommended for deletion in favour of existing PdM routines, i.e. vibration diagnostics, NDT, etc.

It is readily apparent that the majority of MOVs associated with the HPSI System (and the Core Cooling Systems as a whole) are routinely overhauled in accordance with OEM recommendations.

However, such recommendations are often based upon a pessimistic assessment of the number of annual valve operations, i.e. it is assumed that many operations will occur. However, when installed into standby or emergency systems, this is clearly not the case. The fundamental function of any given valve is simply to open or close on demand, i.e. to physically move to the desired position. Preservation of this functionality is supported by routine valve movement rather than fixed interval invasive overhaul. To this end, all HPSI MOV overhauls are recommended for deletion in favour of frequent freedom of movement checks, supported by valve functional tests, most of which are performed by default during 3-monthly, annual and 4-yearly system/full injection tests.

The primary failure mode associated with check valves is that of leaking, i.e. failure of the valve to maintain tightness in order to prevent reverse flow whilst allowing full forward flow (this often also applies to MOVs, etc). To this end, a task of functional test is recommended for HPSI-related check valves. It should be noted, however, that as the relevant check valves are tested as a matter of routine during 4-yearly injection tests, assigning these functional tests does not constitute an increased maintenance burden. Assigning these tasks to the relevant check valves simply formalises the process and ensures that the valves will be routinely stroked from fully open to fully closed during the 4-yearly injection test with the results formally recorded.

As with MOVs and check valves, freedom of movement is the primary consideration regarding manual valves. Assigning infrequent valve overhaul, e.g. every 8 years during alternate outages simply cannot guarantee that the valve will not seize between overhauls. Manual valve seizure can cause significant delays when, for example, attempting to isolate a pump for repair or maintenance. Accordingly, 8-yearly overhauls are recommended for deletion in favour of more frequent (1 monthly) freedom of movement checks.

5.2 Low Pressure Safety Injection System

The conclusions for LPSI tanks, MOVs, check valves and manual valves are the same as in case of the HPSI system.

The LPSI System pump is currently subject to both invasive routine overhaul and a PdM-based strategy. Furthermore, the 3-monthly test includes vibration diagnostics on all 3 pumps and pump motor bearing temperatures are constantly provided. Accordingly, the current scheduled overhaul recommended for deletion in favour of existing PdM.

5.3 Core Flooding System

The current strategy of 4-yearly specialist inspections and NDT (as a component of the overall primary circuit inspection programme) are recommended for retention. However, 4-yearly pressure test is recommended for deletion on the grounds that intentional over-pressurisation of a nuclear-safety related vessel is not good engineering practice. Consultation with the relevant engineers from the Sizewell Power Station (where the hydro-accumulators are similar in size and construction to those in use at Mochovce NPP) has revealed that although similar detailed inspections and NDT are performed, pressure testing is not carried out.

The Pressure Relief Valves (PRVs) associated with the hydro-accumulators require a specific strategy. The 8 Core Flooding System PRVs provide protection against possible hydro-accumulator breach should the nitrogen pillow cause an over-pressurisation. If such a breach were to occur owing to failure of the PRV to lift the minimum consequences would be a reactor trip, system drain down with the loss of up to 14 days of generation, and the requirement to replace an extensive quantity of boric acid.

Notwithstanding the commercial importance of these PRVs, the current annual overhaul (assigned to 8 valves) is a high maintenance burden. Furthermore, consultation with the relevant systems engineer has revealed that during approximately 40 overhauls to date, no problems with the PRVs have been encountered. In close co-operation with the Mochovce engineers, the following strategy has been identified. For a trial period of 4 years, each year 2 PRVs (1 each from a different hydro-accumulator) will be tested. Should any defects or degradation in valve performance be noted after 2 years, the current annual frequency will remain in force. Should there be no problems after 2 years, but observed failures after 3 years, the frequency will be set at 2 years. Should there be no problems after 3 years, but observed failures after 4 years, the frequency will be set at 3 years. Should there be no problems after 4 years, the frequency will be set at 4 years. The whole process is presented on Figure 2.

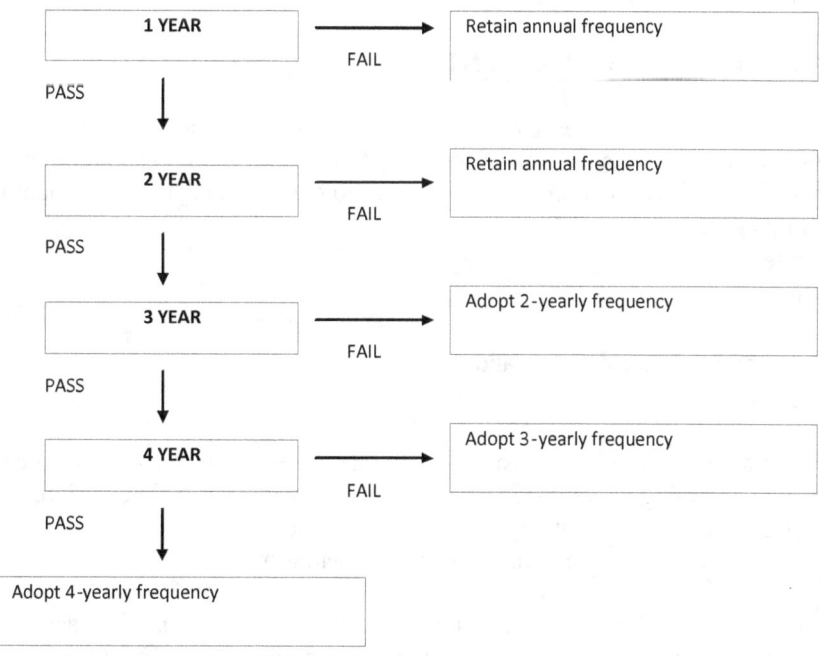

Figure 2: The frequency determination strategy for PRVs

RCM Logic recommends that check valves are subject to a similar trial to that detailed above for the PRVs. However, the trial will be simplified by overhauling 1 valve per train after 4 years. Should no problems be encountered during the second batch overhaul at 8 years, the frequency of overhaul for all 8 valves will remain at 8 years. However, should there be any problems encountered at 8 years, the current 4-yearly frequency will remain in force. During this trial, annual valve tightness checks have been introduced as an extra measure of protection against unforeseen valve failure.

For MOVs a similar trial to that detailed above for check valves is recommended. However, the current 8-yearly pump motor overhaul is recommended for a similar trial with a light maintenance routine replacing full overhaul. As with check valves, an annual valve tightness check is recommended as additional high frequency protection whilst the trial is undertaken.

5.4 Containment spray system

The conclusions for the tanks and pumps are the same as in case of the LPSI system.

The heat exchanger within the Containment Spray System is of primary importance as it not only cools the water re-circulated from the containment sump during HPSI and LPSI System operations, but also provides the means of emergency core cooling following a seismic event. RCM Logic has recommended that the current PM strategy assigned to the heat exchanger should be retained, i.e. continuous leak testing (via integral pumps), permanent measurement of chemical solutions in case of tube leaks and 4-yearly detailed inspections, performed by specialists team and including vessel wall thickness check.

For the majority of MOVs installed within the Confinement Spray System, it was determined that the consequences of failure were of little or no significance. Accordingly, a run-to-failure policy is recommended.

The check valves remain subject routine testing during scheduled system testing. Where necessary, valve testing via air pressure injection should be performed.

As with other manual valves within the Core Cooling Systems, the manual valves associated with the confinement spray system should be subject to monthly freedom of movement checks.

6. THE RESULTS OF RISK ASSESSMENT

The pumps, the motor operated valves and check valves of the HPSI and LPSI system were identified as the most dominant components from the risk point of view. There are the components of the safety systems which are in stand-by operating modes and periodically tested. The unavailability of the components is influenced by:
- failure rate
- testing interval
- repair time
- testing and maintenance duration and
- human errors

The impact of maintenance changes on the failure rates cannot be evaluated before their implementation and receiving experience from operation. Then, after a time period, some trends can be identified in the failure rates. However, the RCM approach has the objectives not increase but to decrease the failure rates by implementation of the new PM tasks.

The analysis recommended changes in the intervals of overhauls or these interactions were deleted. These changes decrease mainly the maintenance duration and impacts of human errors. In some cases the interval of overhauls is changed from 4 to 8 years. It decreases the risk during the refueling outage. The core damage frequency for shutdown operating modes is decreased by about 5%. There are no significant changes in the large early release frequency. However, on the other side, significant financial sources are saved by the proposed changes.

7. CONCLUSIONS

The following conclusions associated with the Mochovce NPP Core Cooling Systems are made:

- Operating experience indicates that the system components are currently over-maintained as the current PM tasks and frequencies are primarily OEM based.
- Mochovce NPP performs sufficient non-invasive PdM monitoring of important components to render invasive overhaul unnecessary.
- The current PM tasks assigned to the Core Cooling Systems have to be replaced with new one detailed in the analysis.
- The modern ultrasound techniques should be employed for the purposes of performing valve tightness checks.
- Several plant items analyzed are recommended for 'run-to-failure' as their failure present no safety, commercial or economic consequences.
- The proposed changes minimize the maintenance activities and reduce the shutdown risk.

References

[1] Smith, A. M., Hinchcliffe, G. R.: RCM Gateway to World Class Maintenance, Elsevier, USA, ISBN 0-7506-7461-X, 2004

[2] M. van der Borst, Schoonakker, H.: An Overview of PSA Importance Measures, Reliability Engineering and System Safety, (2001)

[3] Optimization of the Preventive Maintenance Tasks Using RCM, RELKO report 3R0404, Bratislava 2006 (in Slovak)

An Integrated Management for Occupational Safety and Health throughout the Plant-Lifecycle

Yukiyasu Shimada[a,*], Teiji Kitajima[b], Tetsuo Fuchino[c], and Kazuhiro Takeda[d]

[a] National Institute of Occupational Safety and Health, Japan, Kiyose, Tokyo, Japan
[b] Tokyo University of Agriculture and Technology, Koganei, Tokyo, Japan
[c] Tokyo Institute of Technology, Meguro, Tokyo, Japan
[d] Shizuoka University, Hamamatsu, Shizuoka, Japan

Abstract: The main purposes of occupational safety and health (OSH) management are to assure safe and healthful working conditions for working men and women and to prevent industrial accidents by the establishment of process safety management (PSM) system in the company level as well as the improvement of safety engineering techniques. Business process model has been developed to systematize the engineering activities and information flow throughout a plant-lifecycle (i.e. from research and development through process/plant design, construction and active manufacturing period, including production and maintenance) of chemical processes. This paper proposes an integrated approach for OSH management based on the business process model of engineering activities. This approach consists of three level hierarchical PSM; 1) PSM framework at enterprise-level, 2) HSE (Occupational Health, Process Safety, and Work Environment Protection) management activities at middle-management-level, and 3) SQDC-conscious tasks at manufacturing-site-level. Hierarchical integration of the PSM at each level makes it possible to realize the consistent and collaborative OSH management.

Keywords: Occupational Safety and Health Management, Process Safety Management System, Business Process Model, HSE Management, SQDC (Safety, Quality, Delivery, Cost)

1. INTRODUCTION

There has been a recent surge in the number of disasters and incidents in occurring in the process industry (e.g. the petrochemical, chemical, food and pharmaceutical industries). The reasons include defects in process-safety management (PSM); inadequate safety management systems in companies; inadequate knowledge among managers and insufficient information about the tasks undertaken and resultant erroneous operation and/or misjudgment; no standardization for the PSM activity; and other engineering factors. Expecting that PSM will reduce the hazards and likelihood of disasters, OSHA in USA emphasizes PSM and requires that companies establish PSM systems and improve safety engineering techniques[1]. Existing PSM guidelines, OSHA/PSM, Seveso II Directive[2], OHSAS180001[3], and others, establish only minimum elements for safety management. The Ministry of Labor, Health and Welfare of Japan released "Guidelines on Occupational Safety and Health Management Systems" in 1999 (amendment, 2006)[4]. Purposes of occupational safety and health (OSH) management are to assure safe and healthful working conditions for workers and to prevent industrial accidents by the improvement of each company's PSM system as well as safety engineering techniques.

This paper proposes an integrated approach for OSH management based on business process model of engineering activities. Hierarchical PSM consist of PSM framework at enterprise-level, HSE (Occupational Health, Process Safety, and Work Environment Protection) management activities at middle-management-level, and SQDC (HSE, Quality, Delivery, and Cost)-conscious tasks at manufacturing-site-level. Hierarchical integration of the PSM makes it possible to realize the consistent and collaborative OSH management.

---

* shimada@s.jniosh.go.jp

2. BUSINESS PROCESS MODEL FOR THE PLANT-LIFECYCLE ENGINEERING

2.1 Basis of Business Process Model

In order to achieve a systematic PSM, a model-based engineering framework is needed so that information can be used to inform all stages of the plant-lifecycle (i.e. from research and development through process/plant design, construction and active manufacturing period, including production and maintenance). Constantly updated and revised data and information must be shared at each engineering stage in a transparent way in order to examine the impacts of safety decisions of all activities of the chemical process plant.

Among the challenges by the safety division of the Society of Chemical Engineer Japan, IDEF0 (Integrated DEfinition for Functional model standard, Type-zero)[5] is adopted as a description format to develop the business process model. IDEF0 is a well-known standardized method for enterprise-resource planning or business-process (re)engineering. Figure 1 shows the basis of the IDEF0 format. The rectangle represents an 'activity (function)', and the arrows describe information. The information is classified into four categories: 'Input' which is changed by the activity, 'Control' which constrains the activity, 'Output' which is the result of the activity, and 'Mechanism' which includes the resources of the activity. The information is collectively termed ICOM (Input, Control, Output, and Mechanism). Each activity can be further developed hierarchically to detail sub-activities as needed. Development of a business process model using the IDEF0 format enables function-based discussions.

Figure 1 Basis of IDEF0 Format

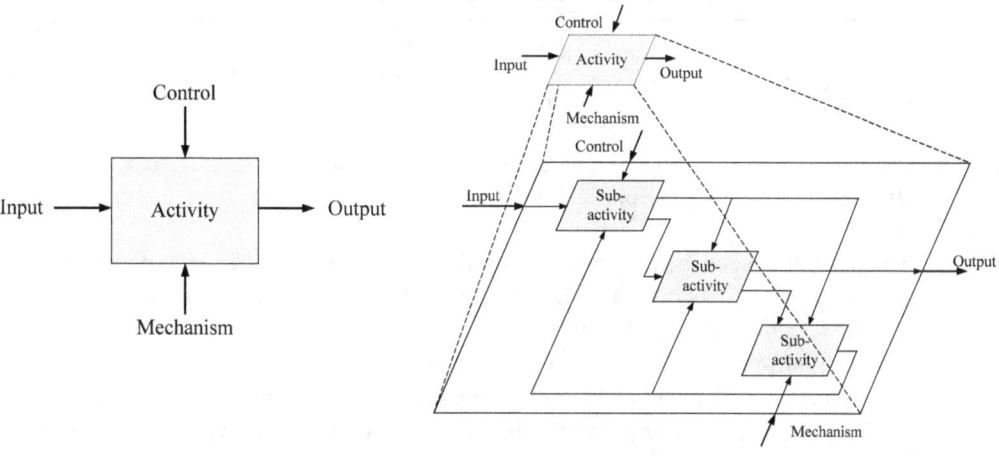

2.2 Template for Business Process Modelling

The PIEBASE (Process Industry Executive for achieving Business Advantage using Standards for data Exchange) was an international consortium to achieve a common strategy and vision for the delivery and use of internationally accepted standards for information sharing and exchange (ISO-STEP), and developed a business process model to represent the core business activity of the chemical process industry[6]. The PIEBASE model uses a template approach across all principal activities. This template consists of three steps: (1) manage, (2) do, and (3) provide resources. The purpose of the PIEBASE model is to provide a common understanding of the engineering and information requirements of processes throughout the plant-lifecycle. However, the activities in the model were defined to reflect current practices.

On the other hands, as shown in Figure 2, a template for business process modelling (BPM-template) of plant-lifecycle engineering (Plant-LCE) has been proposed to generalize the modelling in IDEF0 format and enable discussion of integrating each business process model for Plant-LCE[7]. This BPM-template consists of two functions: "Performance in the form of a PDCA-cycle[8]" and "Resource provision".

Figure 2: BPM-template for Business Process Modelling

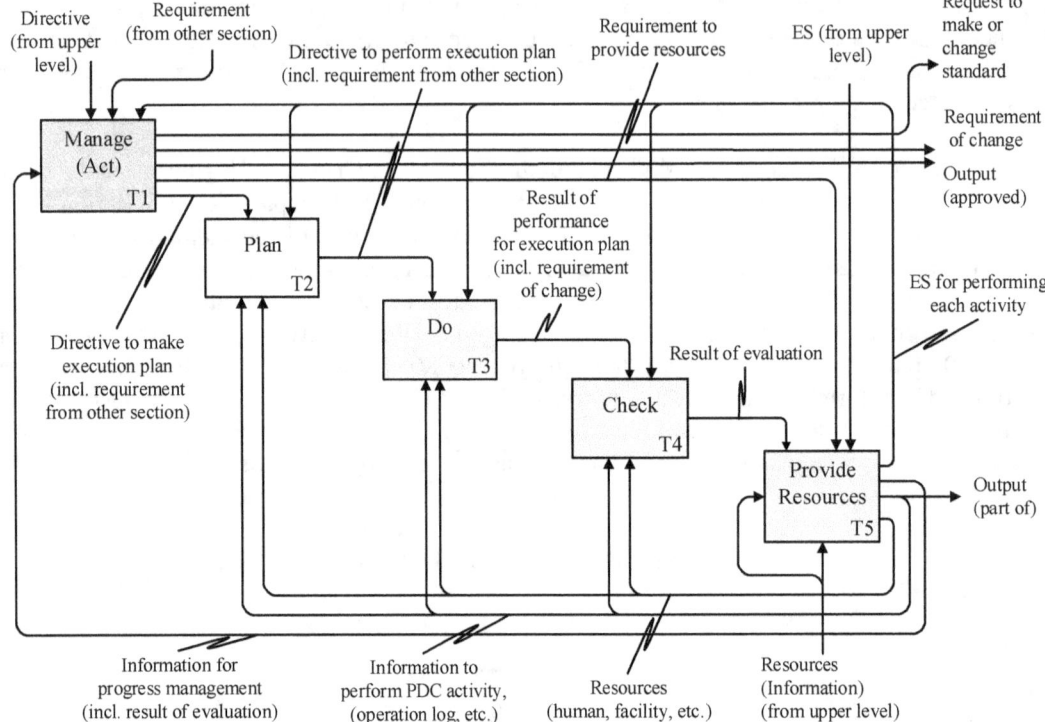

(1) Performance in the form of PDCA-cycle: Each activity should be carried out according to engineering standards (or ESs; e.g., technical standards and control standards) complying with laws and regulations.

- The 'Manage' activity manages the progress of overall activities within the same plane, including the requirement of resource provision, the improvement of engineering standard, and the decision making of the next action for change requirement.
- The purpose of the 'Plan' activity is to make an executable plan for a given specific directive.
- The 'Do' activity executes a plan and yields requirements for administrative defect factors, if any.
- The 'Check' activity evaluates the results and the performance of the previous activities to support these goals: a) performance and results for the directive and the plan, b) compliance with engineering standard, c) sufficient provision of resources, and d) validity of the engineering standard itself.

(2) Resource provision: 'Provide Resources' activity provides the resources to support and control 'Plan', 'Do', 'Check', and 'Act' activities. These resources include: a) educated and trained people and organizations; b) facilities and equipment, tools, and methods for supporting activities; c) information to perform PDC activities and for progress management; and d) engineering standards for controlling each activity, which are given from the activity of the upper plane.

This BPM-template enables the development of a business process model to perform activity planning, execution, evaluation, and improvement at each sub-activity plane. That is, the model based on the proposed BPM-template shows implementation in the form of the PDCA-cycle and the uniform management of engineering standards with the provision of just enough resources. Furthermore, the developed model can clarify the purpose, the contents, and the relevant ICOM of individual activity and provide a framework of logical PSM.

2.3 Business Process Model for Plant-LCE

Business process model should be seen as a 'to-be' model that represents the logical business process. The following points are required for a referenceable model.

- The definitions of business functions and the scope of them must be clarified before starting the development of a model.
- Activities that develop technologies and activities that use technologies for engineering functions should be clearly distinguished.
- Activities must be categorized as 'Plan', 'Do', 'Check (Evaluate)', or 'Manage (Act)' activities in order to develop a model that constitutes an activity framework in the form of a PDCA-cycle.

Furthermore, two points must be kept in mind so as not to create a business process model that only represents a specific company's activities.

- The model should be considered separately from the company by assigning tasks based upon on organizational structure not specific workers in the specific company. That is, tasks should not be based on the question of "who should do them?", but rather on "what has to be done?"
- Specific activities in an individual company should not be the focus of the model. Widely-used and generalized structures of activities and information flow related to the activities should be developed.

Activities that are performed at actual companies (plant engineering companies, plant operation companies, etc.) have been compiled and examined, and business process models have been developed based on the BPM-template. Figure 3 shows a business process model reflecting the activities of Plant-LCE. 'Do' activities of this model are comprised of activities of process and plant design, construction, and manufacturing (production and maintenance) stages[9].

Figure 3: Business Process Model Reflecting the Activities of Plant-LCE

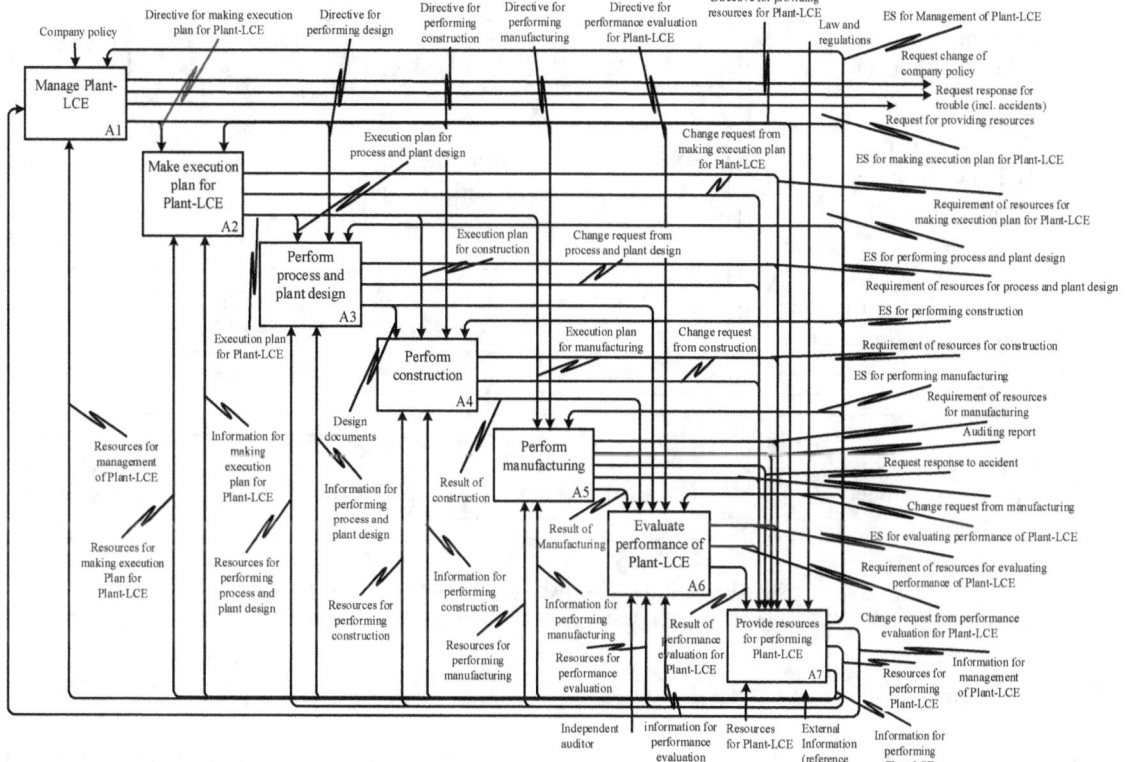

The developed business process models systematically show the universal activities and information flow at each engineering stage. These models can clarify the business process in the form of the PDCA-cycle throughout the plant-lifecycle as well as at each engineering stage, and the resource provision to share the process-safety information, to ensure consistency of engineering standards, etc. A logical and consistent business flow for each company can be developed by referring to these models. A specific business flow shows the framework, activities, information, etc. that are needed in order to prevent malfunctions and accidents, and is useful for the development of an environment for a systematic PSM.

3. INTEGRATED APPROACH FOR OCCUPATIONAL SAFETY AND HEALTH (OSH) MANAGEMENT

3.1. Hierarchical PSM

Systematized PSM can be established based on the essential requirements of the PSM activity, that are management, planning, execution, evaluation, and improvement (PDCA-cycle) and the resource provision. These are embedded clearly in the business process model of engineering activities. This paper proposes an integrated approach for OSH management based on the business process model of engineering activities. Figure 4 shows the structure of hierarchical integration of the PSM. PSMs at enterprise-level, middle-management-level, and manufacturing-site-level are systematized respectively to realize the consistent and collaborative OSH management.

Figure 4: Structure of Hierarchical Integration of the PSM

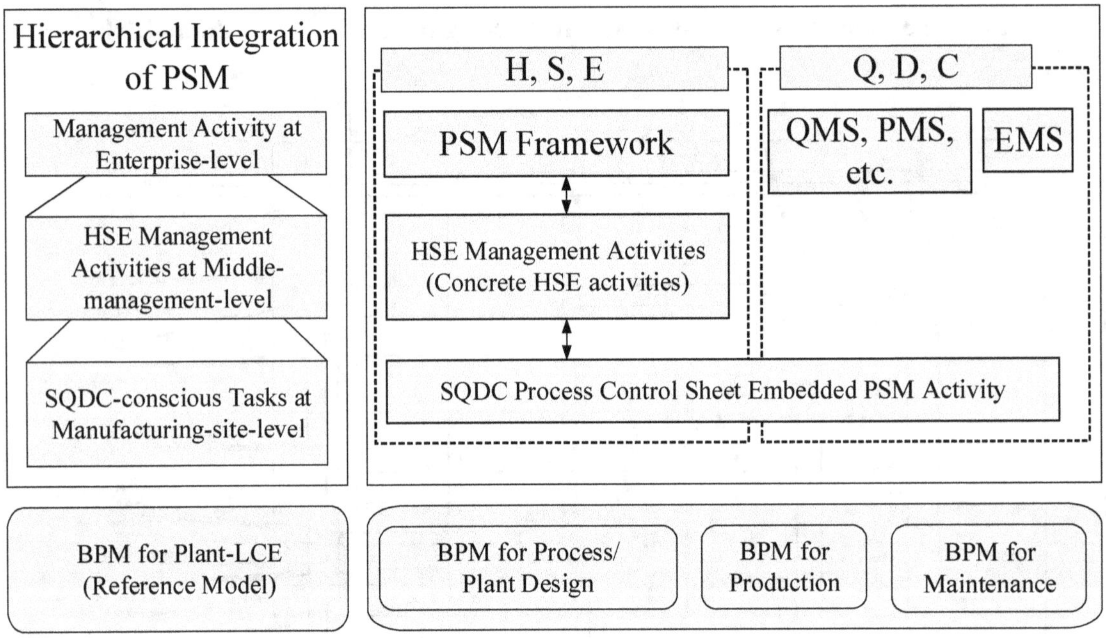

QMS: Quality Management System, PMS: Production Management System,
EMS: Environment Management System, BPM: Business Process Model

3.2. PSM Framework at Enterprise-Level

Shimada et al. has developed business process model for PSM by extracting the essential activities and the information for maintaining the process safety based on the business process model of the engineering activities[10]. For the PSM strategy reflecting corporate philosophy and policy for the social demand, OSH management system is created so as to clarify the implementation of PSM activities in the form of PDCA-cycle and the provision of resources (human resources, facilities and equipments, information, etc.) needed to accomplish the activities.

Business process model for PSM has been summarized by embedded structure of PDCA-cycle at enterprise-level and plant-site-level respectively and a comprehensive PSM framework has been structured based on the model as shown in Figure 5. It has been confirmed that the position of the PSM functions (e.g. 14 elements in OSHA/PSM) can be defined as the concrete engineering activities throughout the plant-lifecycle[10]. This confirmation makes it possible to specify how each PSM element should function in the usual engineering activities. Proposed PSM framework can be applied to improvement of company-specific PSM system to match a business's configuration. Features of proposed PSM framework are follows.

- Relation of PSM activities between enterprise-level and plant-site-level can be clearly specified.
- Improvement requirement against the defect factor on the result of implementing the PSM activities can be clarified as management system in the form of PDCA-cycle.

Figure 5: Framework for Overall PSM Activities

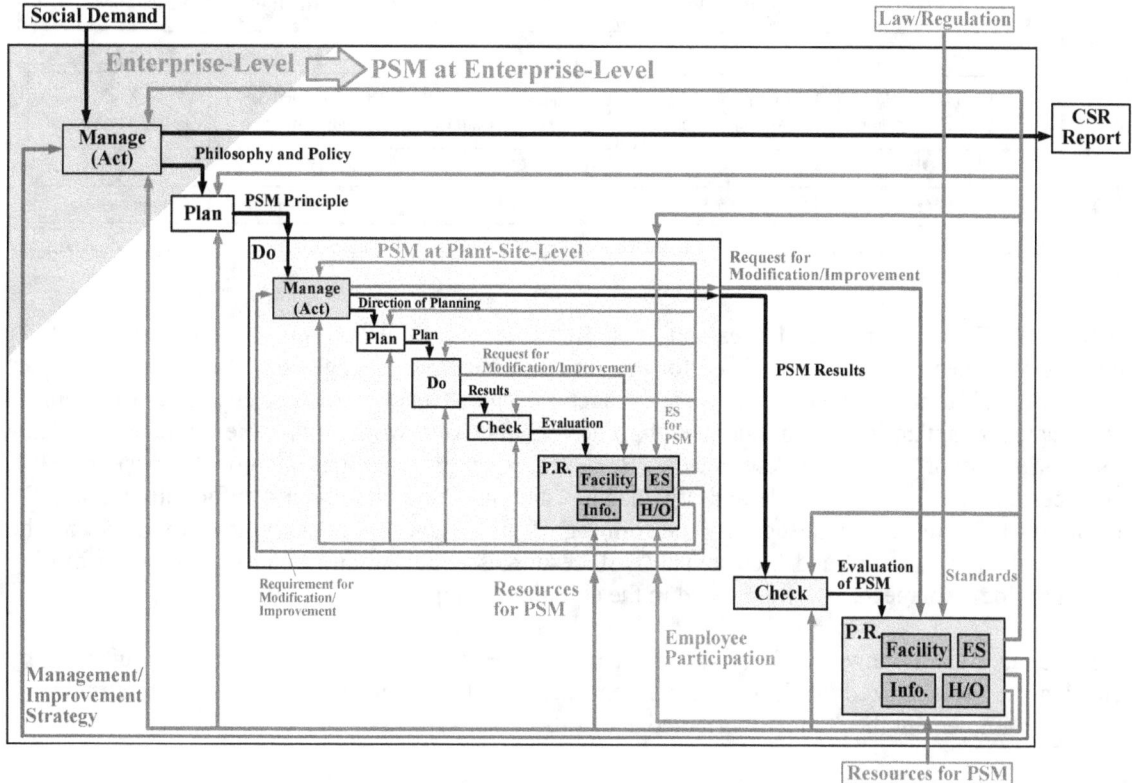

3.3. HSE Management Activities at Middle-Management-Level

In this paper, not only process safety but also occupational health and work environment protection are considered as HSE management activities at middle-management-level. HSE management activities have been systematized to embody PSM activities at middle-management-level under the PSM framework[11,12]. Figure 6 shows systematized HSE management throughout the plant-lifecycle. For each engineering stage, HSE manager demonstrates the HSE target setting and tactics decision according to the HSE strategy presented from the enterprise-level. HSE management activities should be also improved in the form of PDCA-cycle, that is formulating the HSE execution plan for the HSE target and tactics, implementing the plan, evaluating the implementation of the plan, and improving where necessary, with provision of just enough resources to perform each activity. Each HSE management activities are adapted in the form of PDCA-cycle with resource provision.

Figure 6: HSE Management Activities throughout the Plant-LCE

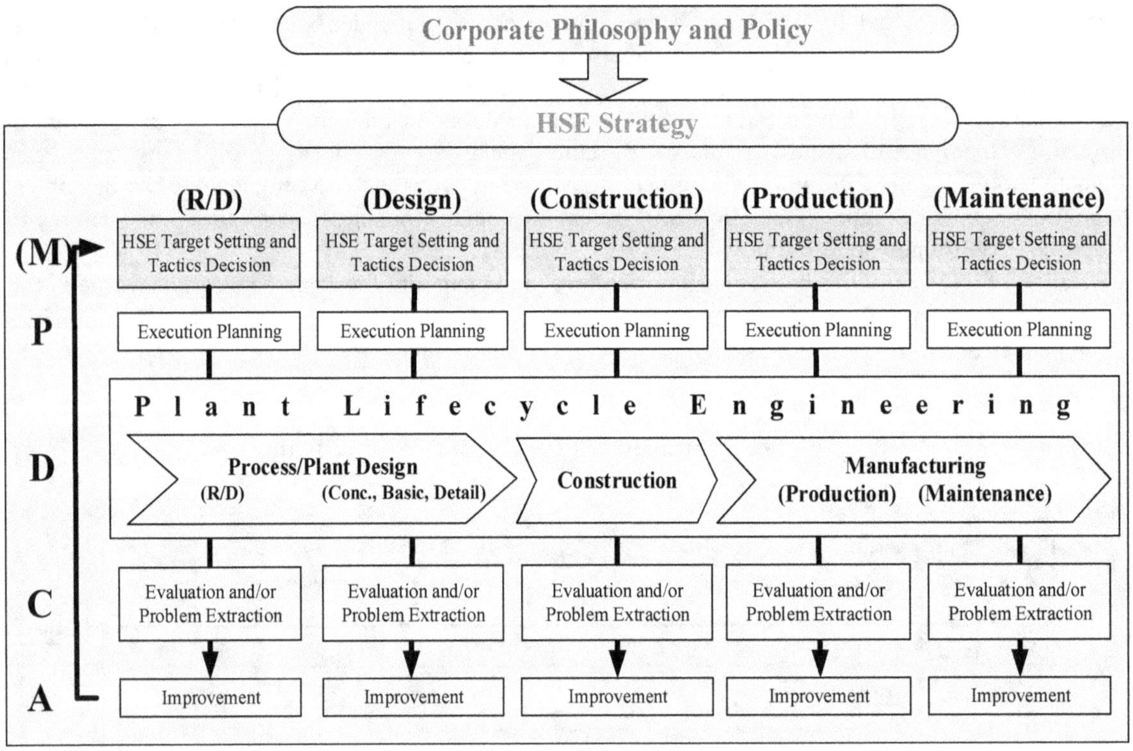

Concrete HSE management activities and related HSE resources at each engineering stage of plant-lifecycle have been analyzed and listed for the systematized HSE management. Table 1 shows a part of list of the HSE management activities at production stage. 'Do' activities are classified according to H/S/E categories, that is occupational health, process safety, and work environment protection. Table 2 shows a part of list of the resources for HSE management activities at production stage. HSE resources are clarified into people and organization, facilities and equipments, information, and HSE standards. HSE resources needed for performing Plan-Do-Check-Act activities are listed in the columns of Input, Control and Mechanism. HSE resources which should be provided by performing "HSE provide resource" activity are listed in the column of Output.

HSE manager can review the HSE management activities which should be performed currently by referring the systematized HSE management and the list of concrete HSE management activities. It can lead to implement the company-oriented and 'to-be' PSM system.

3.4. SQDC-Conscious Tasks at Manufacturing-Site-Level

A method of production management using SQDC process control sheet[12] is proposed as one of the PSM activities at manufacturing-site-level. SQDC process control sheet is developed by the addition of the viewpoints of SQDC for the production with respect to existing QC (Quality Control) process chart which has been used to maintain and improve the quality of products in the manufacturing. In this sheet, engineering standards such as standard operating procedure (SOP), rules are associated with each process as reference information to perform the HSE management activities. Figure 7 shows an example of SQDC process control sheet. A sheet is used for summarizing the process briefly to stimulate the PSM activities at the manufacturing-site. SQDC process control sheet helps field workers execute daily production tasks to prevent any possible troubles and/or industrial accidents as well as improve the quality and the productivity of the product at manufacturing-site. Field manager can also control their sure task executions for PSM according to this sheet.

Table 1: A Part of List of HSE Management Activities at Production Stage

	Activity	Contents
Plan	Execution Planning	▶ Formulation of HSE execution plan - HSE execution plan for HSE targets and tactics at production stage
Do	Engineering for Occupational Health	▶ Investigation of related laws and regulations - Related laws and regulations with occupational health
	Engineering for Process Safety	▶ Investigation of related laws and regulations - Related laws and regulations with process safety - Hazard data of materials, MSDS, etc. ▶ Hazard identification - Operational risk, hazard data of process and materials, MSDS, etc. ▶ Safety countermeasure for safety operation - Confirmation access for operation - Takeover operational information - Monitoring safety under operation - Maintenance request for equipment trouble ▶ Documentation for safety operation management - Education document on operation history - Information of accident or trouble ▶ Education and training for safety operation management - Education and training for safety routine or non-routine operation on education plan
	Engineering for Work Environment Protection	▶ Investigation of related laws and regulations - Related laws and regulations with work environment protection
Check	Evaluation and/or Problem Extraction	▶ Evaluation of implementation of HSE execution plan - Check points for compliance with laws and regulations - Evaluation points for HSE activities performance at production
Act	Improvement	▶ Improvement, where necessary - Review for audit results, defected factors of production, etc. - Modification of HSE targets and tactics of production, etc.

Table 2: A Part of List of Resources for HSE Management Activities at Production Stage

	Category	Input, Control, Mechanism	Output
HSE Resources	People and Organization	- License for related laws and regulations - Document for HSE education and training - Equipment for HSE education and training	- Results of medical check - Results of mental health care - Results of education and training
	Facilities and Equipments	- Fire control or fire-fighting equipment, etc. - Block & blowdown facility, Safety valve, etc. - Drafter, Emergency power source, etc.	*Same as on the left*
	Information	- Hazard data and SDS - Design base information - Construction plan under operation - Maintenance plan and result - Responsible care support data	- Plant operation data (SQEA monitoring data) - Results of hazard identification on HSE - Incident or trouble reports - Equipment trouble reports
	HSE Standard	- HSE department tactics and HSE plan - Applied laws and regulations - Standard for safety operation, MOC, emergency response, etc. - Plan of education and training	*Same as on the left*

Figure 7: An Example of SQDC Process Control Sheet

Name	DOP			Pub. No.			Version 1	2013-03-18	AA
Symb.	○ Process ◇ Inspection						Revision 1	2013-04-19	BB
	○ Transport ▽ Holding						Revision 2	2013-05-19	BB

	Authorization	Audit	Development

Process flow diagram R.M. (Sub-process / Main process)	Process name	Equip./ Fac.No.	Managerial item	Control point	Freq.	Remark (Ref. info.)	Worker	Management method and record sheet	Manager	Operation standard (Work standard)	Precaution — Claim (Operation (process) trouble / Occupational accident)
	0) Planning & scheduling		Quantity, Delivery time, Order, Basic unit		Every order		Planing Sec.	Production plan and schedule	Planing Sec. Chief	Production planning and scheduling rule	
	1) Raw material acceptance										
	Acceptance	T-001 ~T-012	Product (Quantity, Lot) / **Leak**	RM accept. control standard, **Safety standard**	Every accept.	Check gas-detector	Procure. Sec.	R.M. management sheet	Procure. Sec. Chief	Acceptance rule	
	2) Reaction preparation	R-101									
	Tank check	T-001, T-002	Product (Quantity (level))	> Lower storage level limit	Every batch		Production Sec. (A group)		Production Sec. Chief	Storage rule	
	Line-up	T-007, T-012	Line setting	Valve check list		Technical standard, P&ID, etc.		Reaction record sheet			
	N2 substitution		N2 Press., O2 conc., Leak	1 bar (3 times), O2=N/D						SOP	
	3) Raw material Feed	R-101									
	Fresh 2EH Feed	<- T-001	F2EH (Quantity, Rate)	1 ton: 4m3/hr	Every batch	**Emergency response standard, Technical standard**	Production Sec. (A group)	Reaction record sheet	ditto	SOP	**Abnormal reaction**
	Agitation		Agitation rate	Mode: MILD (35 < rpm < 45)							**Local heating**
	Steam heating		Steam temp.	145 C < T < 155 C							**Leak and ignition due to overload press. to elimination equip. (Abnormal press.)**
	4) 1st esterificatin reaction	R-101	Reaction temp. / Reaction time	150 C / 5 hr							
	FAU Feed	<- T-002	FAU (Quantity, Rate)	Stop agitation at feeding / 1.5 ton: 9m3/hr	Every batch	Emergency response standard,	Production Sec. (A group)	Reaction record sheet	ditto	SOP	**Leak and ignition due to overload press. to elimination equip. (Abnormal press.)**

Process flow diagram:
① Raw material → ▽② Reaction preparation → ③ Feed → ④ Reaction 1 → ...

Figure 8 shows a PDCA-cycle of SQDC-conscious tasks execution using SQDC process control sheet. SQDC process control sheet is developed for process arrangement (Plan). SQDC-conscious tasks are implemented using the SQDC process control sheet (Do). Results of implementation are evaluated using the SQDC process control sheet and the SQDC-check-sheet, which is developed separately (Check). For the evaluation results, process, activities and engineering standard such as work instructions are reviewed and improved (Act). This PDCA-cycle activity leads to improvement of PSM at manufacturing-site-level.

Figure 8: A PDCA-cycle of SQDC-conscious Tasks using SQDC Process Control Sheet

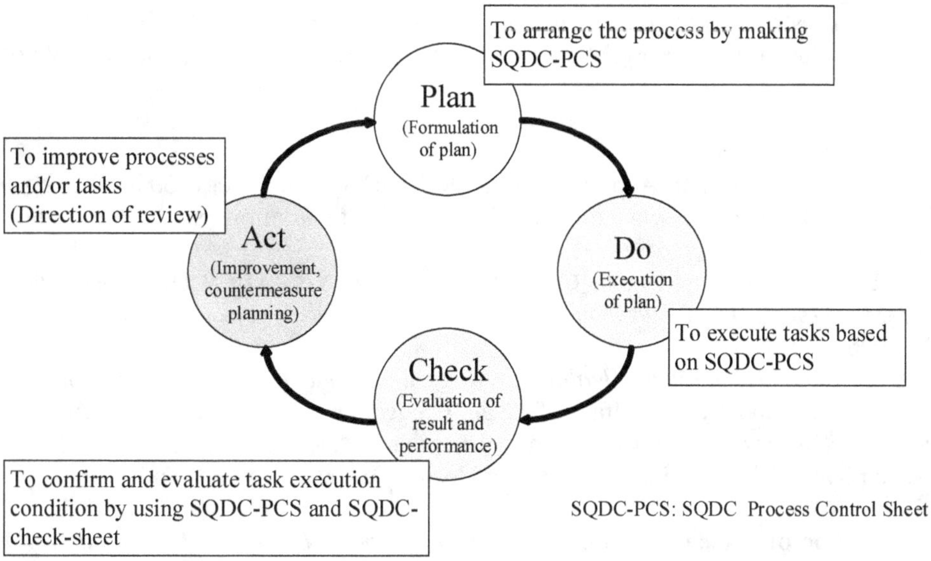

Following positive effects are expected for implementing the SQDC-conscious tasks using the SQDC process control sheet.

- Field workers are not required to perform new and additional tasks, but required to implement steadily the SQDC managerial features written in the SQDC process control sheet. This becomes possible to perform each process without omission and duplicate works for viewpoints of safety as well as quality and productivity. It can lead to cost reduction.
- Field workers can execute their tasks confirming and understanding not only how they can do it (know-how) but also why they should do it (know-why).
- To ensure the consistency of manual, SOP, etc. written in 'Standard' section and to provide them as resources for implementation of each process facilitate the improvement of processes and tasks at manufacturing-site-level and the prevention of the variability of the tasks between field workers, mismanagement, error of check, etc.
- Field manager can manage the production which satisfies the requirements for OSH management.

4. CONCLUSION

This paper proposes an integrated approach for OSH management based on the business process model of engineering activities. PSMs at enterprise-level, middle-management-level, and manufacturing-site-level are integrated in a hierarchical way. The approach is based on formulating the execution plan for specified objectives (Plan), implementing the activities as planned in compliance with engineering standard (Do), and evaluating the result of implementation of activities (Check). If some defect factors happen, countermeasures are considered and improved at next activities and tasks (Act). Furthermore, each activity of 'Plan', 'Do', 'Check', and 'Act' can be performed steadily by providing the resources (human resources, facilities and equipments, information, etc.) needed to perform each activity and complying with the engineering standards.

Proposed hierarchical integration of the PSM does not treat the PSM activities as special and additional tasks, but views the consistency, the affinity, the accommodativeness with production activities as the basis of PSM activity. PSM activities at each level make it possible to realize the consistent and collaborative OSH management.

Acknowledgements

The authors are grateful to the following process safety experts for their cooperation in discussion: Prof. Y.Naka (Tokyo Inst. of Tech.), Mr. O.Usui (Mitsui Chemical), Mr. T.Kawabata (NPO Life Science Technology Spread Center), Dr. H.Saito (Saito MOT Labo.), Mr. H.Sumida (Toyo Engineering Corp.), Mr. K.Bito (Kaneka Corp.), Mr. H.Motoyama (Daikin Industries), Mr. N.Yamamuro (Zeon Corp.), Mr. S.Yori (Mitsubishi Corp.), and Dr. Y.Kikuchi (Univ. of Tokyo).

References

[1] Occupational Safety and Health Administration, 29CFR 1910.119, *Process Safety Management of Highly Hazardous Chemicals*, Occupational Safety and Health Administration, Washington, D.C., USA, (1992).

[2] The Council of the European Union, *Council Directive 96/82/EC of 9 December 1996 on the Control of Major-Accident Hazards involving Dangerous Substances*, Official Journal L 010, pp.13-33, (1996).

[3] OHSAS 18001:2007, *Occupational Health and safety management systems - Requirements*, (2007).

[4] Ministry of Health, Labour and Welfare of Japan, *Guidelines on Occupational Safety and Health Management Systems*, Ministry of Labour Notification 53, Amend.,113, <www.jisha.or.jp/english/pdf/Ministry_of_Labor,Health_OSHMS_Guideline.pdf> accessed 06.03.2014, (2006).

[5] National Institution of Standards and Technology, *Integration Definition for Function Modeling, Federal Information Processing Standards Publication*, 183, http://www.idef.com/pdf/idef0.pdf <accessed 06.03.2014>, (1993).

[6] PIEBASE, *PIEBASE (Process Industries Executive for Achieving Business Advantage using Standards for Data Exchange) Activity Model Executive Summary*, http://www.posc.org/piebase/, (1998).

[7] Y.Shimada, T.Kitajima and K. Takeda, *Practical Framework for Process Safety Management based on Plant Lifecycle Engineering*, Proc. of 3rd International Conference on Integrity, Reliability & Failures, IRF'2009, pp.501-502, Porto, Portugal, (2009).

[8] W.E.Deming , *Out of the Crisis*, The MIT Press, Cambridge, U.S.A., (2000).

[9] Y.Shimada, T.Kitajima, T.Fuchino, and K.Takeda, *Disaster Management Based on Business Process Model through the Plant Lifecycle, Approaches to Managing Disaster - Assessing Hazards, Emergencies and Disaster Impacts*, Chap.2, pp.19-40, INTECH, (2012), Rijeka, Croatia.

[10] Y.Shimada, T.Kitajima, and H.Sumida, *Framework for Process Safety Management based on Engineering Activity through Plant Lifecycle*, Mary Kay O'Connor Process Safety Center International Symposium 2010, (2010).

[11] Y.Shimada, M.Kumasaki, T.Kitajima, K.Takeda, T.Fuchino, and Y.Naka, *Reference Model for Safety Conscious Production Management in Chemical Processes*, 13th International Symposium on Loss Prevention and Safety Promotion in the Process Industries, 1, pp.629-632, (2010).

[12] National Institute of Occupational Safety and Health, Japan, *Promotion of Occupational Safety and Health Management with an Integrated Approach of Safety Management and Production Activities*, JNIOSH-TD-No.1 (in Japanese), (2011).

End User Involvement in the Development of Procedures and Safety Management Systems

Thomas Wold*[a] and Karin Laumann

Department of Psychology, Norwegian University of Science and Technology (NTNU). Trondheim, Norway.

Abstract: IT-based Safety Management Systems contains procedures, safety standards, checklists and descriptions on how different tasks should be performed, and are usually designed at an executive level in the organization, and then communicated to the lower level in the organization where they are being applied. This paper presents data collected from qualitative interviews with executives and operators from two companies in the gas and petroleum industry. The executives generally regard Safety Management Systems as important tools for all work in hazardous environments, while the operators weren't that enthusiastic. How can end user involvement in the development phase of procedures and Safety Management System improve use? A central argument is that Human Factors must be involved as early as possible in the development phase, and that operators need to understand the purpose of the management system in order to use it as intended. The informants that had been involved in the development of the procedures at least to some extent, felt an ownership to the management system, while the ones who hadn't been involved at all felt no ownership to the management system, and did not see the purpose of it.

Keywords: Safety Management System, Procedures, Communication, Human Factors

1. Introduction

In industries operating in hazardous environment there has been a development towards controlling the daily workflow through various forms of management system. IT-based Safety Management Systems contains procedures, safety standards, checklists and descriptions on how different tasks should be performed. Safety standards and work procedures are often, but not always, designed at an executive level in the organization, and then communicated to the lower level in the organization where they are being applied. The purpose of using IT-based Safety Management System is to code and share best practices, create corporate knowledge directories and to create solid knowledge networks within the organization. Executives generally regard Safety Management Systems as important tools for all work in hazardous environments, while the operators aren't always that enthusiastic. The questions investigated in this paper are:

1) How was Human Factors involved in the design of a safety management system for one petroleum company?
2) How does use of Human Factors knowledge increase the operators' satisfaction with and use of the safety management system?

---

[a] Thomas Wold: Thomas.Wold@svt.ntnu.no

1

2. Theory

2.1. Safety Management Systems

Safety management refers to the actual practices, roles and functions associated with remaining safe [1]. Safety management is "the policies, strategies, procedures and activities implemented or followed by the management of an organization targeting safety of their employees" [2, p. 283]. A Safety Management System is hence a formalized way of dealing with these practices, roles, policies and procedures. Safety Management Systems can be seen as "a socio-technical system for knowledge transfer [...], through documented experiences, best practices, and expert references" [3, p. 2]. Safety Management Systems are integrated mechanisms in the organizations, and the purpose is to control the hazards that can affect workers' health and safety [4, 5], to maintain risk within an acceptable range in the operations of any organization [6, 7], and to help the organization meet the regulatory requirements (4, 7, 8, 9]. This is not only a matter of coordinating between tasks, but also the accumulation and diffusion of organizational experience, and to turn tacit knowledge into explicit and shared knowledge [10]. There is also a general agreement that Safety Management Systems is a means to change safety management from being reactive to being proactive [11], and anticipating hazardous situations before they occur, and not just acting after an accident has occurred, or phrased differently; to protect against human error [8, 12, 13, 14]. There is also the matter of defining legal responsibility if incidents should occur [8].

Safety Management Systems in professional organizations has in several cases been less successful. One of the reasons is that in the development and implementation phases the emphasis has been mainly on the technical requirements, ignoring the social and cultural facets of knowledge management [5, 15]. It is also because designers tend to focus primarily on the technology and its features, and forget to look at the use of the system from the human point of view [16].

2.2. Human Factors

According to the European Committee for Standardization human beings should be considered as "the main factor and an integral part of the system" when work systems are being designed (17, p.3). This includes the work processes and the work environment. Several researchers stress the importance of giving high priority to the action of human beings when procedures are being designed [18]. When experts design systems, they are not always able to predict what kind of difficulties other people will experience when using their system. The experts are so familiar with their own system, they know very well how it works and how to control it, that it comes natural for them, but they often forget that the users don't necessarily have the same familiarity with the system. The users don't have the same mental models as the designers, and it may also be that they don't interact with the system frequently enough to develop the same mental models [16]. Another factor is that experts are often distributed across various locations which makes is difficult for them to have general knowledge of the installations and what impact different components might have on each other [19]. A related factor is that onshore workers sometimes lack awareness of offshore processes; they don't have the same direct sensory experience as offshore workers, and they don't have the tacit knowledge and impressions gained while working on the installation [19].

Well intended efforts to promote safety may marginalize the local and system specific knowledge inherent in the organization, and safety professionals must be aware of this. Since employees close to the work are the best qualified persons to make suggestions for improvements, they can be consulted before making final decisions, especially for those decisions that affect the employees. A successful approach on many occasions has been to incorporate users as actual members from beginning to end, although one must be aware that users working with design teams might become so familiar with how the designers think, and so familiar with the system they are designing, that the same problem arises. Therefor it is advisable to bring in a different group of users for the various stages of usability testing [16]. This empowerment of workers provides them with authority, responsibility and accountability for required decisions and ensures that both employees and managements are involved in setting goals and objectives [2]. There is also evidence to suggest that a greater level of commitment and adherence

2

to procedures can be achieved by keeping procedures few and simple and by emphasizing broad and direct worker participation in the process of implementing the procedures [8].

An important part of the design sequence is to analyze the typical user in order to ensure that their needs and the demands of the work situation are understood. One need to clarify the following; who users are, what main functions are to be performed by the system, what are the environmental conditions under which the system will be used, and what are the user's preferences or requirements for the system [16]. This particularly applies to the informations-processing characteristics of the system. Insights from human factors can be very useful here, as it relates to the interaction between human and the system. The precise boundaries of the discipline of human factors cannot be tightly defined but are closely related to ergonomics, engineering psychology, and cognitive engineering [16]. Human factors have often been concerned with the physical aspects of work, but the scope also includes cognitive thinking and knowledge-related aspects and mental interactions with the system [16, 19, 20]. Human factors revolves around the central importance of the user, and the goal of human factors is to enhance performance, increase safety and increase user satisfaction. This includes the communication and cognitive processes involved in using the system. If human factors methods and principles are applied as early as possible in the development of a system; in predesign analysis, technical design, and final test and evaluation, many of the human factors deficiencies will be avoided before they are inflicted on systems design [16].

The European Committee for Standardization state that ergonomic effort should be greatest at the beginning of the design process, as it is here the most important decisions that have consequences in the design are made [17]. This goes for human factors methods and principals as well, which should be applied in all stages of the design: predesign analysis, technical design and final test an evaluation [16]. Hence, human factors should be applied at an early phase in determining how the Safety Management System and the procedures should be developed, and how the staff training should be. Human factors principles are too often either left out entirely, or brought in too late in the development process when the product design is already completed and handed to a human factors expert. This only places everyone at odds with each other [16]. Rather, human factors must be a part of the process from the very beginning of the planning and development of procedures and Safety Management Systems in order to get a balanced development of the technical and human aspects [16, 20]. To achieve this it is important that workers are allowed to be involved in the design process [17].

2.3. Communication
Any Safety Management System, no matter how it is constructed, is communication. It can be convenient to pretend that this is one-way communication, but it's not, because the user interprets the information in the Safety Management System and turns it into knowledge, adding his or her prior experience. This is also part of the communication process which must be addressed as part of the user analysis, to consider the cognitive characteristics of the user. The cognitive strengths of humans must be emphasized, but also how operators feel and interact with operations and management and designed objects [21]. For example, a Norwegian study showed that the workers often thought it was difficult to find the relevant governing documentation within the safety management system, so they needed to use more effort and time in order to find what they needed [22]. This makes it unnecessary difficult for the workers to find the information they need in order to fulfill their tasks and to make the necessary decisions. Not only shall machines be designed to suit the physical abilities of the expected user, but instructions and procedures shall be designed to fit their mental abilities; the cognitive, informational end emotional processes in the human being [21].

If this is not done successfully, we have several studies indicating that workers will deviate from the procedures if they know a better way of doing it [13, 23, 26]. When workers deviate from procedures, one must either figure out how to ensure compliance, or see if they might have a good reason for deviating from the procedures. Have they for instance actually found a better/safer way of doing the job than the procedures prescribe? Safety Management Systems are living systems and should always open for evaluation, adjustment and changes. A Safety Management System is never completely finalized in its making. Ideally it should always be developed on the basis of new experience, in order

3

to replicate success. Three ingredients are particularly important for a successful match between procedures and practice: There should be feedback from the lower to the upper tiers of the organization, the adjustment of procedures should be based on the views of those directly involved, and the time interval between worker feedback and implementing changes should be as short as possible [8]. In existing system one must study the interaction between human and system to find to identify various problems and deficiencies [16].

2.4. Tacit knowledge

In any organization there will always be tacit knowledge, and much effort is made to turn tacit knowledge into explicit and shared knowledge, and to make invisible work processes visible and transparent. If those who actually perform the work are the only ones who knows how it is done, the ability to account for this invisible work and the tacit knowledge that accompanies it, can strengthen the organization's performance significantly [10]. However, tacit knowledge can be so complex that it is difficult to articulate in a way that makes sense, and many professions demand a certain experience in order to be able to make complex considerations [27]. This is not to say that tacit knowledge needs to remain tacit. With Choo's definition of tacit knowledge as "the personal knowledge that is learned through extended periods of experiencing and doing a task, during which the individual develops a feel for and a capacity to make intuitive judgments about the successful execution of the activity" [28], it is clear that this type of knowledge can also be made explicit and brought forward to other workers who lack the experience, which the management systems is an attempt to systematize. This way the separating lines between tacit and explicit knowledge will be moved, so that knowledge that was tacit yesterday is explicit today [27]. This can be done, at least to some degree. It's easier said than done, but it naturally involves the workers in a dynamic communication with the managers and the safety experts.

3. Method

3.1. Subject

The empirical data is collected by conducting qualitative in-depth interviews with 27 employees in two different companies in the Norwegian oil and gas production sector, hereby named Company A and Company B. The first three informants from Company A were onshore executives who have had an active role in the development of the company's Safety Management System, and these were interviewed in a preparatory study. In the same company five offshore executives and ten offshore workers, representing different disciplines; mechanics, electricians, logistics and lab technician, were interviewed on board at the oil and gas producing installation. The third round of interviews was in conducted on land, with nine foremen and offshore installations leaders in Company B, a company that provides contract workers to an oil and gas producing company (not Company A).

3.2. Data-collection process

The interviews were conducted in a separate room during the normal working hour of the informants. The interviews followed a semi-structured interview guide, where certain topics were planned in advance, but also allowing for the informant to bring new topics to the table, and also allowing for the structure for each interview to be different according to how the informant associated the various topics. The interview started off by letting the informants tell about their routines for an ordinary work day and to describe the Safety Management System and the purpose of it. The questions then became more detailed about which procedures they used and in which situations, how they learnt about the Safety Management System, about the user friendliness of the system, shortcomings and advantages. Each interview lasted for 30-45 minutes, with a few exceptions. The interviews with the offshore executives generally lasted longer than the interviews with the workers. The interviews were conducted and transcribed by the first author of this paper.

3.3. Data analysis

The data was analyzed using a thematic analysis. Thematic analysis offers a theoretically-flexible approach to analyze the major themes to be found in interviews (or other qualitative data) [29, 30, 31].

4

The first step was familiarizing with the data by transcribing the audio interview files. The transcribed material was then fed into the software program NVIVO 10 and coded into many categories and sub-categories (nodes is the term used in NVIVO 10). This first round of coding was rather broad, where each interview segment could be coded in several categories, where relevant surrounding data was kept to keep track of the context. After the initial coding some categories were developed into themes while other categories were too small to qualify as themes. The themes were not necessarily the same as the topics in the interview guide, some were of course, but new themes also arrived during the interviews, and some themes emerged when looking at certain keywords mentioned by some of the informants, and by looking at in which context they mentioned these words. After re-reading the coding of each theme, some themes collapsed into each other whilst other themes were broken down into separate themes. The initial codes were partly derived from the interview guides, but several new and sometimes unexpected codes emerged from the interview material. What eventually became themes was mainly guided by the interview material, and not theory driven.

4. Results and discussion

4.1. Purpose

One important basic factor in using a management system is to understand the purpose of it, and the basis of the procedure and its intended higher-level goals. The informants who were able to say something about the background for the development of the procedures and the purpose of the Safety Management System, had a much better use of it than those who couldn't.

In Company A, the Safety Management System was developed by an external consulting company. The consultants had a few meetings with the top management in Company A, but end users weren't involved in testing until it was almost time to launch the management system. The head of the designers said that user friendliness was not given high priority.

> A3: Not really. User friendliness is... well, it is a prerequisite for the management system as a
>
> whole that the user can click his way through a browser.

They did not involve Human Factors in the development of the management system, but had mainly a technical focus. This was partly due to financial and time restrictions, but also because the procedures had already been developed separately by a different company.

> A3: We met a forest of procedures that had been developed by another company. And we had
>
> to make a superstructure that should match all those procedures. And what you discover
>
> when you start to adjust it, is that it doesn't fit.

It will be better to have a more coherent process when developing the procedures and the management system, and to involve Human Factors and end users in the development. The executives in Company A acknowledged that the workers didn't have any sense of ownership to the management system and that this was a problem. The only informants in Company A that felt ownership to the management system were onshore executives who had been involved in the development of the system. The operators in Company A were not able to explain the purpose of the Safety Management System, and hence, they didn't see the point in using it. They acknowledged that it was necessary to have a management system, but couldn't explain why. They said that it had something to do with safety, but saw their own experience and competence as more important for safe conduct.

5

A11: We are perfectly able to do the job without having to sit for an hour reading documents before we start. They have to explain why we should go through all these documents. (…) Why should we sit here and read for an hour when we can just go out and do the job?

When the operators in Company A tried to explain the purpose of the Safety Management System they focused more on responsibility than on safety, and saw the Safety Management System as a way for the management to cover their backs in case of accidents. This has an effect on how they understand the Safety Management System as less important, and negatively influences their motivation for use. In contrast, foremen and offshore installation managers in both companies saw the procedures and the Safety Management System as a collection of best practice principles guided by many years of collective industrial experience. The informants in Company B often worked as contract workers for a large gas and oil producing company, using the management system of the hiring company. But Company B also had their own management system, a smaller entity where several of the informants had been involved in the redevelopment of. This has given them a better understanding of the basic idea of having a management system as a storing and categorising of experience to form a knowledge foundation to evaluate the line of action for a new task.

B24: It's the best of [the company's] 40 years of experience in oil production. (…) What they have gathered there, is the best praxis. How to perform a task and how to relate to HSE and everything we're in touch with. But it is never elaborative. In the end it's still we that have to put the final piece to the puzzle, because it is a lot of good stuff in there, but it can't tell you everything. It doesn't tell you what the weather's going to be like that day, for example. You still have to think.

Antonsen has shown how seamen often interpret attempts to govern work by formal rules as a negation of the seamen`s professional expertise. This will no doubt affect their respect for and their motivation for using the formal procedures. In addition, formal procedures have their origin in onshore organizations, like regulatory authorities and oil companies. This is outside the seaman community, or the ones doing the practical work, and this influences how the seamen interpret the formal safety management. The seamen in Antonsen`s study saw the procedures as based on the theoretical knowledge of some "office worker", and not as based on the practical knowledge possessed by competent seamen, and for them this undermines the legitimacy of formal procedures [12]. By involving the workers in the development of the procedures they will not see it as a negation of their own competence, but rather as an appraisal of their experience and competence, and it will increase ownership to the procedures.

4.2. Language
One of the more specific complaints the informants in Company A had concerned the language used in the Safety Management System. They would prefer it to be in Norwegian and not English. They also though it was a bit "academic" English, with some difficult words and grammar they were unfamiliar with. Both operators and offshore executives mentioned this.

A10: I think Norwegians have pretty good competence in the English language, but it seems to me as if those who use English a lot use a lot of words that we are not familiar with, the common people. I think they do it to impress. They write it wrong. That's what I think.

User tests and user involvement will no doubt help to avoid misunderstandings caused by unfamiliar grammar and vocabulary.

6

4.3. Adaption

In order to operate safely it is necessary to be able to adapt the written procedures to local and immediate circumstances [8, 13, 23]. The informants in Company B pointed to the general purpose and the basic idea of a management system, as a storing and categorization of experience that forms a basis that should be used when evaluation how a specific task should be carried out. Note that they express that individual evaluation is still necessary.

> B26: It's a bit like the Bible, you know. (…) You get an answer, but you have to interpret that answer. It's not very unambiguous. (…) Some places it is very unambiguous, but other places it might be a bit uncertain, and you can experience that they interpret it differently on different installations.

In contrast, the operators in Company A said that they had to do their own evaluations as to how to perform a certain task when the written procedures were useless, instead of letting the procedures form a basis for the decision making process. Informants from both companies valued the workers experience and competence, but in a slightly different way. The informants in Company A saw experience as necessary to compensate for flaws in the procedures were, while the informants in Company B saw experience as necessary to use the procedures as a basis for their evaluation on how a job should be performed.

The informants in Company B saw experience as an important ingredient in cases where the management system didn't give elaborative information. They expressed the opinion that there will always be some situations where the procedures are not entirely elaborate or where they are not entirely in accordance with reality, and in these cases they must adapt the procedures based on their own experience and competence.

> B27: It doesn't always say in the management system, but it's a bit like based on experience and such, so we say to each other "shouldn't we rather do this and that to be a hundred percent sure", you know. That's how it is. Based on experience, really.

With management systems and procedures there will always be a question how detailed they should be and how strict the guidelines should be. In this respect it is important that the management system gives unambiguous information as to where the procedures must be followed to the letter and where there is room for adaptions [13]. Anyhow, the definitive responsibility for safe conduct still lies on the operators who perform the task.

4.4. Informal procedures

Sociological studies of work very often reveal that workers tend to create their own informal work procedures that can be very different from formal procedures, and the existence of informal procedures that guide decisions and actions are a central part of the popular definition of organizational culture as 'the way we do things around here' [8]. This particularly happens when the formal procedures get to bothersome to deal with. In Company A some of the workers had created their own solutions that they found easier and more convenient to use than the Safety Management System. One of their solutions was to simply print out a stack of check lists and keep them in a pile, so they don't have to go into the Safety Management System every time they need a checklist. An obvious drawback with such a solution is that they won't get the updates if there should be any changes on the checklists or procedures. The informants themselves acknowledged this drawback, but still found it to be the more practical solution. Another solution of the workers' own design was a specially made web page with links directly to all the documents they used on a regular basis. There is a potential here to pick up the experience made by the workers, and utilize them in the ever ongoing updating and development of

7

the management system. This is also in accordance with the general guideline that the operator should feel that they have retained control over the system [17].

5. Sum up and conclusion

Procedures and Safety Management Systems are usually developed by management experts who are not involved at the operational level. A key challenge here is to involve the workers in the development of the procedures and Safety Management System. These must be constructed so that they increase ownership of work, and not decrease ownership of work. One should utilize the competence and experience of the workers when developing the procedures. The workers should also have the opportunity to give feedback on how useful the procedures and the management system are. The communication that a Safety Management System constitutes is mainly a linear communication from the upper tiers of the organization to the lower, but one should not forget the cognitive process that takes place when workers interpret and adapt the given procedures. There should also be feedback travelling the other direction as well, so that that adjustments or procedures can be based on the views of those directly involved. It is also important that the time interval between worker feedback and implementing changes is as short as possible.

Broad and direct worker participation in the process of implementing the procedures has been shown to lead to a greater level of commitment and adherence to procedures. Employee involvement will be rewarded by an increased feeling of ownership to the procedures. It will have a positive effect on the employee's use of the procedures when they get the sense that the procedures have originated from themselves, and not from some pencil pusher in an office. However, there are limitations as to how much every individual employee can be involved in the development of the procedures, with respect to money and time, and with respect to what kind of competence is needed in order to develop good procedures and a thoroughly thought-through Safety Management System.

Human Factors-analysis should be a part of the development of procedures and Safety Management Systems to assure user friendliness. Human factors revolves around the central importance of the user, and the goal of human factors is to enhance performance, increase safety and increase user satisfaction. This includes the communication and cognitive processes involved in using the Safety Management System. If human factors methods and principles are applied as early as possible in the development of a Safety Management System, in predesign analysis, technical design, and final test and evaluation, many of the human factors deficiencies will be avoided before they are inflicted on systems design.

References

[1] B. Kirwan. *"Safety management assessment and task analysis—a missing link?"* In: Hale, A., Baram, M. (Eds.), Safety Management: The Challenge of Change. Elsevier, Oxford, pp. 67–92, (1998).

[2] M. N. Vinodkumar and M. Bhasi. *" Safety management practices and safety behaviour: Assessing the mediating role of safety knowledge and motivation"*, Accident Analysis and Prevention 42, pp 2082–2093, (2010).

[3] D. Norheim and R. Fjellheim. *"AKSIO – Active knowledge management in the petroleum industry"*, Research report, (2007).

[4] C. Chen and S Chen. *"Scale development of safety management system evaluation for the airline industry"*. Accident Analysis & Prevention: Volume 47, pp 177–181 (2012).

8

[5] M.N. Vinodkumar and M. Bhasi. *"A study on the impact of management system certification on safety management"*, Safety Science 49, pp 498–507, (2011).

[6] J. Santos-Reyes and A. N. Beard. *"A SSMS model with application to the oil and gas industry"*, Journal of Loss Prevention in the Process Industries. Volume 22, Issue 6, pp 958–970, (2009).

[7] E. M. Koursi, El., S. Mitra and G. Bearfield. *"Harmonising safety management systems in the European railway sector"*, Safety Science Monitor. Vol 11, Issue2, (2007).

[8] S. Antonsen, P. Almklov and J. Fenstad. «*Reducing the gap between procedures and practice – Lessons from a successful safety intervention"*, Safety Science Monitor Vol 12 Issue 1, (2008).

[9] A. R. Hale, B. H. J. Heming, J. Catfhey and B. Kirwan. *Modelling of safety management systems.* Safety Science Vol. 26, No. l/2: 121-140, (1997).

[10] Haavik, T. *"Making drilling operations visible: the role of articulation work for organizational safety"*, Cogn Tech Work 12:285-295, (2010).

[11] J. J. H. Liou, L. Yen and G. W. Tzeng. *"Building an effective safety management system for airlines"*, Journal of Air Transport Management. Volume 14, Issue 1: 20–26, (2008).

[12] S. Antonsen. *"Safety culture: theory, method and improvement"*, Ashgate, 2009, Farnham.

[13] S. Dekker. *"Failure to adapt or adaptions that fail: contrasting models on procedures and safety"*, Applied Ergonomics, 34, pp. 233-238, (2003).

[14] Y. Dien. *"Safety and application of procedures, or how do 'they' have to use operating procedures in nuclear power plants"*, Safety Science, 29, pp.179–187, (1998).

[15] M. Alavi and D. E. Leidner. *"Review: Knowledge Management and Knowledge Management Systems: Conceptual Foundations and Research Issues"*, MIS Quarterly, Vol. 25, No. 1, pp. 107-136, (2001).

[16] C. D. Wickens, J. D. Lee, Y. Liu and S. E. G. Becker. *"An Introduction to Human Factors Engineering"*, Pearson Education, 2004, New Jersey.

[17] European Committee for Standardization. *"Ergonomic principles in the design of work systems"*, EN ISO 6385, (2004).

[18] T. Deaco, P. R. Amyotte, F.I. Khan & S. MacKinnon. *"A framework for human error analysis of offshore evacuations"*, Safety Science, 51, pp. 319–327, (2013).

[19] S. Andersen. *"Using Human and Organizational Factors to Handle the Risk of a Major Accident in Integrated Operations"*. in E. Albrechtsen and D. Besnard, Denis (eds.), *"Oil and Gas, Technology and Humans : Assessing the Human Factors of Technological Change"*. Ashgate, 2013, Burlington.

[20] N. A. Stanton, P. M. Salmon, G. H. Walker, C. Baber and D. P. Jenkins. *"Human Factors Methods. A Practical Guide for Engineering and Design"*, Ashgate Publishing Group, 2005, Abingdon, Oxon, GBR.

[21] European Committee for Standardization. *"Ergonomic design of control centres. Part 1: Principles for the design of control centres"* ISO 11064-1, (2000).

[22] Ø. Dahl. Safety compliance in a highly regulated environment: *"A case study of workers' knowledge of rules and procedures within the petroleum industry"*, Safety Science, 60, pp. 185-195, (2013).

[23] T. Reiman. 2011. *"Understanding maintenance work in safety-critical organisations – managing the performance variability"*, Theoretical Issues in Ergonomics Science, 12 (4), pp. 339-366, (2011).

[24] E. Hollnagel. *"The ETTO principle: efficiency-thoroughness trade-off"*, Ashgate, 2009, Farnham.

[25] E. Hollnagel. *"Barriers and accident prevention"*, Aldershot: Ashgate, 2004, Aldershot.

[26] N. McDonald. *"Organisational resilience and industrial risk"*, In: E. Hollnagel, D.D. Woods, and N. Leveson, (eds), *"Resilience engineering: concepts and precepts"*, Ashgate, 205–221. 2006, Aldershot.

[27] P. Sohlberg. *"Kunskapens former: vetenskapsteori och forskningsmetod"*, Liber, 2009, Stockholm.

[28] C. W. Choo. *"Environmental scanning as information seeking and organizational learning"*, Information Research, 7(1), (2001).

[29] D. Howitt. *"Introduction to qualitative methods in psychology"*, Prentice Hall, 2001, Harlow.

[30] V. Braun and V. Clarke. *"Using thematic analysis in psychology"*, Qualitative Research in Psychology, 3 (2), Pp. 77-101, (2006).

[31] J. Aronson. *"A Pragmatic View of Thematic Analysis"*, The Qualitative Report, Volume 2, Number 1, (1994).

[32] J. M. O'Hara, J. C. Higgins, W. F. Stubler and J. Kramer. *"Computer-Based Procedure Systems: Technical Basis and Human Factors Review Guidance"*, Brookhaven National Laboratory, 2000, Upton.

10

Identifying Requirements for Effective Human-Automation Teamwork

Jeffrey C. Joe[a*], John O'Hara[b], Heather D. Medema[a], and Johanna H. Oxstrand[a]

[a] Idaho National Laboratory, Idaho Falls, ID, USA
[b] Brookhaven National Laboratory, Upton, NY, USA

Abstract: Previous studies have shown that poorly designed human-automation collaboration, such as poorly designed communication protocols, often leads to problems for the human operators, such as: lack of vigilance, complacency, and loss of skills. These problems often lead to suboptimal system performance. To address this situation, a considerable amount of research has been conducted to improve human-automation collaboration and to make automation function better as a "team player." Much of this research is based on an understanding of what it means to be a good team player from the perspective of a human team. However, the research is often based on a simplified view of human teams and teamwork. In this study, we sought to better understand the capabilities and limitations of automation from the standpoint of human teams. We first examined human teams to identify the principles for effective teamwork. We next reviewed the research on integrating automation agents and human agents into mixed agent teams to identify the limitations of automation agents to conform to teamwork principles. This research resulted in insights that can lead to more effective human-automation collaboration by enabling a more realistic set of requirements to be developed based on the strengths and limitations of all agents.

Keywords: Team Performance, Human-Automation Teams, Human Factors Issues with Automation.

1. INTRODUCTION

As the role of automation expands in new and advanced systems, one goal of research is to make automation a team player [1, 2, 3]. One motivation behind this effort is the finding that humans relate to automation in similar ways to the way they relate to human teammates [4]. However, this approach typically uses incomplete models of human teamwork in the design of human-automation teams. Furthermore, we assert that automation agents cannot currently behave and interact with humans in the same way other humans can. These assertions are logical inferences from past research that has found that human-automation teams, that are designed based on an incomplete understanding of what makes good human teams effective, often lead to problems for operators, such as:

- Undesirable changes in the overall role of personnel
- Difficulty understanding automation
- Poor monitoring, lack of vigilance, and complacency
- Out-of-the-loop unfamiliarity and situation awareness
- Workload to interact with automation and when transitioning to greater manual control
- Loss of skills for performing tasks automation typically performs
- New types of human error

We use the term "multi-agent" teams to refer to teams having both human and automation agents who work cooperatively to accomplish tasks and plant functions. Designing automation to be a good "team players" has typically based on an implicit notion of what it means to be a team player and how members of a team should perform to function successfully. General concepts of team characteristics and behavior are employed, such as trust, goal and intention sharing, cooperation, and redundant responsibilities (especially in the case of adaptive automation where shifting of responsibility is a hallmark of the approach). These concepts are based loosely on a sense of how human teams perform.

---

[*]Corresponding Author: Jeffrey.Joe@inl.gov

However, the concept of multi-agent teamwork relies considerably on simplifications and popular notions about what is needed to foster teamwork, which do not always transfer well to human-automated agent teams. That is, the work to define how automation should behave (be designed) to be a team player has not been based on the recognition that there are some fundamental differences between human behavior and how automation has been programmed to behave, and that there are different models of human teamwork, each with its own set of member responsibilities and behaviors [5]. As a result, prior work on identifying how automation can be a team player is fragmented and incomplete at best.

Further, even if a belief about how human teams behave is at the core of this research, it does not sufficiently address the fact that automation agents are not humans, and cannot completely fulfill the role of a crewmember. Nor can we expect that it will fully behave as a human member of a team will. For example, automation agents cannot assume responsibility. Automation can be given the authority to act, but humans always maintain responsibility [6]. As another example, automation is not "concerned" about the consequences of its actions, nor is it as able to innovate as human crews will do when things do not go as planned.

The objective of this research is to identify the principles for effective human-automation collaboration that are appropriate in a commercial nuclear power environment involving a team of agents, specifically one where the anticipated use and/or level of automation is high. Once identified, we can elaborate on the specific requirements of human and automation team members, and can use those requirements to develop guidance for what automation characteristics and attributes are needed to be good "team players."

This paper will present some of our findings related to identifying the important characteristics of human teams, identifying the approaches taken to create multi-agent teams, and our preliminary approach to the formulation of general principles for effective human-automation teams.

2. METHODOLOGY

Our methodology is based primarily on an analysis of the literature. We first examined the research on what are the characteristics of effective teams working in complex systems, with an emphasis on the commercial the nuclear domain. Based on this research, we identified the general principles for effective teamwork. We next examined the research on mixed-agent team, specifically human-automation teams. We examined this literature from two perspectives: First we reviewed the research examining the characteristics of mixed-agent teams. Then we examine the research focused on defining principles of human-automation teamwork. This literature provided us with some understanding of the necessary characteristics of mixed agent team as well as an understanding of the limitations and issues. Then we examined how human-automation teams are different from human teams. To do this we compared the principles of effective human teams to what we know about the characteristics and issues associated with mixed-agent teams. Finally we began the process of identifying the principles for successful automation agent participation in teams. We are currently in the process of developing these principles, which will be further developed into design requirements as the research progresses.

3. RESULTS

3.1. Characteristics of Human Teams

3.1.1. Research on Human Team Effectiveness

The literature on teams composed only of human members is varied in terms of the genres that have studied this subject, and in terms of the depth of analysis that has been on performed to understand how to improve human team performance. Following a brief discussion about research on teamwork in general, we will focus on teamwork in nuclear power plant (NPP) crews.

A considerable amount of research in the business leadership and human factors literatures has been conducted on teams and team performance. The human factors literature has studied team cognition and team collaboration in particular. One popular book from the business leadership literature [7], presents "five dysfunctions" that can undermine team performance: (1) absence of trust, (2) fear of conflict, (3) lack of commitment, (4) avoidance of accountability, and (5) inattention to results. All of these factors make intuitive sense and are backed by other research studies. If team members do not trust one another, are afraid of conflict, lack commitment, etc., the likely outcome is suboptimal team performance. The human factors research on cognitive models of team and teamwork is based on the premise that in addition to the factors that [7] identified, how the team collectively thinks and coordinates cognitive and physical efforts are important elements to team success, but that it is also challenging for humans to do this without guidance and practice. A number of models of team have been developed from the human factors perspective to help elucidate this premise. Perhaps the most well known model of teams is the Crew Resource Management model [8], but there have been a number of other important and insightful contributions including the Teamwork Model [9], work on a meta-cognitive and macro-cognitive model of team collaboration [10], team sensemaking [11], and the Mutual Belief Model [12]. One key insight this literature has revealed is that it is important for team members to know what other people on the team are thinking and doing (i.e., situation awareness about team processes), along with knowing what is going on in the system and surrounding context (i.e., global situation awareness). This research also generally shows that people on teams tend to think that maintaining good global situation awareness is sufficient to also having good situation awareness about team processes. That is, individuals on teams tend to discount the importance of having good situation awareness specifically focused on what other teammates are thinking and doing, and believe that their awareness of team processes can be adequately maintained by having good global situation awareness. Furthermore, research findings have shown that people's habits, assumptions, complacency, and reliance on established conduct of operations standards tend to contribute to people losing situation awareness of other team members. The result of this typically leads to sub-optimal team performance, and sometimes leads specifically to teamwork errors, such as failing to do peer checks or independent verifications, which can subsequently lead to system-wide failures. Helpful reviews this line of human factors research can be found in [13] and [14].

In today's commercial NPPs, several reasons dictate the need for teams, including the distribution of workload, and the physical layout of the control room where tasks are completed via dozens of control boards. Thus, the successful operations of commercial NPPs are accomplished through the coordinated activity of multi-person teams or crews, which has lead to specialized training for NPP crews, and increased focus and rigor on the requirements for becoming a licensed operator. The Institute of Nuclear Power Operations (INPO) is one of the main industry drivers for NPP control room operator training and licensing in that they provide input on the requirements and standards for operator training and licensing. According to [15], along with the expected requirements that operators have the fundamental knowledge of technical topics such as nuclear physics and reactor theory, electrical science, and chemistry, there are guidelines on teamwork, and related interpersonal issues such as control room supervision and operational decision-making [16]. These INPO reports and guidelines cover a variety of teamwork issues including, but not limited to, communications, interaction among team members of different personality types, self- and peer-checking, and constructive conflict management [15]. INPO expects licensed reactor operators to be well versed in these aspects of teamwork as they recognize that teamwork is an important facet of the ability for the crew to function effectively, particularly in novel or emergency situations. According to [15], INPO recommends that teamwork training not only be taught in the classroom, but "should be continually reinforced during day-to-day work and training activities." (pg. 28).

There are also a number of studies of NPP crews that provide additional insights into the specific challenges crews face. Research has shown that factors such as: task characteristics, team member characteristics, and team dynamics affect how successfully the crew performs [17]. Other researchers has shown that successful teams monitor each other's activities, back each other up, actively identify errors, and question improper procedures [18]. Another study showed that communication protocols, communication content, and communications errors in NPP crews vary depending on the philosophy underlying the control room's design and the implementation of instrumentation and controls technologies [19]. Research by [20] studied how characteristics of communication among NPP crews during simulated emergencies affected crew performance, and found that characteristics such as: (1) a tightly coupled communication structure (which is an indicator of good team cohesion), (2) increasing the amount/density of communication to increase team situation awareness (e.g., crew members speaking up when observing changes in the system state), and (3) increasing the thoroughness of communication to make shared understanding more explicit (e.g., greater adherence to three-way communication practices) improved overall crew performance during simulated emergency scenarios. Clearly, important human factors aspects of teamwork in NPPs include having common, coordinated goals, maintaining shared situation-awareness, engaging in open and effective communication, and cooperative planning. In summary, existing NPPs are highly complex systems that require teams of human operators to be well trained on both the technical aspects of operations as well as the 'soft-skills' of effective teaming. This literature reviewed showed that in order for NPP operators to perform well as a team, they need to be both technically proficient, and well trained on how to manage the coordination of both their cognitive processing and physical efforts.

3.1.2. General Principles for Effective Human Teams

Based on all of the literature reviewed, we have derived a set of general principles for effective human teamwork. They are summarized below.

Belief in the Concept of Team: Effective human teams have individuals that believe in (or they are told that their compensation depends on) the idea that there is a mutual benefit to working together. This is also referred to as people having a team orientation versus an individualistic orientation [9].

Effective Communication: High performing teams communicate effectively. Communication is the central behavior team members engage in to function as a team. It is central to exchanging information, establishing team situation awareness, coordinating and regulating individual efforts, and building trust among team members.

Team Leadership: Effective teams have good leadership. Leadership by individuals formally appointed as leaders, and informal leadership by other team members, must principled and consistent, and needs to have both transactional and transformational qualities.

Monitoring Individual and Group Performance and Providing Feedback: Effective teams monitor and provide feedback on both overall group performance and individual contributions. This is key to the systematic adjustment of coordinated team activities. In NPPs, these coordinated team activities are needed to make adjustments to system-level process parameters (that are within the span of control of operators) that are necessary to achieve the desired process outcomes (e.g., operate the NPP safely and generate electricity profitably). Monitoring individual contributions to overall performance is also important to mitigating a well-known social psychological phenomenon called social loafing.

Coordination and Assistance: Effective teams are well coordinated and provide assistance to one another. When this happens, performance above and beyond what each individual could achieve on their own can be realized. Included in the principle of coordination is the extent to which team members have overlapping capabilities and are willing and able to seek and provide assistance to one another (i.e., provide redundancy or defense in depth) and increase overall system resiliency.

Awareness of Internal and External Performance Shaping Factors That Affect Team Processes: Effective teams are aware that personality traits and cognitive abilities can vary from person to person. How the team leverages, and not just manages, individual differences to enhance team performance is important. Similarly, external performance shaping factors (PSFs), such as task complexity, time pressure, and information/knowledge uncertainty can affect the team's ability to collaborate, which in turn can affect overall system performance.

Awareness that each individual's mental model is unique and that it is difficult to create shared mental models in teams: Effective teams recognize that everyone has an idiosyncratic mental model of their environment, and that each team member must clearly communicate their mental model to others. That is, humans construct their mental models based on an attentionally constrained ability to detect and process external stimuli, and a less than perfect long-term memory system, which leads to the formation of an idiosyncratic understanding (i.e., mental model) of the situation. Furthermore, humans have difficulty communicating or sharing their mental models with other humans such that they and one or more team members are subsequently 'on the same page of the playbook.'

Generally speaking, most of the issues with human teams stem from fundamental human frailties that define a large part of what it means to be human. Humans have beliefs, motivations, and emotions that affect their performance. Human team performance is affected by numerous factors including: the quality of communication, the extent to which they trust others, the morale of the group, the quality of leadership, and how well coordinated they are in monitoring individual and team performance, providing feedback, and timely assistance. Team performance is also affected by significant between-person differences in values, and mental and physical abilities. External situational factors further affect individual and team performance. And finally, the mental models humans develop of complex situations are idiosyncratic, often inaccurate, and are difficult to communicate to others such that they all have the same understanding. These are frailties that automated agents do not necessarily share.

In summary, the literature reviewed on human teams provides some good insights on the challenges and issues with these teams, and often provides a range of practical and workable solutions. The extent to which these insights and principles translate to human-automation teams, and automated agents in particular, however, will vary depending on how applicable the principle is to multi-agent teams, and specifically to the automated agent.

3.2. Characteristics of Multi-Agent Teams

3.2.1. Research on the Integration of Automation Agents into Human Teams

Klein, Sycara, and colleagues discussed integrating agents and human in teams [21, 22]. They viewed teamwork from the human team perspective of Sycara and Lewis [22]. They identified the four critical dimensions: information exchange, communication, supporting behavior, and team initiative/leadership. They stated that the human-automation interaction requirements differ based on the role of the automation in the team. They identified three different types of contributions automation can make as part of a team: (1) to work on tasks independently of human teammates, (2) to collaborate with human teammates and support human task performance, and (3) to support teamwork processes such as facilitating communication and coordination of human team members.

Sycara and Lewis noted that the greatest obstacle to integrating machine automation agents with human teams is communicating a human's intent. Furthermore, as automation becomes more flexible and autonomous, it becomes more difficult for humans to monitor and evaluate its performance. They identified three key factors in human-automation interaction: Mutual predictability of teammates, team knowledge (shared understanding), and the ability to redirect and adapt to one another. Agent predictability and shared understanding is made more difficult because automation often does not communicate its intent. They suggest that predictability can be fostered by: (1) consistently pairing simple observable actions with inputs, (2) making the causes and rules governing an agent's behavior accessible to the human, and (3) making the purpose, capability, and reliability of the agent known to the human.

Predicitability is related to trust. In the Introduction, we noted that humans relate to automation similarly to how they relate to human teammates. It is not surprising therefore, that operators' trust in automation greatly impacts their use of it [23]. Trust in automation is based on the operator's perception of its reliability and capability. This perception may or may not be consistent with reality.

When the operator's perceptions accurately match the automation's reliability and capabilities, trust is "well-calibrated" and operators use it appropriately. Miscalibrated trust leads to either an overreliance on automation (i.e., misuse) or its underutilization (i.e., disuse). Misuse (i.e., overreliance) on automation is associated with a failure to properly monitor it. This can engender large errors in system performance, (e.g., the system deviates considerably from the desired performance before it is recognized, if it is recognized at all). Disuse (i.e., underutilization) of automation can lead operators to turn off automation or to ignore its potential benefits.

Directability and mutual adaptation are important aspects of teamwork and enable teams to be flexible in different task contexts. Sycara and Sukthankar define directability as "assigning roles and responsibilities to different team members" and mutual adaptation as "how team members alter their roles to fulfill the requirements of the team" [21]. Researchers acknowledge that the most effective agents change their level of initiative, or exhibit adjustable autonomy, in response to the situation. For agents to appropriately exercise initiative, they must have a clear model of the team's goals, member roles and team procedures.

Communication between agents is key to achieving mutual predictability, shared understanding, and an ability to redirect and adapt. Communication structure between automated and human team members, however, is significantly different than typical human communication. While human-to-human communication includes face-to-face communication and nonverbal cues, automation is limited to transmitting and receiving information via specific formats within an array of displays [14]. Suchman attributed problems with modern automated system to the limits of syntax and semantics in automation [24]. The automated system is at the mercy of its programming, specifically, the types of information the system can provide and request. These constraints limit the human teammate, who must anticipate how to acquire and format the information exchanges with the automation agent. This sharing of information is critically different than the "spontaneous, free-flowing nature of human communication" [14]. Good communication, therefore, is a cornerstone of achieving effective human-automation interaction.

Woods et al. also noted that designers often fail to appreciate the need for humans and automaton to interact even when automation is designed to be mostly autonomous [25]. One reason for this is that designers often underestimate the complexities of operational environments.

Fiore et al. suggested that human-agent team interaction introduces a number of issues with respect to both team and individual cognition such as humans having to deal with an increased level of abstraction that may place unique demands on their information processing abilities, and increased difficulties coordinating team members given that differing communication patterns may be necessary to share cues [26]. Fan and Yen made a similar observation. They note that human teams rely on global situation awareness and shared mental models [27]. When automation agents are involved, developing this awareness and shared mental models can create additional costs attributed to communication, resolution of conflict, and social acceptance.

Pritchett's and colleagues identified ways in which automation agents are different from human agents [28, 29, 30]. Human team members will continue to attempt effective performance in unfamiliar circumstances, while automation generally cannot (e.g., behavior when outside boundary conditions). Good teammates anticipate each other's information needs and provide information (e.g., anticipating the needs of a teammate). Automation is limited in this capability. Humans time their interactions based on an awareness of the current situation, such as another teammate's workload (e.g., managing interruptions). Automation can be "clumsy" in this regard and interrupt human teammates at inopportune times. Finally, humans have a sense of responsibility. Automation does not have motivation, or a sense of responsibility.

Steinberg identified two significant limitations in multi-agent, or human-automation teamwork. First, automation cannot fully capture the human operator's intention in performing tasks [3]. Second, automation cannot flexibly adapt to situations that have not been considered by the designer.

Similarly, human operators often do not have a good understanding of how automation functions and find its interactions disruptive. Thus, Steinberg concludes that it may not be realistic, in the near-term at least, to think automation can be a teammate in the sense that another human operator is. Steinberg suggests that to create more effective multi-agent teams, researchers need to think more broadly about human-machine relationships. Other researchers have made similar observations [26].

Automation agents also affect one person's performance with another person. Roth and O'Hara observed the introduction of a computer-based emergency operating procedures (EOPs) system as part of one utility's digital instrumentation and control upgrade of a NPP [31]. This system automated the functions of information acquisition and analysis and gave some support to decision-making, such as retrieving data and assessing its quality, resolving step logic, and tracking the location in the procedure. The crews handled disturbances on a training simulator. Following each scenario, their interviews focused on the impact of the procedures on operations. They found that by introducing the new system, the procedure management workload was reduced to the point that procedure use became a one-person activity. The board operators were far less engaged in this, except to take occasional control actions at the request of the supervisor. Consequently, the operators felt they were out-of-the-loop, had lost situation awareness of EOP activities, and were unsure what to do. Thus, the introduction of the automation impacted teamwork, a finding that was unanticipated by the designers and plant staff.

Wright and Kaber evaluated the effects of automation on the performance and coordination of teams in a complex decision making task when it was applied to different stages of information processing [32]. Two-person crews performed a simulated mission to protect a home base from enemy attack. They found the effects on teamwork differed based on the generic tasks that were automated. The findings suggested that automation of early and intermediate stages of processing may have benefits with respect to teamwork, while automation of decision selection may be more limited, in that its benefit depends on the context in which it is used.

In summary, this research suggests that a strict human teamwork model may not be appropriate when designing human-automation teams. The limitation of current and near term automation agent capabilities and automation-human interaction include:

- Difficulty establishing shared mental models (i.e., shared understanding) of the situation when human and automation agents are part of the same team
- Failure of both agents to know or anticipate the other's intentions, actions, and overall team goals
- Limited flexibility of automation agents to redirect activities and adapt to shifting needs of the team and novel situations (outside the boundary conditions of its programming)
- Limited interaction between human and automation agents when the level of automation is high
- Clumsy and potentially disruptive effects on teamwork, roles, and responsibilities of the introduction of automation into human teams
- Automation agents have no sense of responsibility, conflict resolution, or social acceptance
- Increase in human workload when the requirements of interacting with automation are high
- Poor communication between agents and difficult constraints on the ease with which humans can communicate with automation

3.2.2. General Principles for Effective Human-Automation Teamwork

Many authors used human teamwork as a model to identify the general characteristics of desired human-automation interaction [4, 33, 34, 35]. Based on a review of studies of multi-agent teamwork, O'Hara and Higgins derived several general principles for human-automation interaction applicable to the commercial nuclear industry [23]. These principles are presented in Table 1.

Table 1: General Principles for Supporting Teamwork with Machine Agents

Principle	Definition
Define the purpose of automation	Automation should have a clear purpose, meet an operational need, be well integrated into overall work practices, and be sufficiently flexible to handle anticipated situational variations and adapt to changing personnel needs.
The designer should establish locus of authority	In general, personnel should be in charge of the automation, i.e., be able to redirect, be able to stop it, and assume control when necessary. This does not preclude the automation from initiating actions. Some actions are allocated to automation because they cannot be reliably performed by personnel within time or performance requirements. Further, there may be situations where automation initiates a critical action because personnel have failed to do so.
The designer should optimize the performance of the human-machine team	The allocation of responsibilities between humans and automation should seek to optimize overall integrated team performance. This may involve defining various levels of automation, each with specific responsibilities for all agents and each. It also may involve flexible allocations that change in response to situational demands. Personnel's interactions with automation should support their development of a good understanding of the automation, and the maintenance of their skills needed if automation fails. The HSIs should support a clear mutual understanding of the roles and responsibilities for both human and automation agents.
Personnel should understand the automation	Personnel should understand automation's goals, abilities, and limitations; and be able to predict its actions within various contexts of use. Minimizing automation's complexity will support this objective. While their understanding largely will come from training and experience, the HSI should support that understanding by reinforcing the operators' mental model through the information provided in automation displays. That is, the HSI should accurately represent how automation performs and how it interacts with the plant's functions, systems, and components.
Personnel should trust the automation	Personnel should have appropriately calibrated trust in automation that involves knowing the situations when the automation can be relied on, those which require increased oversight by personnel, and those for which automation's performance is not acceptable. The HSIs should support trust calibration by providing automation's reliability in various contexts of use.
Personnel should maintain situation awareness	The HSIs should provide sufficient information for personnel to monitor and maintain awareness of automation's goals, current status, progress, processes (logic/algorithms, reasoning bases), difficulties, and the current responsibilities. Special attention should be given to changing levels of automation where the responsibilities of agents may change.
The HSI should support interaction and control	Personnel interaction with automation should support their supervisory role: • HSIs should support personnel's interaction with automation at a level commensurate with the automation's characterization, e.g., level, function, flexibility, and its reliability. • Communication functions should enable personnel to access information about automation's processes. Automation should communicate with personnel when necessary, such as when it encounters an obstacle to meeting a goal, or when information is needed from personnel (e.g., information not accessible to automation). Communications from automation should be graded for importance, so as not to be overly intrusive. • Personnel should be able to redirect automation to achieve operational goals and should be able to override automation and assume manual control of all or part of the system.
The HSI should minimize workload	The HSI design should minimize the workload required to interact with automation, e.g., to configure automation and to communicate with it.
The team should manage failures	The multi-agent team should support error tolerance and failure management: • Personnel should monitor the activities of automation to detect automation errors, and to be sufficiently informed to assume control when automation fails. • Automation displays should support operators in determining the locus of failures as being either the automation or the systems with which the automation interfaces. • Automation should monitor personnel activities to minimize human error by informing personnel of potential error-likely situations. • When situations change sufficiently to render automation's performance unacceptable, it should communicate the situation to personnel in a timely way to enable them to become more engaged in automation's current goals and responsibilities.

Note: From O'Hara and Higgins [23]

3.3. Comparison of Human and Multi-Agent Teams

The well-known dynamics of human teams fundamentally change when teams are comprised of human and automated agents. That is not to say that none of the literature on human teams is applicable to the new human-automation team dynamic. Rather, the change in team dynamics means that the applicability of the insights from this literature on the new problem space defined by the pairing of humans and automated agents is not simple and straightforward. A corollary to this is that the problems associated with the differences between humans and automated agents, and the problems these differences further create in teams, are not directly addressed by this literature's insights and solutions. Additional thought must be applied if the valuable insights from the literature are going to add any value or meaningful guidance to designers contemplating the use of human-automated teams to operate their systems.

This section compares how human-automation teams are different from human teams. The approach for this comparison was to use the principles of effective human teams from Section 3.1.2 above as the standard or set of principles for successful team performance, since much of the automation literature presumes that this is what automation should be like in order to be a good "team player."

We can gain additional insight by comparing the seven principles of effective human teams with the limitations we identified in human-automation teams. This is shown in Table 2.

Table 2: Comparison of Human Teaming Principles and Human-Automation Limitations

Human Team Effectiveness Principles	Limitations of Human-Automation Teams
Belief in the concept of team	• An automation agent's behavior or performance does not change with an appeal to a belief in the concept of team. It cannot work more or less than its predetermined programming • Automation agents have no sense of responsibility, conflict resolution, or social acceptance • Potentially disruptive effects on teamwork, roles, and responsibilities of the introduction of automation into human teams
Effective communication	• Poor communication between agents and difficult constraints on the ease with which humans can communicate with automation • Clumsy and disruptive interactions between agents • Limited interaction between human and automation agents when the level of automation is high
Team leadership	• There is no concept of leadership in automation agents. They are indifferent to the quality and style of team leadership • If automation is ever in a leadership position, it is unlikely that the automated leader will effectively convey these leadership 'soft skills' that would improve the morale and confidence of the human subordinates
Monitoring individual and group performance and providing feedback	• Limited ability to both human and automated agents to know or anticipate the other's intentions, actions, and overall team goals
Coordination and assistance	• Limited flexibility on automation agents to knowing when the human needs additional assistance, and redirect activities and adapt to shifting needs of the team and novel situations (outside the boundary conditions of automation)

Human Team Effectiveness Principles	Limitations of Human-Automation Teams
Awareness of internal and external performance shaping factors that affect team processes	• Automation agents do not feel anxiety (i.e., internal PSF), time pressure (i.e., external PSF), or the interaction of PSFs in the same way humans do • Increase in human workload when the requirements of interacting with automation are high
Awareness that each individual's mental model is unique and that it is difficult to create shared mental models in teams	• Difficulty establishing shared mental models (shared understanding) of the situation when human and automation agents are part of the same team

What this comparison shows is that these principles do not translate well to human-automation teams, primarily because automation agents do not have many of the same key 'soft' characteristics that humans do. This is further exacerbated by the fact that humans know that automation does not have these characteristics, frequently try to compensate for this shortcoming, but nevertheless have trouble adopting new and effective ways of teaming with automated agents. In short, this comparison shows that when viewed from the perspective of what constitutes effective human teams, automation may find it difficult to be a "good teammate" in the same sense as a human teammate can be.

3.4. Developing Requirements for Multi-Agent Teams

Prior research has suggested some general principles for multi-agent teams (see Table 1) that can be developed further into preliminary requirements. However, these principles are based on a simplified model of teamwork that does not fully address the complexity of human teams, their processes, or the effects of automation agents on human team processes. Our analysis was based on a comprehensive consideration of the principles for effective human teamwork. This analysis shows that, at least as far a near-term technology is concerned, significant limitations exist in automations capabilities to fulfil the role of a good teammate.

Such a conclusion is not a showstopper. Instead, we view our findings as an opportunity to define reasonable requirements for integrating humans and automation to accomplish work cooperatively and in a manner that recognizes and capitalizes on the strengths and weaknesses/limitations of all agents in the team. The design of automation teammates should address these differences. Where agent capabilities can mimic those of a human teammate, principles for doing so can be developed. For those teammate characteristics that agents cannot mimic, principles for designing alternative approaches for accomplishing the characteristic need to be developed. In short, automation agents affect human team performance and this has to be addressed in the design. Further, operator training and experimentation will need to establish and reinforce a realistic view of automation's role in NPP monitoring and control as part of a multi-agent team.

Defining reasonable requirements is the next phase of our research. Using the general principles in Table 1 as a start, we will develop requirements for human-automation collaboration that realistically considers the capabilities and limitations of all agents on the team.

4. CONCLUSIONS

The human factors literature on human-automation teams has had the implicit belief that the way to make automated agents better team players is to translate the best practices and principles for effective human teams and use them to design human-automation interactions. This analysis shows, however, that many key principles for effective human teams do not translate well to automated agents, primarily because there are inherent differences between humans and automated agents. These differences have had significant consequences on human-automation team performance. Clearly, additional research should be directed to determine how insights from the human team literature can help improve human-automation team performance. The recognition of an automation agent's strengths and weaknesses should also be factored into the design of more effective collaborations.

Furthermore, given the human operators' tendency to interact with automation in ways similar to their human teammates, more emphasis is needed on calibrating the expectations and behaviors of human operators toward a more realistic means of collaborating with their automation teammates.

Acknowledgements

INL is a multi-program laboratory operated by Battelle Energy Alliance LLC, for the United States Department of Energy under Contract DE-AC07-05ID14517. This work of authorship was prepared as an account of work sponsored by an agency of the United States Government. Neither the United States Government, nor any agency thereof, nor any of their employees makes any warranty, express or implied, or assumes any legal liability or responsibility for the accuracy, completeness, or usefulness of any information, apparatus, product, or process disclosed, or represents that its use would not infringe privately-owned rights. The United States Government retains, and the publisher, by accepting the article for publication, acknowledges that the United States Government retains a nonexclusive, paid-up, irrevocable, world-wide license to publish or reproduce the published form of this manuscript, or allow others to do so, for United States Government purposes. The views and opinions of authors expressed herein do not necessarily state or reflect those of the United States government or any agency thereof. The INL issued document number for this paper is: INL/CON-14-31340.

References

[1]   K. Christoffersen, and D. Woods, *"How to Make Automated Systems Team Players,"* In E. Salas (Ed.) *Advances in Human Performance and Cognitive Engineering Research*, Elsevier, 2002, New York, NY.

[2]   S. Land, J. Malin, C. Thronesberry, and D. Schreckenghost, *"Making Intelligent Systems Team Players: A Guide to Developing Intelligent Monitoring Systems,"* (NASA Technical Memorandum 104807), 1995, National Aeronautics and Space Administration, Houston, TX.

[3]   M. Steinberg, *"Moving from Supervisory Control of Autonomous Systems to Human-Machine Teaming,"* 4th Annual Human-Agent-Robot Teamwork Workshop, (2012).

[4]   J. Lee, and K. See, *"Trust in Automation: Designing for Appropriate Reliance,"* Human Factors, 46(1), 50-80, (2004).

[5]   E. Salas, N. Cooke, and R. Rosen, *"On Teams, Teamwork, and Team Performance: Discoveries and Developments,"* Human Factors: The Journal of the Human Factors and Ergonomics Society, 50(3), 540-547, (2008).

[6]   A. Pritchett, *"Reviewing the Role of Cockpit Alerting Systems,"* Human Factors and Aerospace Safety, 1, 5–38, (2001).

[7]   P. Lencioni, *"The Five Dysfunctions of a Team,"* Josey-Bass, 2002, San Francisco, CA.

[8]   R. Helmreich and H. Foushee, *"Why Crew Resource Management? Empirical and Theoretical Bases of Human Factors Training in Aviation,"* In E. Wiener, B. Kanki & R. Helmreich (Eds.) *Cockpit Resource Management.* Academic Press, 1993, San Diego, CA.

[9]   T. Dickinson, and R. McIntyre, *"A Conceptual Framework for Teamwork Measurement,"* In M. Brannick, E. Salas, & Prince, C. (Eds.) *Team Performance Assessment and Measurement.* Lawrence Erlbaum Associates, 1997, Mahwah, NJ.

[10] M. Letsky, N. Warner, S, Fiore, and C. Smith, *"Macrocognition in Teams: Theories and Methodologies,"* Ashgate Publishing Limited, 2008, Hampshire, UK.

[11] G. Klein, S. Wiggins, and C. Dominguez, *"Team Sensemaking,"* Theoretical Issues in Ergonomics Science, 11(4), 304-320, (2010).

[12] Y. Soraji, K. Furuta, T. Kanno, H. Aoyama, S Inoue, D. Karikawa, and M. Takahashi, *"Cognitive Model of Team Cooperation in En-Route Air Traffic Control,"* Cognition, Technology & Work, 14(2), 93-105, (2012).

[13] E. Salas and S. Fiore, *"Team Cognition: Understanding the Factors that Drive Process and Performance,"* American Psychological Association, 2004, Washington, DC.

[14] C. Bowers, E. Salas, E. and F. Jentsch, *"Creating High-Tech Teams: Practical Guidance on Work Performance and Technology,"* American Psychological Association, 2005, Washington, DC.

[15] Institute for Nuclear Power Operations. *"Guideline for Training and Qualification of Licensed Operators,"* (INPO ACAD 10-001), 2010, Atlanta, GA.

[16] Institute for Nuclear Power Operations. *"Control Room Supervision, Operational Decision-Making, and Teamwork,"* (INPO SOER 96-1), 1996, Atlanta, GA.

[17] J. Toquam, J. Macaulay, C. Westra, Y. Fujita, and S. Murphy, *"Assessment of Nuclear Power Plant Crew Performance Variability,"* In M. Brannick, E. Salas, & Prince, C. (Eds.) *Team Performance Assessment and Measurement.* Lawrence Erlbaum Associates, 1997, Mahwah, NJ.

[18] J. O'Hara and E. Roth. *"Operational Concepts, Teamwork, and Technology in Commercial Nuclear Power Stations,"* In C. Bowers, E. Salas, & F. Jentsch (Eds.) *Creating High-Tech Teams: Practical Guidance on Work Performance and Technology.* American Psychological Association, 2005, Washington, DC.

[19] Y. Chung, W. Yoon, and D. Min, *"A Model-Based Framework for the Analysis of Team Communication in Nuclear Power Plants,"* Reliability Engineering & System Safety, 94(6), 1030-1040, (2009).

[20] J. Park, W. Jung, and J. Yang, *"Investigating the Effect of Communication Characteristics on Crew Performance Under the Simulated Emergency Condition of Nuclear Power Plants,"* Reliability Engineering & System Safety, 101, 1-13, (2012).

[21] K. Sycara, and G. Suktghankar, *"Literature Review of Teamwork Models,"* (CMU-RI-TR-06-50), 2006, Carnegie Mellon University, Pittsburgh, PA.

[22] K. Sycara, and M. Lewis, *"Integrating Intelligent Agents into Human Teams,"* In E. Salas & S. Fiore (Eds.) *Team Cognition: Process and Performance at the Inter and Intra-Individual Level.* American Psychological Association, 2004, Washington, DC.

[23] J. O'Hara, and J. Higgins, *"Human-System Interfaces to Automatic Systems: Review Guidance and Technical Basis,"* (BNL Technical Report 91017-2010), 2010, Brookhaven National Laboratory, Upton, NY.

[24] L. Suchman, *"What is Human-Machine Interaction,"* In W. Zacahary, S. Parasuraman, & J. Black (Eds) *Cognition, Computing, and Cooperation.* Ablex Publishers, 1990, New York: NY.

[25] D. Woods, J. Tittle, M. Feil, and A. Roesler, *"Envisioning Human–Robot Coordination in Future Operations,"* IEEE Transactions on Systems, Man, and Cybernetics – Part C: Applications and Reviews, 34(2), 210-218, (2004).

[26] S. Fiore, F. Jentsch, I. Becerra-Fernandez, E. Salas, and N. Finkelstein, *"Integrating Field Data with Laboratory Training Research to Improve the Understanding of Expert Human-Agent Teamwork,"* Proceedings of the 38th Hawaii International Conference on System Sciences, (2005).

[27] X. Fan, and J. Yen, *"Modeling Cognitive Loads for Evolving Shared Mental Models in Human–Agent Collaboration,"* IEEE Transactions on Systems, Man, and Cybernetics – Part B: Cybernetics, 41(2), 354-377, (2011).

[28] K. Feigh, and A. Pritchett, *"Requirements for Effective Function Allocation,"* Journal of Cognitive Engineering and Decision Making, 8(1), 23-32, (2014).

[29] A. Pritchett, S. Kim, and K. Feigh, *"Measuring Human-Automation Function Allocation,"* Journal of Cognitive Engineering and Decision Making, 8(1), 52-77, (2014).

[30] A. Pritchett, S. Kim, and K. Feigh, *"Modeling Human-Automation Function Allocation,"* Journal of Cognitive Engineering and Decision Making, 8(1), 33-51, (2014).

[31] E. Roth, and J. O'Hara, *"Integrating Digital and Conventional Human System Interface Technology: Lessons Learned from a Control Room Modernization Program,"* (NUREG/CR-6749), U.S. Nuclear Regulatory Commission, 2002, Washington, DC.

[32] M. Wright, and D. Kaber, *"Effects of Automation of Information-Processing Functions on Teamwork,"* Human Factors, 47(1), 50-66, (2005).

[33] C. Billings, *"Aviation Automation: The Search for a Human-Centered Approach,"* Lawrence Erlbaum Associates, Inc., 1997, Mahwah, NJ.

[34] R. Parasuraman, and V. Riley, *"Humans and Automation: Use, Misuse, Disuse, Abuse,"* Human Factors, 39(2), 230-253. (1997).

[35] D. Woods, and E. Hollnagel, *"Joint Cognitive Systems,"* CRC Press (Taylor & Francis), Boca Raton, FL.

Characterization of resilience in Nuclear Power Plants

Florah Kamanja[*], and Kim Jonghyun[a]
Kenya Electricity Generating Company, Nairobi, Kenya*
KEPCO International Nuclear Graduate School, Ulsan, South Korea[a]

Abstract: An emergency operation system in a nuclear power plant consist of operators, human-machine interface, procedures, and the interactions among these elements working together to respond to incidents. The complexity of dynamic systems such as nuclear power plants poses a challenge for safety as it can be a source of deviations from normal behavior during system operation. NPP control rooms consist of many elements that result in complex interactions between them. Resilience is the ability of a system to recover from a disturbance, so that it can sustain required operations under both expected and unexpected conditions.

Nuclear power plants must anticipate the operating risks caused by either the hardware, human, or organizational failures in order to be resilient. The ability of NPPs to monitor the current status of the system, anticipate possible problems, react appropriately to events, and learn from past incidents is a measure of success hence the resilience. Although the significance of resilience has been stressed in the literature, there is a lack of adequate literature attempting to analyze system resilience. To achieve a practical an insightful understanding of the EOS resilience complexity, this paper aims at characterizing resilience attributes based on the existing literature.

Keywords: Resilience, emergency operation system (EOS), resilience attributes

1. INTRODUCTION

A nuclear power plant is a safety-critical organization whose main objective is to control hazards and risks that can lead to release of radioactive elements to the environment. There has been a significant improvement of safety designs as well as risk analysis tools and methodologies of nuclear power plants over the past few decades. The first safety design concept in a nuclear power plant was based on defence in depth philosophy which relies to a great extent on multi-level physical barriers and engineered safety features to protect the workers, public and the environment should an accident occur. The next significant safety analysis concept that was introduced was classical probabilistic safety assessment (PSA) which was hardware oriented. Human reliability analysis followed thereafter after it was recognized that human errors contributed to major accidents, e.g. Three Mile Island accident. The history of nuclear power plants illustrates a shift of emphasis in the safety considerations from a technical perspective to human factors and broader issues connected to organization and management [1].

Conventional safety analysis methods such as PSA have several limitations [2,3,4]: 1) they primarily focus on technical dimension, 2) the analysis are linear and sequential, 3) they are dominated by static models, 4) they do not take a systemic view into account, and 5) they focus primarily on why accidents happen and not how success is achieved. Insights from research and failures in complex systems have also demonstrated that safety is an emergent rather than a resultant property of systems, therefore it cannot be predicted by considering only the constituents parts of a system. New approaches to risk analysis for NPPs are needed to complement the conventional approaches [3].

Nuclear power plants being safety critical organizations have had low number of accidents and this has weakened the ability to learn from experience. Thus resilience is needed to increase the system's ability to cope by enhancing anticipation for both expected and unexpected events. The study of a nuclear power plant safety can be further improved by characterizing NPP EOS resilience to gain an understanding of the various resilience attributes.

2. THEORETICAL BACKGROUND

2.1 Resilience Engineering

A resilient system is defined by "its ability effectively to adjust its functioning prior to or following changes and disturbances so that it can continue its functioning after a disruption or major mishap, and in the presence of continuous stresses" [5]. A study of a nuclear power plant emergency operation demonstrated that for system operations to be successful, more than procedure guidance is required and that in some incidents some degree of adaptability from the operators is needed [6,7]. The studies further shows that problems occur when operators fail to adapt plans and procedures to the situation. The adaptive capability to such situations is a measure of system resilience. Another study investigating the possibilities of operating crews to act flexibly in situations where procedures cannot be applied showed that expertise gained from training and teamwork effectiveness is important when the unexpected strikes [8]. A framework was proposed to analyze micro-incidents during nuclear power plant operation [9]. In this framework, micro incidents were defined as complex with four basic properties: singularity, unpredictability, importance, and pertinence to the situation. The findings indicate that to achieve a resilient performance the operators cannot rely only on the formal organizational constructs such as procedures, local adaptation by operators is necessary to solve plant problems.

A mathematical optimization model proposed for measuring resilience categorized resilience characteristics into two: inherent or adaptive. Inherent refers to resilience under normal operating conditions and adaptive refers to the use of a different strategy in crisis situations [10]. Adaptive capacity of a system is not static; the time dimension is important. Recovery time of the system after a disturbance should be taken into account when measuring resilience [11].

2.2 The safe Regulation Model

The safe regulation model shown in Figure 1 was developed by EDF research and development team to explain the impact of organizational factors on the operation of safety-critical systems such as nuclear power plants [5]. Three safe regulation phases are defined in this model; stabilisation, interruption, and stabilisation (post interruption stabilisation).

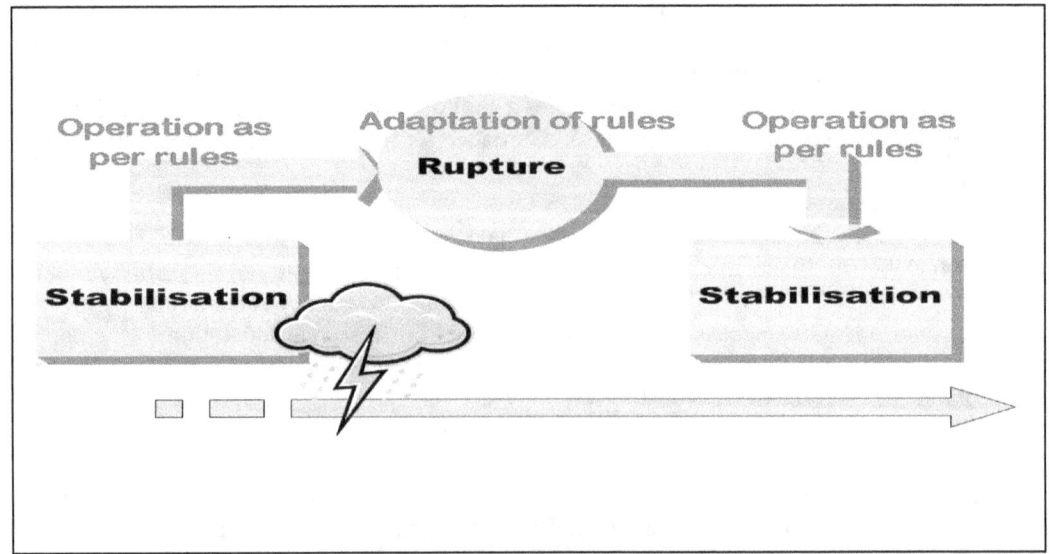

Figure 1: Safe Regulation Model [5]

3. EOS RESILIENCE MODEL (MRS)

EDF Research and Development Human Reliability team suggested five resilience attributes in their model of resilience in situation (Figure 2). The five high level resilience attributes are anticipation, adaptation, collective functioning, robustness, and learning organization.

3.1 Anticipation

A resilient system must be able to anticipate disruptions and their consequences. According to MRS, prescription, human resource, human machine interface, training, safety culture, and experience of the employees contribute to anticipation attribute. Resilient systems gauge their ability to anticipate using the following patterns [12]:

- Ability to recognize that adaptive capacity is falling/deteriorating,
- Ability to recognize that buffers or reserves become exhausted,
- Ability to recognize when to shift priorities across goal tradeoffs,
- Ability to navigate changing interdependencies across roles, activities, levels, and goals, and;
- Ability to recognize the need to learn new ways to adapt.

3.1.1. Prescription

Prescription consists of procedures used by the operators and collective rules such as task allocation and delegation rules. In a nuclear power plant, the operators have procedures to guide them in all modes of plant operation. While it is clear that the procedures give important support in accident operation, the degree of adequacy varies according to the characteristics of the actual situation during the team's operation. Licensees and applicants should ensure that all operators receive training on the use of EOPs prior to their implementation [13]. If an accident deviates from the procedures, the operators need to use their expertise in assessing the situation and eventually adjust the operation.

Emergency Operating Procedures (EOPs) are plant procedures that direct operators' actions necessary to mitigate the consequences of transients and accidents that have exceeded the set limits [13]. Procedures can be symptom based or event based; emergency operating procedures are mostly symptom based. Event-oriented EOPs require that the operator diagnose the specific event causing the transient or accident in order to mitigate the consequences of that transient or accident [13]. Symptom-based EOPs provide the operator with guidance on how to verify the adequacy of critical safety functions and how to restore and maintain these functions when they are degraded. Symptom-based emergency operating procedures are written in a way that the operator need not diagnose an event, such as a LOCA, to maintain a plant in a safe condition. The following limitations are related to strict application of procedures [14].

- Procedures do not take into account the individuals' variability in experience, attitude, and perceptions of the risk activity,
- It's not possible to guarantee the correct use of procedure during an emergency,
- The application conditions are not always well defined due to uncertainties: contingencies can turn the procedure inadequate depending on the actual conditions of the plant,
- In many cases procedures are developed by the system designers in a country different from where the system will be installed; different social aspects such as culture, and the language of the country that the plant has to operate may lead to wrong use of the procedure, and;
- Procedures generally refer to ideal situations, previously modeled by system designers which in most cases differ from the actual situations.

3.1.2. Human resource

Human resource refers to the way that the organization hires and assigns tasks to personnel. The organization should ensure that the main control room (MCR) team size is adequate to completely handle all of the scenarios under normal, abnormal, and emergency conditions. Team size should be determined with respect to both staffing requirements originating from the size of the task, as well as teamwork requirements originating from task complexity and uncertainty [15].

Task analysis can also be used to determine the staffing level by identifying the specific tasks needed to accomplish human actions, the information, control, and task support required to complete those tasks [16]. A task analysis report has detailed narratives of what personnel have to do including analyzing the alarms, information, controls, and task support needed to accomplish the task. The task analysis forms the baseline data upon which to allocate roles and responsibilities [17].

Table 1: Task Considerations [16]

Topic	Example
Alerts	- Alarms and warnings
Information	- Parameters (units, precision, and accuracy) - Feedback needed to indicate adequacy of actions taken
Decision-making	- Decision type (relative, absolute, probabilistic) - Evaluations to be performed
Response	- Actions to be taken - Task frequency and required accuracy - Time available and temporal constraints (task ordering) - Physical position (stand, sit, squat, etc.) - Biomechanics - Movements (lift, push, turn, pull, crank, etc.)
Teamwork and communication	- Coordination needed between the team performing the work - Personnel communication for monitoring information or taking control actions
Workload	- Cognitive - Physical - Overlap of task requirements (serial vs. parallel task elements)
Task support	- Special and protective clothing - Job aids, procedures or reference materials needed - Tools and equipment needed
Workplace factors	- Ingress and egress paths to the worksite - Workspace needed to perform the task - Typical environmental conditions (such as lighting, temp, noise)
Situational and performance shaping factors	- Stress - Time pressure - Extreme environmental conditions - Reduced staffing
Hazard identification	- Identification of hazards involved, e.g. potential personal injury

3.1.3. Human Machine Interface (HMI)

Human machine interface includes alarm system, indicators, controllers, operator support systems, and ergonomics. Human Machine Interfaces (HMI) is the primary mechanism through which personnel interact with the system during plant operation. HMI support the delivery of nuclear plant safety functions through detection, diagnosis, decision-making, and action. Nuclear power plant operation is a safety-critical organization where ultimate diagnosis and execution of tasks decisions lies with the operators. Thus it is important to provide a reliable decision support through effective supervisory control operator interfaces. Advances in digital technology have resulted to more application of automation for plant control. The systems in use now are advanced and more flexible because the personnel interact with plant at varying levels. Examples of advanced HMI include computer-based procedures, computerized operator support systems, intelligent agents that perform information processing tasks for operators in an autonomous manner, visual displays, and advanced Controls that combine multiple control methods [18]. When HMI performance is degraded, two main scenarios can be envisioned [19]: If the HMI is capable of returning to the initial nominal performance the system is resistant. If the HMI is capable of recovering from a disturbance and stabilizing at another functioning level; the system can be defined is resilient.

3.1.4. Safety culture

The term safety culture was first introduced by IAEA following their analysis of the nuclear power plants accidents at Chernobyl, Ukraine in 1986 [20]. The identification of poor safety culture, as a contributing factor to accident, led to a large number of studies investigating safety culture in many high hazard industries. There is no universally accepted definition for safety culture, however, majority of research studies commonly describe it as including norms, rules, and behaviours that are presented with respect to safety, as well as characteristics, beliefs, and values that are exhibited in an organization [21]. Safety culture is that assembly of characteristics and attitudes in organizations and individuals establish that, as an overriding priority, nuclear plant safety issues receive the attention warranted by their significance [22]. The most important safety culture attributes are; communication, learning culture, management commitment to safety, problem identification, roles and responsibilities, and technical knowledge [23].

3.1.5. Training

Training refers to the knowledge and experience imparted to the personnel by the organization. The content, scheduling, and the frequency of the training should be considered when establishing a training program. Training operators is important in ensuring safe and reliable operation of nuclear power plants. Training programs enable the plant personnel have the knowledge, skills, and abilities needed to perform their roles and responsibilities.

Teams can learn from the repetitive simulations of various types of environments that may be confronted in a naturalistic setting. Simulation can establish effective learning environments that enhances team problem solving expertise [24]. The number of serious accidents recorded in NPPs is low therefore training through simulation provides the teams with skills to handle emergency scenarios. Simulator studies are of great importance in testing the applicability of the procedures.

Table 2: Implications of simulation for training (25)

Skill	Implications for Training	Source
Information Processing skills Encoding Storage Retrieval	- Simulation can provide a shared environment which fosters similar templates - Similar experiences within a simulation can enable team members to have consistent knowledge organizations that lead to the development of common goals	[26] [27]
Situation awareness Cue recognition Template recognition	- Simulation is an excellent environment in which to receive practice - Simulation can provide many more trials than would be possible in a natural setting - Simulation can highlight specific patterns	[24] [28] [29]

Problem solving skills Domain specific skills	- Simulation can provide a safe setting to practice problem solving in complex dynamic environments - By providing multiple practice opportunities, simulation can accelerate team member proficiency - Opportunities for feedback can be established in simulation environments	[24] [30] [28] [27]
Monitoring Detecting faults Metacognition	- Simulation can be used to provide examples of normal system states. - Team members are able to practice self-regulating behaviors within a simulation environment	[24] [31] [32]

3.2. Adaptation

Adaptation is the ability to detect deviations from expected or unexpected paths and to readjust operation accordingly [33]. A resilient system responds to regular and irregular threats in a robust, yet flexible, manner. Emergency operation system stability therefore relies on dynamic and adaptive strategies to unanticipated situations. System verification strategies and reconfiguration approaches contribute to the adaptive capability of an EOS. Reconfiguration process involves stopping wrong rules, selection of adequate procedure, crew negotiation to adapt new rules, and validation of the new rules by a person with in-situation delegation of control regulation [5].

A theoretical framework for team adaptation shows that high performing teams adapt to the following: (a) decision making strategy, and (b) behaviour and organizational structure to the demands of the situation in order to achieve effective team performance [34]. Cooperation and trust are required to enable the team members engage in adaptive behaviour. Team trust among team members is increasingly recognized as important in applied research, especially because of the interdependence required in dynamic tasks [35]. Trust is important especially during validation of rules during system reconfiguration because the team share and commits to ideas in a decision-making and help each other in solving problems.

3.3. Robustness

Robustness of an EOS is the ability to carry out the required operation strategies and monitor them to ensure they are correctly applied. Tasks execution strategy and system control affect the robustness capability of the system. Execution involves information selection and the related operator actions. Operator actions can be done in series or in parallel depending on the procedure instructions. The operators need feedback information about the actual state of the controlled process for situation awareness purpose and satisfy their safety management objectives [19]. Human machine interface system supports interaction between the operators and the environment to aid in detection and interpretations of the plant process and this enhances system robustness.

The operator obtains information directly from the process system or processed information through the HMI. The tasks included in the information acquisition are collecting process parameters data, grouping the information, noting the necessary information, and recognizing required parameter values [36]. Diagnosis of the plant condition is done by either human operator or automation. Once the diagnosis has been performed, the operators select the response guided by the emergency operating procedures.

Execution of the various control tasks are done either by human operator or automation. The implementation of the selected response can also be done by automation with the consent of the human operator [36].

Table 3: The information required for monitoring or verifying automation activities [36]

Process Stage	Information	Example
Information acquisition	The process of raw measurements from process system, in terms of how the raw measurements are being processed	The calculation of the difference between SG level of each SG is shown, including the readings from all channels of the interested parameters
Plant Diagnosis	The process of how the diagnosis is made	The logic or steps of how the diagnosis result is achieved, for example for a loss of feedwater, the corresponding parameters and the set points, such as steam flow, SG level, etc. are shown, as well as the criteria of coming to the conclusion during diagnosis
Response Selection	The basis of response selection (the criteria that resulted in the particular response to be chosen)	The diagnosis result and the goal that needs to be achieved to deal with the diagnosis are shown
Response Implementation	The basis of implementation	The reason for the implementation, for example when automation reduces the feedwater flow through economizer valve, the HSI should be able to show the reason why the automation acts such way, along with the diagnosis result and the selected procedure to support the basis of implementation

3.4. Collective functioning

Nuclear power plant control room crew performs the plant operational tasks collectively. The operating tasks includes monitoring the system, detecting and receiving information, interpreting and assessing situations, diagnosing symptoms, making decisions, and task execution. The resilience of complex systems such as NPPs emerges in the core of team coordination and cooperation processes [37]. Communication and collective management of the situation determines the team collective effort. Communication is an important means of exchanging information between individuals during a group activity which is a prerequisite for good teamwork by establishing a shared mental model [38]. Communication is a cornerstone for teamwork and it becomes very critical especially during abnormal and emergency conditions. Communication influences attitudes, behaviours, and builds commitment and ownership [39]. The importance of communication has been stressed in reports of previous major incidents. In nuclear industry, a study in Japan showed that about 13% of incidents involving human error were caused by written communication and about 5% were caused by verbal communication [40]. A study carried out in Germany showed that 10% of 232 operational events were caused by communication problems [41]. Standardized communications among operators is fundamental as it can increase the sureness of the communications and reduce the possibilities of confusion, misunderstanding, or errors [42]. In a nuclear power plant MCR, the amount of conversation is significantly reduced by using computer-based procedures instead of conventional paper-based procedures, because information can be shared easily through computer screens [43].

Collective management of the situation involves spontaneous sharing of information among team members, co-ordination for action or diagnostics, validation of information or action with someone, collaboration, co-operation, close inter-monitoring of activity, and recap of rule points to be applied.

Teamwork defines how operators interact with each other in order to exchange information, coordinate actions, and maintain social order [24]. Team work deficiencies have attributed to incidents in nuclear power plants for example radioactive release accident from the Biblis nuclear power plant, Germany, in 1987 [44]. Team coordination is an important characteristic of teamwork and is the process of planning, scheduling, integrating, and allocating resources and responsibilities for coordinated tasks.The taxonomy of team skills shown in Table 4 provides a framework to identify the team skills required to be developed and aid in investigating teamwork mishaps [45].

Table 4: Nuclear team skills taxonomy [45]

Category	Elements	Definition
Building situation awareness	Develop understanding	Analyzing and sharing the information in order to develop an accurate model of the problem or task
	Anticipation	Forward planning to identify and discuss contingency strategies
	Maintain overview	Retaining abroad picture of a task or situation without becoming involved in the details
	Performance monitoring	Observing the activities and performance of other team members
Team focused decision making	Analytical decision making	Gathering and integrating information from team members, selecting the best solution, and evaluating the consequences
	Procedure following	Following written procedures.
	Intuitive decision making	Associating cues in the environment to appropriate corrective actions and making a decision
	Initiative	Using judgment to make decisions and carry out tasks without needing to be told what to do
Communication	Assertiveness	Communicating ideas and observations in a manner which is persuasive to other team members
	Information exchange	Exchanging information clearly and accurately between team members
Coordination	Adaptability	Reacting flexibly to changing requirements of a task or situation
	Supporting behavior	Giving help to other team members in situations in which it was thought they need assistance
	Team workload management	Prioritizing and coordinating tasks and resources
Collaboration	Leadership	Directing and coordinating the activities of, and motivating other team members, assessing team performance, and establishing a positive atmosphere
	Cooperation	Two or more team members working together on a task which requires meaningful task interdependence without any leadership
	Followership	Cooperating in the accomplishment of a task as directed by a senior team member

3.5. Learning organization

A Learning Organization is an organization that continuously monitors its environment for changes, and learns from and adapts to these changes. Organizations are seen as learning through processes that create new knowledge or modify existing knowledge [46]. The effectiveness of learning from experience depends on which events or experiences are taken into account, as well as on how the events are analyzed and evaluated [8]. Learning orientation can lead to a favorable culture for innovation, behavior improvement, and capability of individuals so that the organization can effectively respond to changes in its environment [47]. A learning organization creates an atmosphere where workers freely report concerns and the management responds to these concerns appropriately.

Learning organization is determined by factors such as knowledge management, telling stories by actors, in-situation learning, simulations, and learning from internal and external events. Learning from accidents is to extract, put together, analyze, and also to communicate and bring back knowledge on accidents and near-accidents, from discovery to course of event, damage, and cause to all who need this information" (as defined by Swedish Centre for Lessons Learned from Accidents). The six basic quality criteria for experience feedback are; initial reporting, selection methodology, investigation, dissemination of results, preventive measures, and evaluation [48].

Simulation produces situations as close as possible to the future reality and aims to create future work scenarios by using future operating means (interfaces, procedures, and operators) to impart experience and emergency coping skills to operators [49]. Learning is also a direct in-situation feedback, and the team plays an essential role in this by mutual assistance and cooperation mechanism in case of an incident [5]. During incidents, the less experienced members learn from their more experienced colleagues enabling the team to cement collective experience.

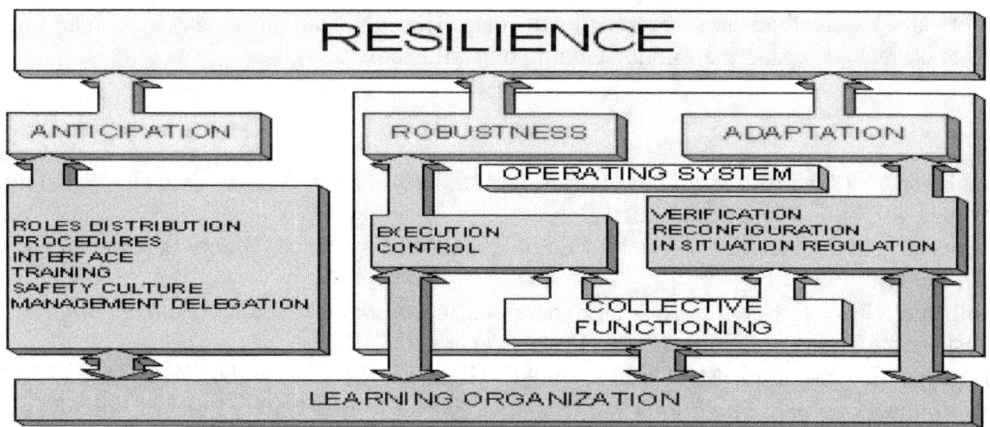

Figure 2: Model of resilience in situation [50]

4. Discussion

This paper is based on resilience engineering, which a new paradigm for safety management of complex systems. Traditional safety approaches such as PSA are inclined toward how accidents happen but not how success is achieved, they are linear and sequential, and they treat systems as static, and focus mainly on technical aspects. Safety is not being free of acceptable risks but the ability to succeed during expected and unexpected conditions. Resilience approaches address this limitation by treating safety as an emergent property of a system and evaluating the system's ability to adjust to its functioning status after disturbance. The nuclear domain being a strongly procedure guided environment is characterized by small number of accidents; hence flexibility during emergency situations that fall outside the procedures is critical for system success. Resilience engineering takes into account not only the technical dimension of the system but also human and organizational factors.

Resilience attributes are likely to deteriorate with time due to changes in the system environment. Although incident reduction measures may initially improve system safety, the absence of incidents decreases situational awareness of the system [51]. For ultra-safe systems such as nuclear power plants, continued elimination of errors, incidents, and breakdowns may lead to decrease in safety and hence the resilience [52]. Safety improvement programs can be expensive and often do not show immediate results leading to less emphasis on safety and adjustment of the goals of the safety program. A history of operations without incidents often leads to growing complacency which also results in decreased safety goals. To assure continued preparation to operators against expected and unexpected events, a certain level of incidents is tolerable. Responding to incidents provides the organization with the desire to adapt and therefore increases the resilience of the system. It is also the responsibility of the organization to evaluate the EOS resilience periodically to assure system success during expected and unexpected conditions.

5. Conclusion

This paper analyzes and characterizes resilience attributes to improve understanding of the EOS resilience dynamics of complex systems such as nuclear power plants. By improving understanding of resilience attributes, it may provide insights to new resilient strategies that the management can adapt

.

The main conclusion is that EOS resilience analysis approach can supplement traditional safety approaches to help in addressing their inherent limitations. This approach focuses on how success is achieved in a dynamic environment such as a nuclear power plant. With rapid growth in technology, large socio-technical systems such as nuclear power plants have become so complex that the established safety analyses methods have become inadequate. The characterization may help managers and employees to correct or expand their understanding on resilience.

Bibliography
[1]   B.Wahlstrom. *"Organisational Learning-Reflection from the Nuclear Industry"*, Safety Science,Volume 49, Issue 1, pp. 65-74 (2011).
[2]   N. Leveson. "A New Accident Model for Engineering Safer systems", Safety Science, Vol. 42, pp. 237-27, (2004).
[3]   E. Hollnagel. *"How Resilient Is Your Organization?An Introduction to the Resilience Analysis Grid"*, Sustainable Transformation, Hal-00613986, Version 1, (2011).
[4]   H. Bouloiz, E. Garbolino, M. Tkiouat, and F. Guarneri. *"A System Dynamics Model for Behavioral Analysis of Safety Conditions in a Chemical Storage Unit"*, Safety Science,Vol. 58,, pp. pp. 32–40, (2013).
[5]   E. Hollnagel, C.P. Nemeth, and S. Dekker. **"Resilience Engineering Perspectives, Remaining Sensitive to the Possibility of Failure"**, Ashgate Publishing Limited, 2008, UK.
[6]   D.D. Woods. *"Some results on operator performance in emergency events"*, Institute of Chemical Engineering Symposium Series 90, (1984).
[7]   D.D. Woods, J. O'Brien, and L.F. Hanes. " *Handbook of human factors/ergonomics"*, Wiley, 1987, New York
[8]   P. Gustavsson. *"Resilience and Procedure use in the Training of Nuclear Power Plant Operating Crews"*, dissertation, (2011).
[9]   V. R. P. Carvalho, I. L. Dos Santos, O. J. Gomes and R. S. M. Borges, *"Micro incident analysis framework to assess safety and resilience in the operation of safe critical systems:A case study in a nuclear power plant"*, Journal of loss prevention in the process industries,Volume 21, Issue 3 , pp. 277–286, (2008).
[10] A. Rose. and S. Liao, *"Modeling regional economic resilience to disasters: A computble general equilibrium analysis of water service disruptions"*, Journal of Regional Science, vol. 45, pp 75-112, (2005).
[11] S. Carpenter, B. Walker, J.M. Anderies and N. Abel. *"From Metaphor to Measurement:*

Resilience of What to What?", Ecosystems, vol. 4, pp. 765–781 (2001).

[12] Woods, D. D. "Resilience Engineering in Practice: A Guidebook", Ashgate Publishing Limited, 2011,UK.

[13] USNRC. *"Guidelines for Preparation of Emergency Operating Procedures"*, NUREG 0899.

[14] V.R.P. De Carvalho. "Ergonomic field studies in a nuclear power plant control room", Progress in Nuclear Energy 48, pp 51-69 (2006).

[15] M. Hoegl, K.P. Parboteeah, and H.G. Gemuenden."*When teamwork really matters: team innovativeness as a moderator of the teamwork–performance relationship in software development projects",* Journal of Engineering and Technology Management, vol 20, pp. 281–302, (2003).

[16] USNRC, *"Human Factors Engineering Program Review Model"* NUREG 0711, Rev 3, (2012).

[17] J. Stokes, K. Rich, and T. Foord. *"A human factors approach to the optimisation of staffing in the process industry",* Institution of Chemical Engineers Symposium, Series 151. pp 502-515, (2006).

[18] A. Hall. *"Human Machine Interface"*, T/AST/059, Issue 1, (2010).

[19] A.B. Ouedraogo, S. Enjalbert, and F. Vanderhaegen. *"How to learn from the resilience of Human–Machine Systems?",* Engineering Applications of Artificial Intelligence, vol 26, ppm 24-34, (2013).

[20] S. Gadd, and A.M. Collins. *"Safety Culture:A Review of the Literature"*, Report by the Health and Safety Laboratory. UK : Sheffield University,(2002)

[21] D.L. Potter. *"Research report organizational culture and safety:integrating for a safe workplace",* (2003) http://www.debpotter.com/admin/files/files.

[22] International Nuclear Safety Advisory Group, *"Safety Culture",* Safety Series No.75-INSAG-4, (INSAG. 1991).

[23] L.E. Alexander. *"Safety Culture in Nuclear Power Industry:Attributes for Regulatory Assessment"*, Dissertation, (2004).

[24] G. Klein and C. Zsambok *" Naturalistic decision making"*, Erlbaum, 1997, Mahwah, NJ.

[25] R.L. Oser, J.W. Gualtieri, J.A. Cannon-Bowers, and E.Salas. *"Training team problem solving skills: an event-based approach",* Computers in Human Behavior, vol 15,pp 441-462, (1999).

[26] J. A. Cannon-Bowers, E. Salas, and S. Converse. *"Cognitive psychology and team training: training shared mental models of complex systems",* Human Factors Society Bulletin, 33 (12), pp. 1-4, (1990).

[27] R.L. Oser, J.A. Cannon-Bowers, D.J. Dwyer, and H. Miller. "A*n event based approach for training: enhancing the utility of joint service simulations".*65th Military Operations Research Society Symposium, (1997).

[28] J. Orasanu, R. Calderwood, C. Zsambok and G. Klein. *"Decision making in action: models and methods",* Ablex, 1995, New Jersey.

[29] D. F. Noble, C. Grosz, and D. Boehm-Davis. *"Rules, scheme, and decision making,"* Engineering Research Associates, Vienna.

[30] D. Dorner, and H. Schaub. *"Errors in planning and decision making and the nature of human information processing",* Applied Psychology: an International Review, vol 43 (4), pp.433-453, (1994).

[31] M.S. Cohen, J. Freeman, S. Wolf, and L. Militello. *"Training metacognitive skills in Naval combat decision making",* Arlington, (1995).

[32] J. Orasanu. *"Shared mental models and crew decision making",* Technical Report No. 46, Princeton University, Cognitive Sciences Laboratory, (1990).

[33] E. Salas, D.E. Sims, and C.S. Burke. *"Is there a big five in team work?",* Small Group Research, vol 36, pp 555-599, (2005).

[34] E.E. Entin. *"Adaptive Team Coordination",* Human Factor, (1999).

[35] G. R. Jones and J. M. George. *"The experience and evolution of trust:Implications for cooperation and teamwork",* Academy of Management Review, vol 23, pp. 531–54, (1998).

[36] N. Anuar, and J.H. Kim. *"A direct methodology to establish design requirements for human - system interface (HSI) of automatic systems in nuclear power plants",* Annals of Nuclear Energy,.

vol 63, pp 326-338, (2014).

[37] N. Leveson. and D. D. Woods. *"Resilience engineering:Concepts and precepts",* Aldershot, Ashgate, 2006, UK.

[38] M. Hoegl and H. G. Gemuenden. *"Teamwork quality and the success of innovative projects: a theoretical concept and empirical evidence",* Organization Science, vol 12(4), pp. 435-449, (2001).

[39] A.M. Vecchio-Sadus. *"Enhancing Safety Culture Through Effective Communication",* Safety Science Monitor, vol 11(3), (2007).

[40] Y. Hirotsu, K. Suzuki, M. Kojima, and K. Takano. *"Multivariate analysis of human error incidents occurring at nuclear power plants: several occurrence patterns of observed human errors",* Cognition, Technology & Work, vol 3, pp. 82-91, (2001).

[41] R. von Meltzer and T. Dietrich. *"Communication in High Risk Environment",* Linguistische Berichte : Sonderheft 12, pp. 155-179, (2003).

[42] C.M. Kim, J. Park, W. Jung, H. Kim, and J.Y. Kim. *"Development of a standard communication protocol for an emergency situation management in nuclear power plants",* Annals of Nuclear Energy, Vol 37, Issue 6, pp. 888–893, (2010).

[43] D. H. Min, Y.H. Chung, and W.C Yoon. *"Comparative analysis of communication at main control rooms of nuclear power plants",* IFAC/IFIP/IFORS/IEA symposium, (2004).

[44] B. Wilper, and T. Ovale. *"Reliability and Safety in Hazardous Systems",* Lawrence Erlbaum Associates, 1993, New Jersey.

[45] P. O'Connor, A. O'Dea, R. Flin, and S. Belton. *"Identifying the team skills required by nuclear power plant operations personnel",* Operational Performance Information System for Nuclear Power Plants, Int. J. Ind. Eng. 38, pp.1028–1037, (2008).

[46] C.W. Phang, A. Kankanhalli, and C. Ang. *"Investigating organizational learning in eGovernment projects: A multi-theoretic approach",* Journal of Strategic Information Systems 17, pp. 99-123, (2008).

[47] P. Murray and K. Donegan. *"Empirical linkages between firm competencies and organisational learning",* Learning Organization, vol 10 (1), pp. 51-62, (2003).

[48] A. Lindberg, S.O. Hansson, and C. Rollenhagen *"Learning from accidents – What more do we need to know?",* Safety Science, vol 48, pp.714–721, (2010).

[49] J. Labarthe, and C. De La Garza. *"The Human Factors Evaluation Program of a Control Room:The French Approach"* , (2010).

[50] J. Kim, L. Podofillini, and V. N. Dang. *"Characterization of Emergency Operation System (EOS) of Nuclear Power Plants:Feedback to EDF Model and Intial Apllication to Swiss EOS",* LEA 009305, rev 0, (2009).

[51] M. Marais, H.J. Saleh, and G.N. Leveson. *"Archetypes for Organizational safety",* Safety Science 44, pp.565-582, (2006).

[52] R. Amalberti. *"The paradoxes of almost totally safe transportation systems",* safety science, vol 37, pp 109-126, (2001).

RECENT INSIGHTS FROM THE INTERNATIONAL COMMON CAUSE FAILURE DATA EXCHANGE (ICDE) PROJECT

Albert Kreuser[a], Gunnar Johanson[b]

[a] Gesellschaft für Anlagen- und
Reaktorsicherheit(GRS) mbH, Cologne, GERMANY
[b] ES konsult, Solna, SWEDEN

Abstract:

Common-cause failure (CCF) events can significantly impact the availability of safety systems of nuclear power plants. In recognition of this, the international CCF data exchange (ICDE) project was initiated in 1994. The objectives of the ICDE project are: to provide a framework for a multinational co-operation; to collect and analyze CCF events over the long term so as to better understand such events, their causes, and their prevention; to generate qualitative insights into the root causes of CCF events which can then be used to derive approaches or mechanisms for their prevention or for mitigating their consequences; to establish a mechanism for the efficient feedback of experience gained in connection with CCF phenomena, including the development of defenses against their occurrence, such as indicators for risk based inspections; and to record event attributes to facilitate quantification of CCF frequencies when so decided by the member countries of the Project. Until January 2014, 1346 ICDE events had been analyzed and reported in public OECD/NEA reports
This paper presents recent activities and lessons learnt from data collection on Control Rod Drive Assemblies and Heat Exchangers and on cross-component analysis on events which were due to external factors.

Key Words: Common cause failure, CCF, ICDE

1   INTRODUCTION

Common-cause-failure (CCF) events can significantly impact the availability of safety systems of nuclear power plants. In recognition of this, CCF data are systematically being collected and analysed in most countries. A serious obstacle to the use of national qualitative and quantitative data collections by other countries is that the criteria and interpretations applied in the collection and analysis of events and data differ among the various countries. A further impediment is that descriptions of reported events and their root causes and coupling factors, which are important to the assessment of the events, are usually written in the native language of the countries where the events were observed.

To overcome these obstacles, the preparation for the international common cause data exchange (ICDE) project was initiated in August of 1994. Since April 1998, the OECD/NEA has formally operated the project. Phase II had an agreement period that covered years 2000-2002, phase III covered the period 2002-2005, phase IV covered years 2005-2008 , phase V covered 2008-2011, and phase VI covers 2011-2014. Member countries under the Phase VI Agreement of OECD/NEA and the organisations representing them in the project are: Canada (CNSC), Finland (STUK), France (IRSN), Germany (GRS), Japan (JNES), Korea (KAERI), Spain (CSN), Sweden (SSM), Switzerland (ENSI), United Kingdom (HSE), United States (NRC) and Czech Republic (UJV).

The objective of this paper is to give generic information about the ICDE activities and the lessons learnt from recent analysis of CCF events in the ICDE database of Control Rod Drive Assemblies and Heat Exchangers and a cross-component analysis of common-cause-failure events which were due to external factors.

2   ICDE OBJECTIVES AND OPERATING STRUCTURE

The objectives of the ICDE project (denoted later as the Project) are:

- to provide a framework for a multinational co-operation;

- to collect and analyze CCF events over the long term so as to better understand such events, their causes, and their prevention;

- to generate qualitative insights into the root causes of CCF events which can then be used to derive approaches or mechanisms for their prevention or for mitigating their consequences;

- to establish a mechanism for the efficient feedback of experience gained in connection with CCF phenomena, including the development of defenses against their occurrence, such as indicators for risk based inspections; and

- to record event attributes to facilitate quantification of CCF frequencies when so decided by the member countries of the Project.

The ICDE Steering Group (SG) controls the Project with assistance from the NEA project secretary and the Operating Agent. The Operating Agent is responsible for the database and consistency analysis. The NEA Secretariat is responsible for administering the project on behalf of the OECD. The ICDE operating structure and documents related to it are depicted in Figure 1.

Running an international project requires funding and consequently the participating countries make yearly an agreed ICDE contribution to the NEA for reimbursement of the costs of the Operating Agent and the OECD NEA Secretariat. In addition, each participant bears all other costs, like the ones for data collection and national analysis, associated with participation in the Project. These costs are generally much higher that the costs of running the Operating Agent. Moreover, the in-kind principle is followed in the data exchange in that each country gets the dataset corresponding to its own data sent to the Operating Agent. Thus, just participating and paying the fees does not lead to directly receiving any data without a member's own data collection and submittal effort.

The SG meets twice a year on average. Its responsibilities include the following types of decisions: to secure the financial (by approval of budget and accounts) and technical resources necessary to carry out the project; to nominate the ICDE project chairman; to define the information flow (public information and confidentiality); to approve the accession of new members; to nominate project task leaders (lead countries) and key persons for the Steering Group tasks; to define the priority of the task activities; to monitor the development of the project and task activities; to monitor the work of the Operating Agent & quality assurance and to prepare the three-year legal agreements "Terms and Conditions", see Figure 1, for project operation. The ICDE experience tells that such a legal agreement completed by internal operating rules and summary presentations are vital prerequisites of mutual understanding and a functioning framework for a long-term internal co-operation with many countries involved.

An agreement and an Operating Agent do not alone guarantee good quality results, but data collection and analysis has to be organized at national level. In most countries, the data exchange is carried out through the regulatory bodies. They often delegate this to other organizations, since arriving at the information required by ICDE requires access to plant maintenance data. That data is normally proprietary. Consequently, the ICDE database is only available for signatory organizations and restricted by proprietary rights. The only possibility to get access to the working material is to actively take part in the data exchange.

OECD/NEA is responsible for administering the project according to OECD rules. This means secretarial and administrative services. Issuing publicly available ICDE reports, calling for member contributions/fees, paying expenses incurred in connection with the Operating Agent activities and keeping the financial accounts of the Project are examples of these activities. NEA appoints the Project Secretary from amongst its administrators.

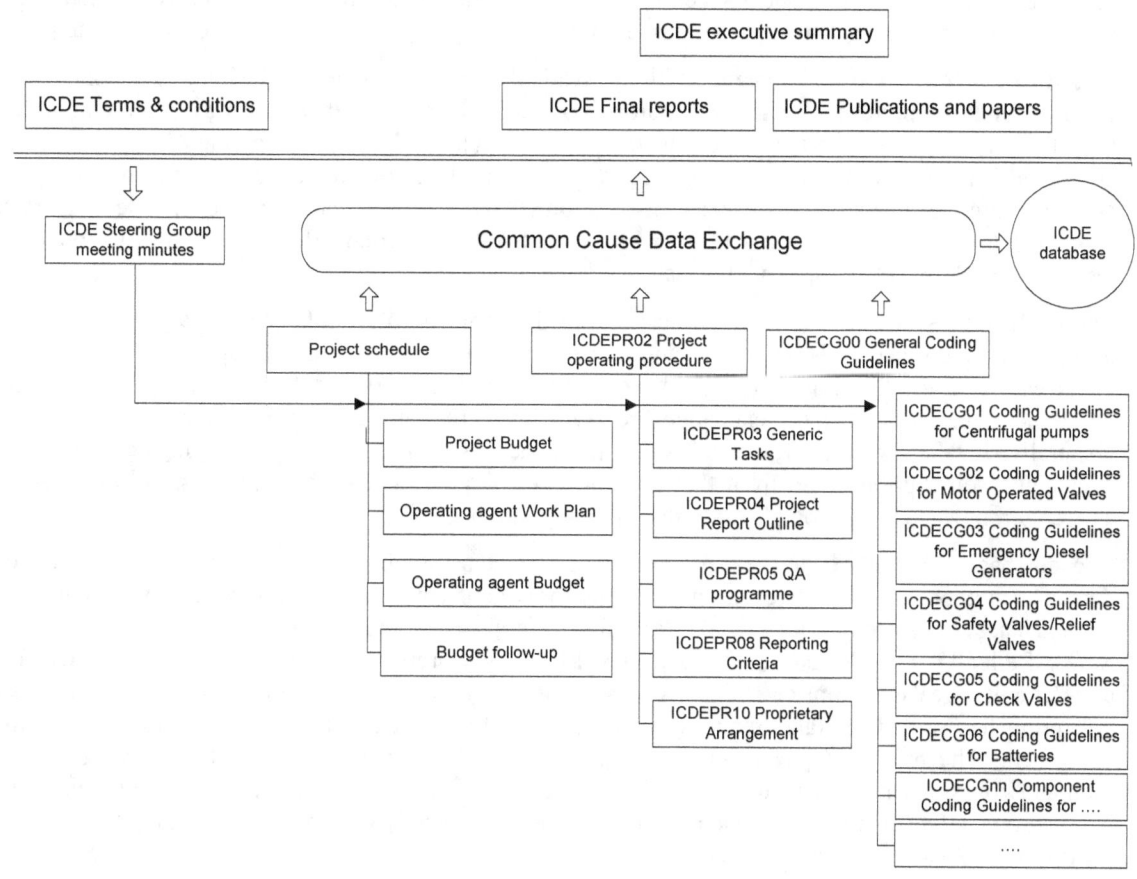

Figure 1. Operating procedure documents overview.

To assure consistency of the data contributed by the national coordinators, the project operates through an Operating Agent. The Operating Agent verifies whether the information provided by members complies with the ICDE Coding Guidelines. It also verifies the correctness of the data included in the database jointly with the national coordinators. In addition the Operating Agent operates, and develops if necessary, the ICDE databank. ES Konsult in Sweden is currently running the Operating Agent.

3   TECHNICAL SCOPE OF THE ICDE ACTIVITIES

The ICDE operates with a clear separation of the collection and analysis activities. In the first stage of the project, emphasis has been on the collection of data. The analysis results mostly in qualitative CCF information. It may be used for the assessment of 1) the effectiveness of defenses against CCF events and 2) the importance of CCF events in the PSA framework. The qualitative insights on CCF events generated are made public as CSNI reports. The member countries are free to use the data in their quantitative and PSA related analyses.

It is intended to include in ICDE the key components of the main safety systems. The data collection and qualitative analysis result in a quality assured database with consistency verification performed within the project. The responsibilities of participants in technical work, document control and quality assurance procedures as well as all other matters dealing with work procedures are described in a special ICDE Quality Assurance Program and the ICDE operating procedures.

The ICDE activity defines the formats for collection of CCF events in order to achieve a consistent database. This task includes the development and revision of a set of coding guidelines describing the classification, methods and documentation requirements necessary for the ICDE database(s). Based on the generic guidelines, component specific guidelines are developed for all analyzed component types as the Project progresses. These guidelines are made publicly available as a CSNI technical note [1].

The ICDE Steering Group prepares publicly available reports containing insights and conclusions from the analysis performed whenever major steps (i.e. analysis of a dataset for a certain component type like check valves) of the Project have been completed. The ICDE Steering Group assists the appointed lead person in reviewing the reports. Following this, an external review is provided by the NEA Committee on Safety of Nuclear Installation (CSNI). ICDE reporting also includes papers to suitable international conferences like PSAM and PSA, and journals. The intention is to make the lessons learnt known to a large nuclear safety audience.

The ICDE time schedules define the milestones of data collection tasks for each analyzed component group. The time schedule is reassessed and revised at each ICDE Steering Group meeting. The work starts with drafting the guidelines, getting comments, making a trial data collection, approving the guidelines, making the data exchange, resolving the remaining problem cases and reporting. Generally, it takes between 1,5 and 2 years from the first guideline draft to commencing the data exchange itself. Furthermore, from that point it takes about 2-3 years to approving the final report. Thereafter, new exchange rounds (database updating) are possible.

The database contains general information about event attributes like root cause, coupling factor, detection method and corrective action taken. As for the current phase V (January 2014), data analysis and exchange have been performed for Centrifugal Pumps, Diesel Generators, Motor-operated Valves, Safety Relief Valves, Check Valves, Batteries, Level Measurement Components, Switching Devices and Circuit Breakers, control rod drive assemblies and heat exchangers. Also, first round data collection has started on fans, main steam isolation valves and digital instrumentation and control equipment. The breakdown of resulted ICDE events in the database, i.e. events involving at least incipient common cause characteristics, of various components is shown in Table 1. Special emphasis is given on CCF events in which each component fails completely due to the same cause and within a short time interval. These events are called "Complete CCF".

Public **final reports** for Centrifugal Pump, Diesel Generators, Motor-operated valves, Safety & Relief Valves, Check Valves, Batteries, Level Measurement Components, Switching Devices and Circuit Breakers and Control Rod Drive Assemblies have been issued in the NEA CSNI series [2]-[11], a report for Heat Exchangers is in preparation, (see also: http://www.nea.fr/html/nsd/docs/indexcsni.html). Guidelines for fans, main steam isolation valves and digital instrumentation and control equipment have been approved. Also, an updated report on Centrifugal Pumps has been issued.

4 STATUS OF ICDE DATA COLLECTION

Until January 2014, 1346 ICDE events had been analyzed and reported. The total number of event records collected in the database for the analyzed component types is 1712. The breakdown into the various components is shown in Table 1. The third column shows the numbers of events in which each redundant component failed completely due to the same cause and within a short time interval.

Table 1 Number of ICDE events

Component	No. of events in component report	No. of ICDE events with complete CCF in report	No. of events in database (January 2014)	Data amount change since component report
Centrifugal pumps	353^{i2} (125^{i1})	42^{i2} (19^{i1})	384	9%
Diesels	106	17	223	110%
MOVs	86	5	167	94%
SRVs	149	14	261	75%
Check valves	94	7	116	23%
Batteries	50	3	77	54%
Breakers	104	6	106	2%
Level measurement	146	6	154	5%
Control Rod Drive Assemblies	169	0	171	1%
Heat Exchanger	46	4	53	15%
Total	1303	104	1712	31%

[i1] Issue 1 presented in year 1999

[i2] Issue 2 published 2013

5 ANALYSIS OF CCF EVENTS OF CONTROL ROD DRIVE ASSEMBLIES

This study was performed on a set of 169 common-cause failure (CCF) events of Control Rod Drive Assemblies (CRDA) derived from the ICDE-database.

These events were examined by tabulating the data and observing trends. Once trends were identified, individual events were reviewed for insights.

The data span a period from 1980 through 2003. The data are not necessarily complete for each country through this period. Besides verbal descriptions, these data includes coded information like failure mode, root cause, coupling factor, detection method, corrective action, observed population (OP) size, degree of failure and affected subsystem.

The most frequently occurring **failure mode** of control rod drive assemblies was 'Failure to completely insert for gravity insertion systems (FCI-G)' with 50 percent of the events.

The analysis of the 169 events of the database reveals that there are two complete CCF events for CRDAs. These events did not affect the scram function of the CRDAs, but they affected other CRDA safety functions. For these events the exposed populations associated with these safety functions were only a portion of the total observed populations of CRDAs for the respective plants. However, the entire exposed population associated with the safety function was completely failed. One of these events involved a backup insertion system containing 29 CRDAs. The total observed population at the plant is 57 CRDAs.

The most likely **root cause** is 'state of other component' (44 percent). This is consistent with CRDA architecture which implies a high interaction between control rods and fuel assemblies. Fuel assemblies can be deformed by irradiation, thermal, mechanical and hydraulic loading and jam control rods. Another important root cause is 'design, manufacture or construction inadequacy'; it accounts for 25 percent of CCF events.

The dominant **corrective action** is 'design modifications' (58 percent). The major parts of the components which are modified are fuel assemblies, axial seals of drive shaft and anti-rotation screws.

In looking for further qualitative engineering aspects the events were analysed with respect to failure symptoms and failure cause categories which are defined as follows:

Failure Symptom: An observed deviation from the normal condition or state of a component, indicating degradation or loss of the ability to perform its mission.

Failure Symptom Categories: Are component-type-specific groupings of similar failure symptom aspects.

Failure Cause Categories: A list of potential deficiencies in operation and in design, construction and manufacturing which rendered possible a CCF event to occur.

One failure symptom category was identified as dominant in the data: 'Movability problems due to deformation of core internals / fuel assemblies'. Most of the movability problems are caused by 'deficiency in design of hardware', and these design deficiencies involve deformations of the core and fuel assemblies.

Deficiencies in operation contributed to 21 percent of the failure causes. Each of the failure cause categories 'Deficient procedures for maintenance and/or testing' (13 events), 'Insufficient attention to aging of piece parts' (10 events) and 'Operator performance error during maintenance/test activities' (13 events) has a significant contribution to the total deficiencies in operation.

Design, construction and manufacturing deficiencies contributed to 79 percent of the failures causes, mainly due to failure cause category 'Deficiency in design of hardware'. Most of these failures were caused by core or fuel assembly deformations due to irradiation, thermal, mechanical and hydraulic loading, and their mutual interaction.

One additional conclusion is that some CCF events may be qualified as 'generic' for a specific plant series or CRDA design. That is, the same CCF mechanism has been observed in events occurring at plants with similar CRDA designs. Two examples of this are revealed in the database:

1) Many of the events coded with root cause category 'State of other components' involve CRDAs that failed to completely insert due to fuel assemblies that may have deformed due to creep induced by irradiation, thermal, mechanical and hydraulic loading, and their mutual interaction. There are 69 events in the database that match this description. These events occurred at a series of similar plants.

2) A number of events in the database involve degradation of a sub-component found in hydraulic-driven CRDA designs. Degradation of the seating material used in some SCRAM solenoid pilot valves has been found to slow the actuation of the valves and ultimately result in high rod insertion times. This failure mechanism appears in 26 ICDE events. The events occurred at plants with similar CRDA designs during the 1980s and early 1990s.

These problems have been addressed by licensees by communicating operating experience and/or in using a generic modification of the CRDAs or other components. These events highlight the importance of having a reliable design for the CRDA component, its sub-components, and those components that interface or interact with the CRDAs. These events also demonstrate that CCF phenomena can appear across a series of plants with similar CRDA designs. Communication of operating experience with CCF phenomena is important to ensure that plants can implement the appropriate defences and controls to prevent significant impacts on plant safety.

6 ANALYSIS OF CCF EVENTS OF HEAT EXCHANGERS

This study was performed on a set of ICDE events related to heat exchangers. Organisations from Canada, Germany, Japan, Spain, Sweden and the United States contributed to the exchange.

46 ICDE events, exhibiting at least some degree of dependency, and spanning a period from 1987 through 2007, were examined in the study.

The **failure mode** relevant from a PSA point of perspective is the failure mode HT-General (Failure of heat transfer) representing 100% of the events.

Degree of impairment: 4 of the events (8,7%) are complete CCFs (all redundant components had failed in a short time interval and for the same cause) while 1 event is defined as partial CCF (at least two, but not all completely failed components). The majority of the events (78%) have low impairment vectors, i.e. less than two components that have completely failed. Because of the small number of complete CCFs, the statistical significance of any result concerning complete CCFs should be handled carefully.

Dominant **root causes** were "abnormal environmental stress", "procedure inadequacy" and "design, manufacture or construction inadequacy", accounting for in total 37 events (83% of the events) with 1/3 in each of these root cause classes.

The **coupling factors** are strongly dominated by "environmental internal" (28%). However, if coupling factors are combined into top-level categories of environmental, hardware and operational, there is no dominant group.

Detection modes: 26 events (57%) were detected during test and maintenance activities, i.e. the equipment failure was discovered during the performance of a scheduled test or during maintenance activities. Only 7 events (15%) were revealed by demand events. Furthermore, 3 of the 4 complete CCFs were revealed by "test during operation". These results imply that the employed procedures and practices for detecting common-cause failures have been effective.

Concerning **corrective actions**, design related actions make up only 23% of the corrective actions, although "deficiencies in design, hardware and manufacturing" were involved in 65% of the events.

The identification of the relationship of **failure symptom categories** and **failure cause categories** was based on the verbal event descriptions and further engineering analysis for all of the ICDE events.

Heat exchangers are passive components operated in different systems and environmental conditions. In the majority of the events, dependencies occur in systems with an aggressive environment affecting heat exchanger internals as tubes, plates, chambers in multiple trains and components. Observed failures have also lead to leaks and impeded flow due to corrosions (corrosion, erosion) and dirt accumulation (pitting, fouling). There are also direct human/operator related faults causing dependencies of heat exchanger trains, e.g. by faulty alignment of valve configuration and wrong maintenance procedures and/or –practices.

The failure symptom aspect analysis reveals that there are three strong manifestations leading to flow problems: fouling, foreign objects or dirt accumulation impedes flow. If grouping the failure symptom aspects, there are two groups completely dominating:

- Impeding flow problems, 56 % of the events
- Erosion/corrosion problems leading to internal leak, 40 % of the events

Deficiencies in design, construction and manufacturing contribute 65% of the failure causes, the majority due to failure cause category "Deficiency in design of hardware". The other 35% of failure causes are deficiency in operation, mainly due to failure cause category "Deficient procedures for maintenance and/or testing". Among the four complete CCFs, 3 of 4 were due to deficiencies in design, construction and manufacturing.

The study shows that there are several test interval lengths practiced in the member countries.

A more frequent testing and –maintenance practice would be a powerful approach to reduce failures on less important failures, e.g. as shorter test intervals, more frequent cleaning, faster change of degraded heat exchangers, improved instrumentation of in/out flows and water temperatures, improved maintenance and/or testing instructions.

7   ANALYSIS OF CCF EVENTS DUE TO EXTERNAL FACTORS

In the light of the Fukushima accident, a cross-component study was performed on a set of common-cause failure events due to external factors, "External events", meaning that not only storms and hurricanes are included but also high outdoor temperatures and excessive algae growth. The events were derived from the ICDE database, where a brainstorming exercise performed by the Operating Agent on how to identify interesting events resulted in finding 52 events related to the topic out of 1600 ICDE events in total. The study is based on a workshop performed during an ICDE Steering Group meeting in April 2012, were the scope of "external events" were analysed in work groups. During the workshop additionally nine events were pointed out as not external events and therefore outside the workshop scope, i.e. this study includes the assessment of 43 ICDE events.

The majority of the events include centrifugal pumps (40%), followed by diesels (30%). The most common failure modes for pumps respectively diesels are failure to run (FR) and demand was the main way of detecting external problems (37%).The high number of demand events suggests that these type of "external failures" may be difficult to detect in periodic movement tests.

Additional engineering insights about the events were achieved by identifying the observed failure mechanisms. Table 2 lists representative failure mechanisms sorted by component type

Table 2 Representative Failure Mechanisms sorted by component type

Component type	Occurred failure mechanisms
Battery	- Potential loss of function during earthquake due to cracks in battery casings
Centrifugal Pumps	- Freezing led to blocking by ice of suction lines of service water pumps - Heavy seaweed in combination with low tide caused lack of water - Excessive sand and shellfish in sea water led to wear of pump impeller - Extremely low level of sea water was not considered in design - Algae growth in diesel fuel tank led to failure of operation of diesel driven pumps
Diesels	- Sludge in sea water reduced cooling capacity - Excessive sand and shellfish in sea water led to clogging of heat exchangers
Heat Exchanger	- High temperatures led to fast growth of clams and mussels with subsequent clogging of heat exchangers - Very high water level in combination with highly polluted water (foliage and grass) led to clogging of heat exchangers
Safety and Relief Valves	- Diaphragms installed in the air supply regulators of safety relief valves were dry and cracked due to long term high temperature environment leading to failure to open of the valves

The identified areas of improvements and lessons learnt can be divided into two subcategories - human/operational respectively hardware related improvements. Both "increased monitoring" and "improve cleaning of strainers" was concluded as important improvements for events involving pumps, diesels and heat exchangers. In addition, there were especially three events where the surveillance procedure was identified as a successful defence. All three events involve slow processes where excessive sand or shellfish in the sea water causing wear of the pump's impeller or clogging in the heat exchanger. Due to the slowly developing failure, it was possible to detect the event with differential pressure monitoring before degradation of the pump or heat exchanger.

Three diesel events at the same site experiencing the same cause failure mechanism are indication that back fitting of operational experience takes long time sometimes. These events involved sludge in the sea water leading to reduced cooling capacity and therefore too high temperatures of the diesel's cooling water. Here it could be concluded that thorough root cause identification is crucial before continuation of operation to prevent a second failure.

Since many of the events due to external factors involve sea water problems, important hardware improvements involve the construction of the water intake. One diesel event, where sludge in the sea water led to reduced cooling capacity and therefore too high temperatures of the diesel's cooling water, could have been prevented if the water intake was diversified. An example of a diversified water intake could be one surface intake and one deep water intake. Another interesting event was a pump event where both emergency feed water pumps run by diesel engines where degraded due to algae growth in the shared diesel fuel tank. The shared fuel tank is an indication that the separation of redundant pumps is not consequently done.

Two other interesting aspects were found. The first involves correlated hazards, which should be taken into account for better defence. There is one heat exchanger event where very high water level in combination with high amount of pollution in water such as foliage and grass led to clogging of the tubes in the heat exchangers. The second interesting aspect is related to a pump event where it was concluded "slight impairment by chance" because the detection was not via monitoring but by testing during outage.

The results of this analysis may serve as input for an in depth review of the methods and assumptions used in external hazards PSA.

8   DISCUSSION

What can be said is that the ICDE has changed the views to CCFs a great deal. Many insights would not have been possible to identify without a deep plant data collection and combining information from many sources. This paper discusses such insights about CCF events of the component types control rod drive assemblies and heat exchangers as well as insights about CCF event which were due to external factors.

Maybe the most important generic lesson is that it is worth forming specialized data exchange projects like ICDE. This, however, requires firstly the will of several countries to form a critical mass by combining their operating experience efforts, secondly national efforts to collect lower level data than made publicly available as LER or IRS reports, thirdly forming a legal framework to protect this proprietary data and fourthly a long term commitment to consistently continue and develop the activity.

OECD NEA and ES Konsult as the Operating Agent have provided means to run the international dimension of the ICDE, but national efforts are the key to the success of any project relying on operating experience. The success of ICDE has given a birth to several similar types of projects, among which are the OPDE for pipe failure events and OECD-FIRE for NPP fire events.

More information about ICDE may be obtained by visiting the site CSNI report site: http://home.nea.fr/html/nsd/docs/indexcsni.html, the Operating Agent website: www.eskonsult.se/icde or contacting the responsible OECD administrator.

9   REFERENCES

1. NEA/CSNI/R(2011)12 ICDE General Coding Guidelines – Updated version (replacing (2004)4)
2. NEA/CSNI/R(1999)2 ICDE Project Report on Collection and Analysis of Common-cause Failure of Centrifugal Pumps.
3. NEA/CSNI/R(2000)20 ICDE Project Report on Collection and Analysis of Emergency Diesel Generators.
4. NEA/CSNI/R(2001)10 ICDE Project Report on Collection and Analysis of Common-Cause Failures of Motor Operated Valves.
5. NEA/CSNI/R(2002)19 ICDE Report on Collection and Analysis of Common-Cause Failures of Safety and Relief Valves.
6. NEA/CSNI/R(2003)15. ICDE Project Report: Collection and Analysis of Common-Cause Failures of Check Valves
7. NEA/CSNI/R(2003)19 ICDE Project Report on Collection and Analysis of Common-Cause Failures of Batteries
8. NEA/CSNI/R(2008)8 ICDE Project Report: Collection and Analysis of Common-Cause Failures of Level Measurement Components
9. NEA/CSNI/R(2008)1 ICDE Project report: Collection and Analysis of Common-cause Failures of Switching Devices and Circuit Breakers
10. NEA/CSNI/R(2013)2 ICDE Project report: Collection and Analysis of Common-cause Failures of Centrifugal Pumps
11. NEA/CSNI/R(2013)4 ICDE Project report: Collection and Analysis of Common-cause Failures of Control Rod Drive Assemblies

Internal Flooding According to EPRI Guidelines – Detailed Electrical Mapping at Ringhals

Per Nyström[a], Carl Sunde[a], and Cilla Andersson[b]

[a] Risk Pilot, Gothenburg, Sweden
[b] Ringhals AB, Varberg, Sweden

Abstract: Eleven different tasks should be executed according to the EPRI guidelines for performing internal flooding PSA. Task 2 deals with identification of flood sources/mechanisms as well as with Systems, Structures and Components (SSCs). In this task it is briefly mentioned that not only the main components such as pumps and valves can be affected by flooding but also associated components such as circuit breaker, junction boxes and instrumentation and control circuitry are affected. It is fairly easy to locate the main components as well as the impact of flooding on these components. However it is more difficult to make a detailed mapping of the cable routing and the electrical dependencies (at Ringhals called electrical mapping) for the main components. This paper describes how this type of work is being executed and documented at Ringhals NPP in Sweden.

Keywords: Flooding, PSA, Database, Electrical mapping, Ringhals

1. INTRODUCTION

When performing a flooding analysis in accordance with the EPRI guidelines for performing internal flooding eleven different tasks should be considered [1]. Task 1-4 are associated with qualitative evaluation of flood phases such as defining flood areas and identifying flood sources and components, while task 5-10 are quantitative evaluation phases where evaluation of flood areas, that have not been screened out in the qualitative phases are performed. The quantitative phases consider the implementation of the flooding analysis in the PSA-model. The last task described in the guidelines is documentation. This task is an ongoing work and should be considered during each of the 10 tasks mentioned before.

The main concern in this paper is the second task in the EPRI guidelines, [1], which consists of the identification of flood sources/mechanisms and Systems, Structures and Components (SSCs). More specifically this paper will deal with how to identify and analyze electrical components associated with the plant main components at Ringhals NPP in Sweden.

Ringhals NPP is located on the west coast of Sweden approximately 60 km south of Gothenburg. The site consists of four reactors, one ASEA-ATOM BWR and three WESTINGHOUSE PWRs with the oldest one set in operation in the year 1975.

2. MAIN COMPONENTS AND ASSOCIATED ELECTRICAL COMPONENTS

As explained in the second task of the EPRI guidelines, [1], an identification of all interesting components must be performed to be able to perform a flooding analysis. It is also pointed out that not only the main components should be considered but associated dependent electrical components such as circuit breaker, junction boxes and instrumentation and control circuitry as well. The function of a main component is in most cases dependent on several components and failure of any of these may be crucial. The main components are presented in the PSA-model but the associated electrical components are hidden in the electrical mapping. For example a valve has dependencies to power supply and activation which in turn have dependencies to cables and junction boxes.

How the detailed electrical mapping is performed and used in the flooding analysis at Ringhals NPP will be described in chapter 3.

3. METHOD

In chapter 2 it is described how components at a NPP are dependent on different electrical components and a failure of the electrical components can lead to a failure of the main component. In order to keep track of all different electrical components and how they affect the main components it is of major importance to make a detailed mapping of the cable routing and the electrical dependencies (at Ringhals called electrical mapping) for the main components. The aim of this chapter is to describe the different steps when working with electrical dependencies in the PSA-model at Ringhals NPP and how to use this information when performing a flooding analysis.

The electrical dependencies in the PSA-model at Ringhals NPP are based on three different databases that process information about cables and components. The three databases are:

1. Cable database
2. Database of objects and connected micro circuit breaker (Object-MCB)
3. Database that compiles step 1 and 2 with flooding analysis.

The three steps will be described in the following subsections.

3.1. Cable database

The first step in processing information about cables and components consists of a database called cable database where information about all the cables is stored. This database is a living product where information is added whenever a new cable is installed or rerouted. It consists of cable routing (through which rooms the cable runs) and connected objects that can vary from pumps or valves to small cabinets. In Figure 1 an example is given of how the information about cables in the database is presented. The cable normally runs in cable trays that have connection points and this makes it possible to locate the cable in cable drawings and during walk downs. The name of the cable in the example in Figure 1 is 20020Y and it is runs from object 20554RI-03A in room H 1.09 through a number of rooms that can be seen in the right table to object X404 in room H 1.14. This information about cables and connected objects is then used to build up the second database which will be described in the next subsection.

Figure 1: Cable card – routing information

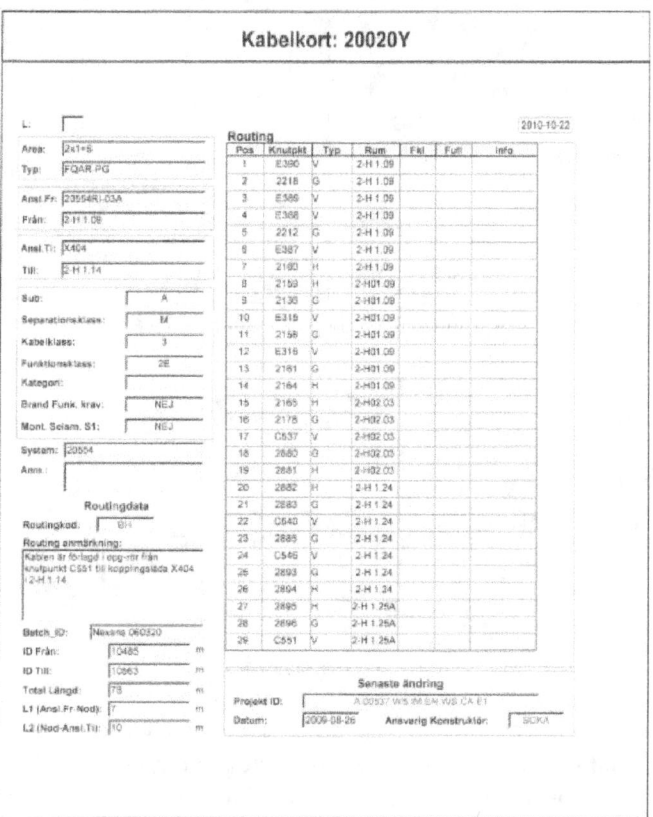

3.2. Database of objects and connected micro circuit breaker

The second step of the electrical mapping consists of making the information about cables compatible with the PSA-model and to identify cables and power supply for objects of interest. Objects in the PSA-model have dependencies to micro circuit breakers (MCB) and analogous for cables. Cables are not presented in the PSA-model explicitly but can be found through their dependencies to the MCBs. The relationship between MCBs, cables and objects are compiled in a second database, called Object-MCB database. This database presents all the room dependencies for the MCBs and connect these dependencies to the objects in the PSA-model. The MCB will represent connected cables and junctions hence the room dependence for the MCB will include all rooms that connected objects pass through. The MCBs with corresponding room dependencies can be found in the PSA-model on each of the interesting objects and therefore the location of objects in the PSA-model and their room dependencies is mapped in a satisfying way. An example of how the components and connected MCBs are presented in the PSA-model can be seen in Figure 2. The example consists of a pump and connected MCBs between the pump and feeding power busbar (power supply). There are two MCBs in the example; EHE300.XXX and EHE400.XXX. In these breakers all the room dependencies from the electrical mapping is collected. Failure of any of these breakers will lead to a failing power supply to the pump, i.e failure of any of the components connected to the MCBs such as cables and junction boxes will lead to a failing power supply to the pump.

Figure 2: Main component and connected MCB:s in the PSA-model

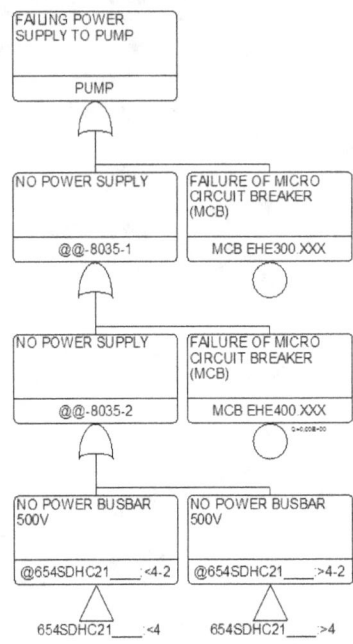

It should also be noted that several objects can be connected to the same MCB and therefore a detailed mapping is of great importance. In the example in Figure 3 three different objects (A/B/C) are connected to the same micro circuit breaker (S) and failure of any of these objects is assumed to trip the MCB and result in failure of all three objects. The objects A, B and C can be located in three different rooms and cables connected to them can be routed through additional rooms.

Figure 3: Dependencies to micro circuit breaker

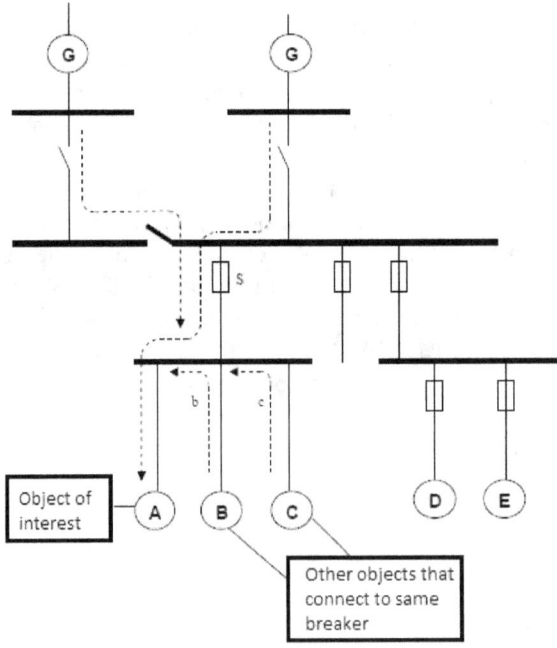

It is crucial to keep track of these dependencies when performing a flooding analysis because when for example a pump fails it's not always the pump that's the reason but a connected object. The connected object may have a direct connection to the pump or maybe a completely different function in which it is just connected to the same MCB.

The third step of the work with electrical dependencies in the PSA-model will be described in the next subsection.

3.3. Detailed electrical mapping and flooding analysis

The third and last step of the electrical mapping consists of compiling information from the detailed electrical mapping performed in step 1 and 2 and adding the different flooding scenarios. This compilation is done in a third database which is also used for easily importing the electrical dependencies into the PSA-model.

All micro circuit breakers (MCB) that are represented in the PSA-model and mapped in the previous steps can be found in this third database with connected objects. In the PSA-model only the main components for example pumps are represented and no cables and cabinets can be found. Cables and cabinets can however be found in this third database and therefore it's a powerful tool when determining the reason for a failing object in case of a flooding. The reason for a failure as previously explained, does not always have to be a failure of the main component but could also be a failure of a connected object, for example a cable or a cabinet. In the PSA-model it will only be indicated which MCB that has failed but by using the database additional information can be obtained. An example of how the connections between a main component in this case a pump and the micro circuit breaker can be seen in Figure 4. The component chain in Figure 4 consists of a pump pictured as a circle with an arrow located in room A and three cabinets located in room A, B and C. The pipe break occurs in a high energy line in room B resulting in dynamic effects on the cable between cabinet X200 and H400. The water does not propagate further but steam spreads into room C and damage cabinet EHE300 where the micro circuit breaker (EHE300.XXX) is located. No steam or water propagates into room A where the pump is located and therefore the component chain between the main component and the MCB is crucial. Otherwise the pump function would not seem to be affected by the pipe break which it actually is. In the PSA-model only the failing MCB is shown (see Figure 2). However by using the database one can conclude that it actually is the components located in room B and C that fails.

Figure 4: Components between pump and MCB and their exposure in case of pipe break

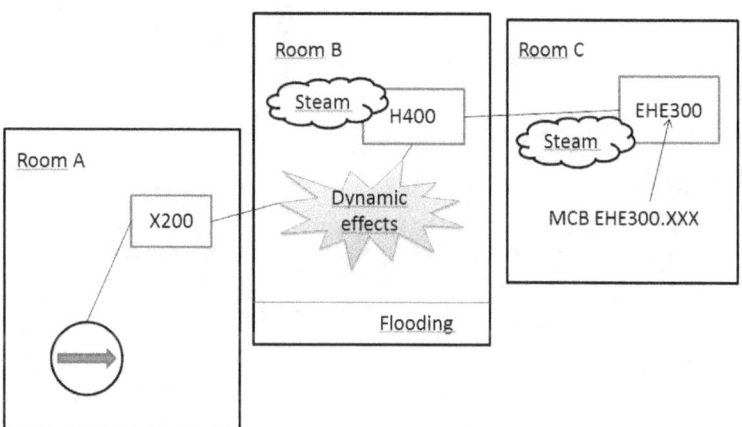

As shown in Figure 4 information about the studied objects are of great interest when performing a flooding analysis, not only in which room they are located but also for example on what level above the floor they are placed and if they are water/steam protected. Therefore all such information that can be of help is collected and compiled in this third database, an example can be seen in Table 1. This information can be of importance when analyzing the consequences of flooding. One example of important information is if an object is above the actual water level or not. If the object is above the actual water level this means that it would not be damaged by water drowning.

Table 1: Collected information about objects

ObjectID	Type	Steam Proof	Water Proof	Level (cm)	Fire Proof	Smoke Proof
303348154.41	Valve	False	False	0	False	False
30334CSAPBA-01.01	Pump	False	False	10	False	False
302950	Cable	True	True	0	False	False

Other example of information that is used when performing a flooding analysis and compiled in the database is flow paths for water, blow aperture for steam and potential dynamic effect sources. The information about water paths and blow apertures consist of where the pipe breaks occurs, flow rate, expected water level along the water path and to which rooms the water or steam propagates. Dynamic effects are analyzed in case of pipe break in high energy lines and only analyzed in the room where the pipe break occurs. A piping system is defined as high-energy line if the operating temperature is equal to or exceeding $93^{O}C$ ($200^{O}F$) or internal operating pressure equal to or exceeding 20 bara (275 psig). An example of a how a pipe break case is presented in the database can be seen in Figure 5.

Figure 5: Pipe break case in database

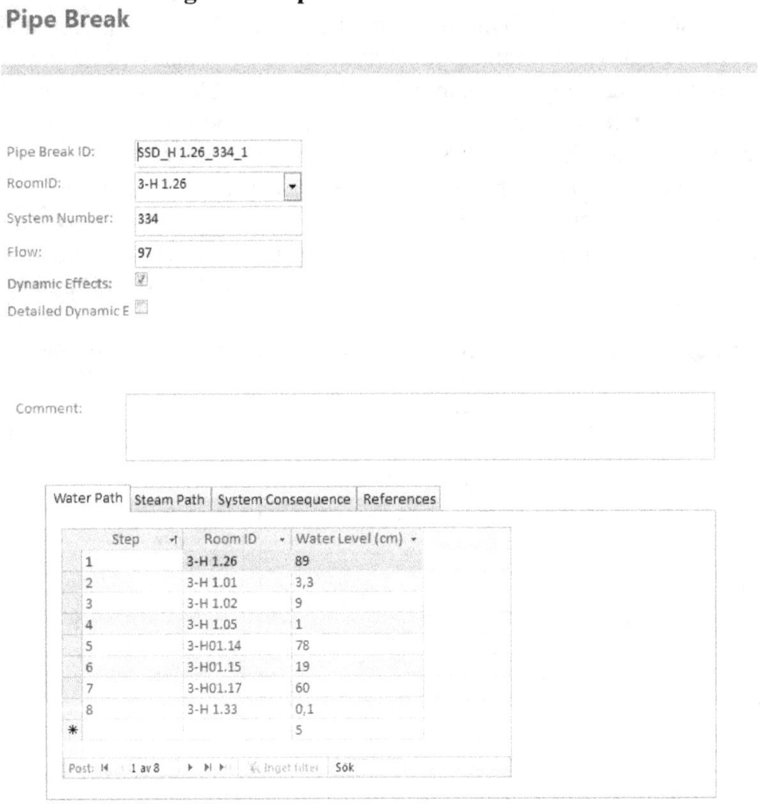

This example represents a pipe break in room H 1.26 in system 334. System 334 is defined as a high energy system in room H 1.26 and therefore dynamic effects are analyzed. The water path for a pipe break in room H 1.26 can be seen in the sheet "Water path" and the water propagates in the example from room H 1.26 after reaching 89 cm to room H 1.01, H 1.02 and further. The sheet "Steam Path" is composed in the same way as "Water Path" and therefore no further explanation is needed. The sheet "System Consequence" can be used in case the pipe break results in failure of the whole system or a part of the system. For example a pipe break in a pipe from one pump is not only a source of flooding but the consequence can also be an unavailable pump. The sheet "References" is used to keep track of all references such as separate analysis of water paths and blow apertures for steam.

After collecting all information that could be useful for defining the pipe break cases a group of analysis cases are created and exported to the PSA-model. The analysis cases have corresponding boundary conditions that specifies the water/steam path and possible dynamic effects or system consequences which is also exported to the PSA-model. By using a database this work is automated and possible human errors in implementing the cases in the PSA-model is depleted. This work responds to the quantitative phases of the EPRI guidelines, [1], for performing a flooding analysis which includes implementation in the PSA-model.

Even after implementing the flooding analysis cases in the PSA-model usage of the database continues. If a deeper study of the analysis cases with reason to failing objects needs to be done a complete list of failing objects is to be found in the database and not in the PSA-model. Therefore the work with the database follows through all the tasks which are considered when performing a flooding analysis in accordance with the EPRI guidelines, [1].

4. CONCLUSION

According to the EPRI guidelines, [1], the second step in performing a flooding analysis is identification of flood sources/mechanisms and Systems, Structures and Components (SSCs). In order to completely perform this task it is of major important not only to identify main components but also associated electrical component. Therefore, it is of major importance to make a detailed electrical mapping in order to perform a complete flooding analysis. At Ringhals NPP three different databases are used for the detailed electrical mapping. By storing information in a database it is easy to maintain and update the electrical mapping and it is also easy to update the PSA-model. Quality assurance of electrical mapping is improved by using a database, this since it is more manageable to review and the database also creates a useful overview over large quantity of information. For example there are about 3000 cables and 800 cabinets handled in the final database discussed above for one of the reactors at Ringhals NPP.

The PSA-model only model the main components such as pumps and valves with room dependencies but there are also other objects that affect the main components. These are hidden in the electrical dependencies in the PSA-model. By using a database that compiles these electrical dependencies other objects such as cables and cabinets that affect the main components can easily be detected. Without this information a flooding analysis would be incomplete and would provide an inaccurate image of an actual pipe break scenario.

Acknowledgements

Thanks to Ringhals AB who has made this paper possible.

References

[1]   EPRI – Guidelines for Performance of Internal Flooding Probabilistic Risk Assessment

NRC Reactor Operating Experience Data

Shawn Walter St. Germain

Idaho National Laboratory, Idaho Falls, USA

Abstract: Idaho National Laboratory (INL) has been providing technical assistance to the U.S. Nuclear Regulatory Commission Division of Risk Analysis in the Office of Nuclear Regulatory Research in the areas of data collection and reliability and risk calculation. INL collects, codes, assures the quality of, and maintains all reactor operating experience data necessary to support the Industry Trends Program and various risk-associated NRC studies requiring reactor operating experience data. The types of data collected under this effort include initiating event data, system reliability data, loss of offsite power data, common cause failure data, fire event data, and shutdown initiating event data. The data sources for this effort primarily consists of Licensee Event Reports (LERS), Event Notifications, and equipment failure reports provided by the Institute for Nuclear Power Operations (INPO). This data is analyzed and results published annually on the NRC website. The data is primarily used to support the NRC's standardized plant analysis risk (SPAR) models but also provides generic industry average values for use by the industry in their individual PSA models. This paper characterizes the types of data collected, the various uses of this data, and the methods of collection, storage and retrieval.

Keywords: Data, Operating Experience.

1. INTRODUCTION

Idaho National Laboratory (INL) provides technical assistance to the US Nuclear Regulatory Commission Division of Risk Analysis in the Office of Nuclear Regulatory Research in the areas of data collection and reliability and risk calculation. The results of this work are generally made publically available on the NRC's website to support various PSA activities. In addition, several specialized databases and data tools are provided to the public. INL also supports the NRC's Industry Trends Program (ITP) in the areas of operating experience data and models. INL maintains a database of operating experience data to support various ongoing studies related to Probabilistic Safety Analysis (PSA) including initiating events, system reliability, loss of offsite power, and common cause failure studies. The primary sources of NRC operating experience data are Licensee Event Reports (LERS), Event Notifications and equipment reliability data collected by the Institute for Nuclear Power Operations (INPO). This paper outlines the NRC's Reactor Operating Experience Data Collection and Analysis Program, including data collection, data sources and some of the uses of this operating experience data.

2. DATA COLLECTION METHODS

The Integrated Data Collection and Coding System (IDCCS) provides the structure to ensure consistent data collection, coding, and quality assurance of reactor operating experience data. IDCCS is a program using a Microsoft™ Access project (.adp) user interface to a SQL Server database (SQL Server 2008 R2). Each record in IDCCS is linked to a source document such as an LER or INPO failure record. Source documents are organized by "events". An event is defined by a plant and date; therefore, while viewing an event, all documents and records for a given plant on a given day are displayed. The use of the event concept helps to prevent creating duplicate records from multiple data sources. The process to collect operating experience data includes several quality controls to maintain data integrity including:

- A user's guide describes each study and provides guidance for filling out each field in the IDCCS.
- The IDCCS program utilizes numerous lookup tables and automated checks to ensure data consistency.
- Records are entered by qualified coding engineers.
- Each record is independently checked by a second qualified coding engineer.
- The IDCCS software randomly selects a sample of records for an independent quality review semi-annually.

Figures 1 and 2 outline the data collection, coding and analysis process including quality assurance activities.

Figure 1: INL Data Collection and Coding Process

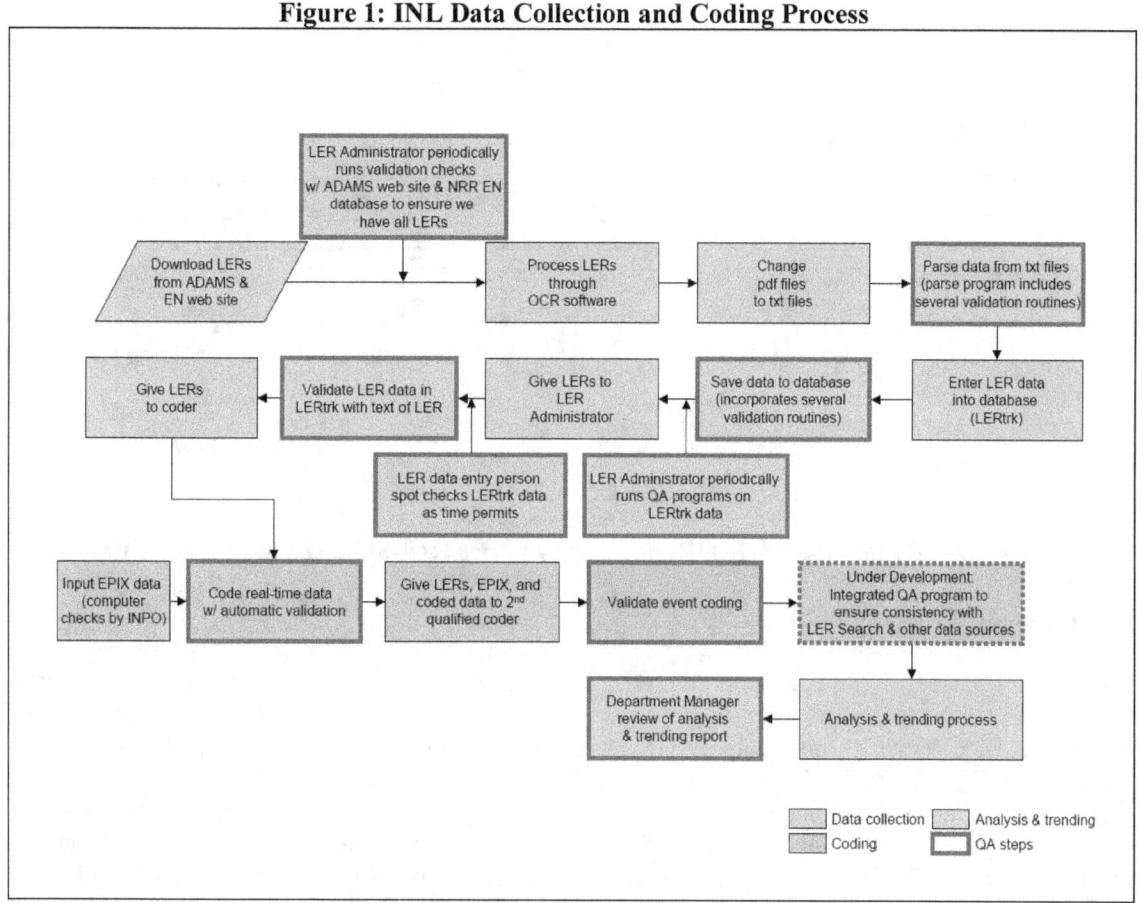

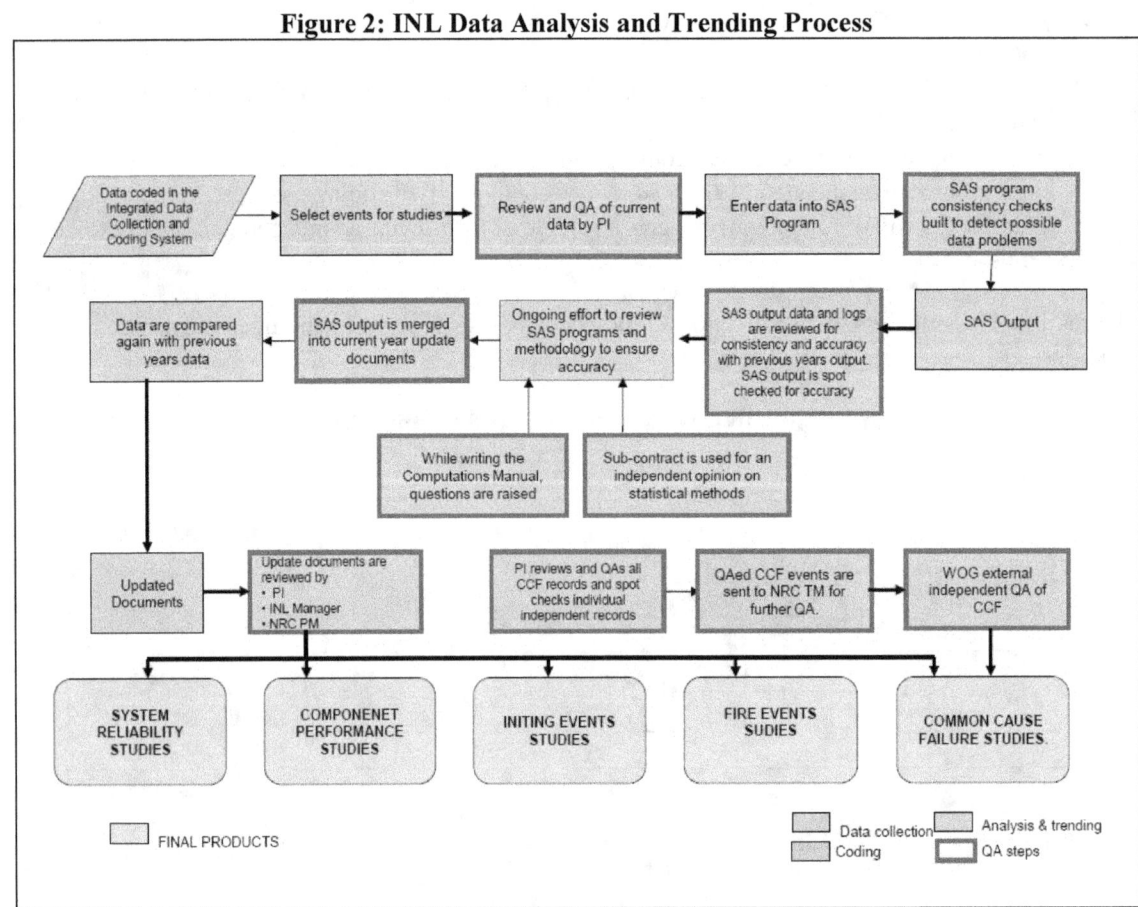

3. PRIMARY NRC REACTOR OPERATING EXPERIENCE DATA SOURCES

3.1. Equipment Reliability Data

The primary source of equipment reliability data is from the Institute of Nuclear Power Operations (INPO). All US nuclear power plants report equipment reliability data via INPOs Consolidated Event System (ICES), formerly the Equipment Performance and Information Exchange (EPIX) System Database. ICES is part of a larger INPO data collection program, Consolidated Date Entry (CDE). ICES supports several regulatory functions, including support for the maintenance rule and the Mitigating Systems Performance Index (MSPI). ICES data are provided to INL quarterly through a contract between INPO and the NRC. INL is allowed to publish industry average equipment performance information, but does not publish individually identifiable failure reports to the public. ICES data supports the industry average parameter estimates, systems studies, component performance studies, and common-cause failure studies. Individual components are referred to as devices in the ICES database. Devices can be Key Components, Sub-Components, or Supporting Components. Figure 3 shows the relationship between the device types.

Figure 1: ICES Device Types

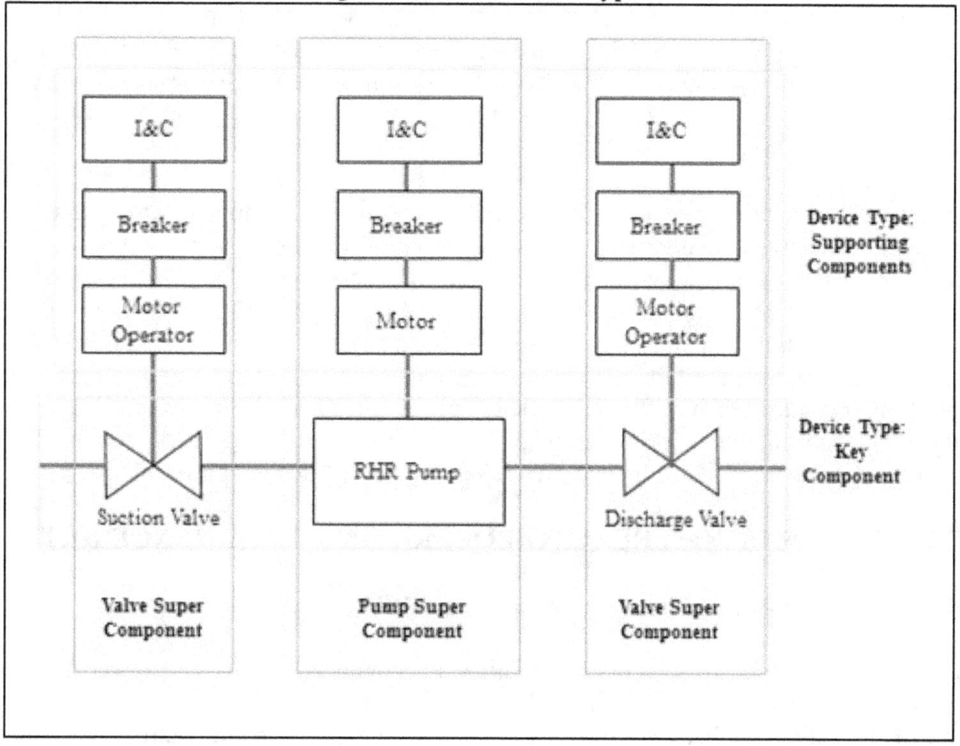

The ICES database contains over one million devices, of which approximately 140,000 are Key Components. INL only codes failure records for select set of devices in the ICES database, queries in IDCCS capture failure records for those devices of interest. Additionally, since the data entered into the ICES database is not specifically for PSA purposes, INL staff code the failure records of interest into the IDCCS database to ensure consistent capture of failure modes and related information.

INL provides a software tool to allow analysts to estimate industry and plant-specific reliability and availability parameters for selected components in risk-important systems for PSA applications. This tool is known as the Reliability and Availability Data System (RADS). RADS contains data and information based on INPOs ICES database. RADS uses failure data from 1997 through present and initiating events from 1987 through present. Since ICES data are proprietary, the NRC only provides the RADS database and the RADS analysis software to nuclear power plant licensees who are members of INPO and NRC staff on request. The reliability parameters estimated by RADS include:

- Probability of failure on demand
- Failure rate during operation
- Maintenance out-of-service unavailability
- Initiating event frequencies
- Time trends in reliability parameters.

INPO members or NRC staff may request access to RADS from a link at http://nrcoe.inl.gov/resultsdb/RADS/. A publically available reliability calculator is available at the same web link. The public calculator is not linked to ICES data.

3.2. Licensee Event Reports

Licensee Event Reports (LERs) are the only source for initiating event data and are the primary source of data for the Loss of Offsite Power study. LERs are downloaded from the NRC's Agency-wide

Documents Access and Management System (ADAMS) weekly. LERs are required by NRC regulation under Title 10 of the Code of Federal Regulations (10 CFR) Part 50.73. Additional guidelines for LER reporting is provided in NUREG 1022, rev. 3. In addition to collecting LERs for coding into the IDCCS, INL maintains a public, searchable database of LERS for the NRC at https://lersearch.inl.gov/LERSearchCriteria.aspx. LER Search contains LERs from 1980 to present.

3.3. Operating Time Data

Operating time data, including reactor critical years, shutdown time and total reactor years are required for a variety of risk analysis studies being conducted by the NRC. The primary sources of operating time data are the Monthly Operating Reports (MORs). MORs are required per NRC Generic Letter 97-02, and are submitted to INPO monthly by each licensee. Quarterly the MOR data is provided to the NRC and then to INL. INL maintains a database of MOR data called MORTRK. Additional outage information is maintained in a database maintained by INL called OUTINFO. Each morning the NRC Operations Center contacts each commercial nuclear power plant and obtains a verbal operations status report. This information is reviewed and a record of each plant outage is created in OUTINFO. These two data sources provide a complete picture of commercial nuclear power plant operating time.

4. PRIMARY USES OF NRC REACTOR OPERATING EXPERIENCE DATA

4.1. Industry Average Probabilistic Risk Assessment (PRA) Parameter Estimates

The NRC maintains risk models covering all U.S. commercial nuclear power plants. These standardized plant analysis risk (SPAR) models support NRC staff in regulatory decision-making, evaluation of inspection findings, and precursor studies. The SPAR models utilize industry average data for initiating event frequencies and component failure rates. The NRC published the results and parameter estimation process to estimate component failure probabilities, component failure rates, maintenance unavailability, and initiating event frequencies in NUREG/CR-6928 [1]. The data used to characterize industry-average performance in NUREG/CR-6928 was typically from the years 1998-2002. Rather than periodically publishing revisions to NUREG/CR-6926, updated parameter estimates are now posted on the NRC's website at http://nrcoe.inl.gov/resultsdb/. The most recent parameter estimates available on the website include industry average performance data collected through 2010.

4.2. Common-Cause Failure Parameter Estimates

INL maintains a database of common-cause failure (CCF) data for the NRC. CCF data collection for the NRC consists of CCF event identification, coding, and parameter estimation. CCF failure events satisfy four criteria:

- Two or more individual components fail, are degraded (including failures during demand or in-service testing), or have deficiencies that would result in component failures if a demand had been received.
- Components fail within a selected period of time such that success of the PSA mission would be uncertain.
- Components fail because of a single shared cause and coupling mechanism.
- Components fail within the established component boundary.

A description of how CCF data are gathered, coded, and analyzed is presented in NUREG/CR-6268, Rev.1 [2]. The NRC provides CCF parameter estimates for industry PSA applications and to support the NRC's SPAR models. The results were originally published in NUREG/CR-5497 [3]. Quantitative results of this data collection effort are now presented annually on the NRC's website at http://nrcoe.inl.gov/resultsdb/ParamEstSpar/. The current report contains CCF data through 2012. INPO members and NRC staff may request access to the Common-Cause Failure Database (CCFDB). The CCFDB is a data collection and analysis system to support CCF parameter estimates. CCF data and analysis tools are also available to INPO members and NRC staff in the NRC Reactor Operating

Experience Data (NROD) website at https://nrod.inl.gov. The NROD website also provides a search feature for IDCCS coded data an enhanced web version of the Reliability and Availability Data System (RADS).

4.3. Loss of Offsite Power Study

Loss of offsite power (LOOP, also referred to as LOSP) events are important contributors to overall risk at nuclear power plants. Some U.S. commercial nuclear power plants attribute over 70% of the overall risk to LOOP events. The LOOP data covers both at power and shutdown operations. LOOP industry frequencies are determined for four LOOP event categories: plant-centered, switchyard-centered, grid-related, and weather related. Trend plots are provided for each LOOP category for both critical operations and shutdown operation. The LOOP report contains important PSA parameters such as bus probabilities and Emergency Diesel Generator (EDG) repair times and non-recovery probabilities. The LOOP report also contains an engineering analysis section that provided quantitative insights into LOOP causes. The NRC published NUREG/CR-5496 to provide updated LOOP model parameters, frequency, and recovery time data [4]. Updated results are published annually on the NRC's website at http://nrcoe.inl.gov/resultsdb/LOSP/. The current report contains results from 1986 to 2012.

4.4. Initiating Events Study

The NRC collects data for all unexpected reactor trips during power operations at commercial nuclear power plants. Each reactor scram is reviewed and categorized according to the initiating event. To be included in the study, an event must meet all the following criteria:

- Include an unplanned reactor trip (not a scheduled reactor trip)
- Sequence of events starts when the reactor is critical and at or above the point of adding heat
- Occurs at a U.S. commercial nuclear power plant
- Is reported by a Licensee Event Report (LER)

The operating data is used to determine trends or patterns of plant performance on a plant type, plant specific or industry-wide basis. The initiating events study is also used to validate the estimates used for PSA initiating event frequencies. NUREG/CR-5750 documents the results of the initial study [5]. Updates to the Initiating Events Study are posted annually on the NRC's website at http://nrcoe.inl.gov/resultsdb/InitEvent/. Initiating Event results through 2012 are currently presented.

4.5. System Studies

System performance evaluations have been made for select safety systems.   For Boiling Water Reactors (BWRs), the available systems are:

- High Pressure Coolant Injection (HPCI) System
- High Pressure Core Spray (HPCS) System
- Isolation Condenser (IC) System
- Reactor Core Isolation Cooling (RCIC) System

For Pressurized Water Reactors (PWRs), the available systems are:

- Auxiliary Feedwater (AFW) System
- High Pressure Safety Injection (HPSI) System

Common available systems are:

- Emergency Power System (EPS)

- Residual Heat Removal System (RHR)

Reactor Protection Systems available for:

- General Electric
- Westinghouse
- Combustion Engineering
- Babcock and Wilcox (B&W)

The system studies provide unreliability results and trends for these given systems.
The initial system studies were documented in a series of NUREGs (NUREG/CR-5500, Volumes 1-11) [6]. Current results are posted on the NRC's website at http://nrcoe.inl.gov/resultsdb/SysStudy/.

4.6. Component Performance Studies

A study of safety-related components used in both PWRs and BWRs in risk important systems is provided. The study provides a risk-based analysis and engineering analysis of trends of operating data for select nuclear power plant components to provide insights into the performance of these components on an industry basis. The analyzed components are:

- Emergency Diesel Generators (EDGs)
- Turbine-Driven Pumps (TDPs)
- Motor-Driven Pumps (MDPs)
- Air-Operated Valves (AOVs)
- Motor-Operated Valves (MOVs)

The initial component performance studies were documented in a series of NUREGs (NUREG-1715, Volumes 1-4) [7]. Current results (through 2012) are posted on the NRC's website at http://nrcoe.inl.gov/resultsdb/CompPerf/.

4.7. Fire Events

The fire study uses operating experience data to characterize the frequency and nature of fire events from U.S. commercial nuclear power plants. The fire study contains data from 1987 to 2012 from various sources including LERs, ENs, EPIX data, Nuclear Plant Reliability Data System (NRPDS), and the National Electric Insurers Limited (NEIL). Some of these data sources have a limited time frame of contribution and fire events are not necessarily required to be reported through LERs and ENs. Due to the inconsistent data sources, the NRC fire events study should not be considered as comprehensive, rather only used for qualitative distribution and trending purposes. For quantitative fire data for use in PSA, the enhanced Fire Events Database (FEDB) developed by the Electric Power Research Institute (EPRI) in cooperation with the NRC should be used. The EPRI FEDB contains a complete and comprehensive data source of fire events at US nuclear power plants from 1990 through 2009 [8].

4.8. Industry Trends Program

The industry trends program was established to collect and monitor industry-wide data to ensure the nuclear industry is maintaining the safety performance of operating plants. The NRC reports annually to Congress the results of the ITP performance indicator trends. The current performance indicators included in the ITP are:

- Automatic Scrams While Critical
- Safety System Actuations (SSA)
- Significant Events

- Safety System Failures (SSF)
- Forced Outage Rate (FOR)
- Equipment Forced Outages per 1000 Commercial Critical Hours (EFO)
- Collective Radiation Exposure (CRE)
- Accident Sequence Precursors (ASP)
- Unplanned Power Changes
- Reactor Coolant System (RCS) Specific Activity
- Reactor Coolant System Leakage
- Drill/Exercise Performance
- Emergency Response Organization (ERO) Drill Participation
- Alert and Notification System Reliability
- Baseline Risk Index for Initiating Events (BRIIE)

A description of the ITP program as well as the results may be found at
http://www.nrc.gov/reactors/operating/oversight/industry-trends.html

5. CONCLUSION

This paper provided a summary of operating experience data collection efforts performed by Idaho National Laboratory for the U.S. Nuclear Regulatory Commission. The results of this ongoing work were generally provided to the public through a series of NUREGs and currently through periodic web reports on the NRC's website. Several databases and data analysis tools are also provided for specific purposes. The operating experience data and data analysis tools provided by the NRC are available to support various PSA modeling needs.

Acknowledgements

This work is supported and funded by the U.S. Nuclear Regulatory Commission.

References

[1]   S.A. Eide et al., "*Industry-Average Performance for Components and Initiating Events at U.S. Commercial Nuclear Power Plants,*" U.S. Nuclear Regulatory Commission, NUREG/CR-6928, 2007.
[2]   T.E. Wierman et al., "Common-Cause Failure Database and Analysis System: Event Data Collection, Classification, and Coding" U.S. Nuclear Regulatory Commission, NUREG/CR-6268, Rev.1, 2007.
[3]   F.M. Marshall et al., "Common-Cause Failure Parameter Estimations," U.S. Nuclear Regulatory Commission, NUREG/CR-5497, 1998.
[4]   C.L. Atwood et al., "Evaluation of Loss of Offsite Power Events at Nuclear Power Plants: 1980 – 1996," U.S. Nuclear Regulatory Commission, NUREG/CR-5496, 1998.
[5]   J.P. Poloski et al., "*Rates of Initiating Events at U.S. Nuclear Power Plants: 1987 – 1995,*" U.S. Nuclear Regulatory Commission, NUREG/CR-5750, 1999.
[6]   J.P. Poloski et al., "Auxiliary/Emergency Feedwater System Reliability, 1987-1995," U.S. Nuclear Regulatory Commission, NUREG/CR-5500, Vol. 1, 1998.
[7]   J.R. Houghton et al., "Component Performance Study – Turbine Driven Pumps, 1987-1998," U.S. Nuclear Regulatory Commission, NUREG-1715, Vol. 1, 2000.
[8]   P. Baranowsky et al., "The Updated Fire Events Database: Description of Content and Fire Event Classification Guidance," EPRI 1025284, 2013.

This manuscript has been authored by Battelle Energy Alliance, LLC under Contract No. DE-AC07-05ID14517 with the U.S. Department of Energy. The United States Government retains and the publisher, by accepting the article for publication, acknowledges that the United States Government retains a nonexclusive, paid-up, irrevocable, world-wide license to publish or reproduce the published form of this manuscript, or allow others to do so, for United States Government purposes.

Component Reliability in the T-Book – The New Approach

Anders Olsson[*a], Erik Persson Sunde[a], and Magnus Gudmundsson[b]

[a] Lloyd's Register Consulting, Stockholm, Sweden
[b] TUD Office, Vattenfall, Stockholm, Sweden

Abstract: T-Book is a reliability data handbook for use in Nordic Nuclear PSAs (Probabilistic Safety Assessments). Due to its ambitious scope, high level of detail, and high QA standard, it has become world-famous, and is frequently used even outside the nuclear field. Since 2008, Lloyd's Register Consulting, on behalf of the Nordic PSA Group (NPSAG) and TUD (the editor of T-Book), has performed a series of projects to enhance and consolidate the process, right from the classification and sampling of data, through parameter assessment, PSA modeling, and up to the final interpretation of results. Two aspects have proven to be of particular interest. Firstly, providing more homogeneous groups of T-Book components, which will have positive impact on PSA in terms of less conservative and more precise parameters, as well as increased consistency in the entire modeling process. Secondly, the benefits of said homogenization need to be weighed against the use of the multi-parametric model for standby components, because these two aspects are not fully compatible. A comprehensive approach, addressing both these aspects, is presented for selected components: pumps, batteries, diesel generators, and motor operated control valves. In this paper, the background and motives for the proposed strategy will be outlined, as well as the "tool box" to put it into practice. The presentation will also include what has been accomplished during 2013, and what is going to be introduced in the new version of the T-Book.

Keywords: Component reliability, T-Book

1. INTRODUCTION

The main objective of the T-Book [1], currently in the 7th edition, is to provide reliability data for the unavailability computations that are made for each component that is considered in the compulsory probabilistic safety assessments (PSA) of nuclear power plants. Safety, reliability and availability of nuclear power plants (NPP) are of paramount concern to employees, power companies, authorities and the public at large. As the use of PSA is wide in the normal safety work at the NPPs, there is a need for easily accessible and reliable failure data.

The failure characteristics presented in the T-Book are primarily based on the failure reports stored in the central database TUD (managed by the TUD Office) and the Licensee Event Reports delivered to the Swedish Radiation Safety Authority (SSM). The TUD database was started already in the middle of the 1970s, quite voluntarily by the Swedish power companies. In 1981, the Finnish power company TVO, operating two reactor units of Swedish design, joined the data collection system. Before the TUD data are statistically treated they are carefully examined with respect to consistency and correctness. The T-Book comprises only critical failures, i.e. failures that stop the function of components or lead to repair. The release history of the T-Book is outlined in table below.

Table 1: T-Book history

Version	Year	Comment
1	1982	Operational statistics from 21 reactor years
2	1985	Operating data covering about 40 reactor years
3	1992	Data up to the operating year 1987 included (108 reactor years)
4	1994	Data up to the operating year 1992 included (178 reactor years)

---

[*] anders.olsson@lr.org

Table 1: T-Book history

Version	Year	Comment
5	2000	Data up to the operating year 1996 included (234 reactor years)
6	2005	Data up to the operating year 2005 included (315 reactor years)
7	2010	Data up to the operating year 2007 included (378 reactor years)

As the amount of data has increased with each successive edition of the T-Book, continuous work has been done to improve the methods for the statistical inference and related tools required to derive the reliability parameters from the operational data in the database.

Already in the initial edition there was a Bayesian reasoning in the description of the uncertainty associated with the failure rates or demand-related failure probabilities. Analytically attractive distributions, gamma distributions for the failure rate, were used to model this uncertainty, and straightforward statistical methods were used to estimate the parameters of these distributions.

2. SUMMARY OF PREVIOUS WORK

As mentioned in the introduction there has been a continuous work to improve the methods and tools used to derive the reliability parameters in the T-Book. Since 2005 a series of such projects have been launched starting with a master thesis [4] with the purpose to study the mathematical model used in the T-Book and to compare it with other models. This has been followed by a series of additional works, in several cases in the form of master thesis's (see references [5], [7] and [8]). A brief summary of this previous work and the conclusions that were derived upon which has led to the work that is currently ongoing is given in the subsections below.

2.1. Study of the statistical methods used in the T-Book

The two-stage Bayesian method that is used in the T-Book today was studied in the master thesis work presented in [4]. The Bayesian method has been specifically developed for the T-Book and is particularly appropriate when data is extremely sparse. One objection against it however, is that it is somewhat non-transparent and therefore it may be worth studying if it could be simplified or even replaced by a simpler alternative, if there is one.

On this basis alternative methods were studied in the thesis work and an alternative Finnish method developed by Jussi Vaurio was identified an interesting alternative to study further. The main advantages with this alternative method would be fast calculations, transparent mathematical expressions and straightforward replication. Moreover, the numerical results presented in the thesis looked quite promising. Hence, though further comparisons were deemed to be necessary, this alternative method was proposed as a simpler and in some respects even better alternative to the method used in the T-Book. Even though the thesis pointed out this specific alternative method the work conducted after the thesis has not been focused on this alternative will is shown further on in this paper.

2.2. Test and analysis of homogeneity

The first master thesis, see [4], was followed by a second master thesis [5] which had the purpose to study homogeneity. The two-stage Bayesian method used in the T-Book comprises an assumption of inhomogeneity among the components of a population which is different compared to how it is done in the German ZEDB framework [9]. If components are assumed to be inhomogeneous it is possible to assign a specific failure rate for each individual component. On the other hand, if the components are assumed to be homogeneous the data can be pooled before a common reliability parameter is derived representing all components in the group. The objective with this thesis work was to design a statistical method for testing the homogeneity of Nordic data with emphasis on their failure rate. A *chi-square goodness-of-fit test* with consideration taken to operation time (or standby time), was implemented and applied on failure event data for the Nordic utilities.

From the tests it was concluded that the failure intensity for continuously operating components for most populations can be considered homogeneous with regard to failure rate. However, the test results also indicated that populations of standby components are to a larger extent inhomogeneous, which might be explained by differences in the data set due to unequal number of demands. It was also concluded that larger populations, i.e. components of all plants, must be considered as more inhomogeneous. Furthermore, it was recommended in the thesis work that a statistical test of homogeneity should be introduced.

2.3. Pros and cons using a using a multi-parametric model

As a result from the two master thesis works mentioned above additional work has been performed, which was presented at PSAM11 in Helsinki related to pros and cons with the $q_0 + \lambda t$ model [3] and evaluation of grouping criteria [2]. The conclusions presented in [3] are briefly summarized here.

One of the conclusions was that the T-Book approach deviates from what is state-of-the-art internationally in using a two-parametric model for assessment of reliability parameters of standby components. The principal merit of the model is that it is intuitively attractive because standby components are naturally associated with two different failure mechanisms. The $q_0 + \lambda t$ model estimates parameters representing these two mechanisms statistically from a joint data set. Thus, the type of failure does not have to be determined beforehand. Instead, the correlation between the two parameters is derived by virtue of the different activation intervals that are present in a group of components. Nevertheless, two major challenges were pointed out related to this model:

- The study shows that pooling of data (i.e. summarizing failures and exposure time, respectively, for components in the same group) has positive impact on PSA in terms of increased precision, decreased conservatisms, and improved conditions for implementation of parameters. However, the $q_0 + \lambda t$ model uses differences between individual components thus implying that data cannot be pooled.
- It was also pointed out that it has been questioned if it is possible to apply a multi-parametric model in an industrial field where data are generally sparse, which is challenging even for single parametric models.

Thus, to keep the $q_0 + \lambda t$ model, it was concluded that it has to be stated that its advantages are overriding the advantages of pooling data, and that the model is well suited for the area of application.

2.4. Evaluation of grouping criteria

The second topic that was addressed in the continuous work, and which is described in [2], was an evaluation of grouping criteria. The results to date, as described above and documented in [5] show that pooling of data (i.e. summarizing failures and exposure time, respectively, for components in the same group) has positive impact on PSA in terms of increased precision, decreased conservatisms, and improved conditions for implementation of parameters. However, pooling of data requires the component groups to be homogeneous, which has not been verified for the T-Book. Thus, alternative grouping criteria have been studied.

One such alternative is the so called function oriented criteria used in the German PSA community (represented by VGB, RiSA and the ZEDB software). This criterion is assumed to result in homogeneous populations, thus strengthening the motives for data pooling. It was then concluded that application of these criteria requires splitting up today's groups based on attributes like water chemistry, operational conditions and a finer division of quantitative variables. It was furthermore recommended that statistical tests should be used to verify homogeneity for the groups obtained.

It is therefore assumed that alternative grouping criteria, together with homogeneity tests, will considerably enhance the quality and usability of reliability parameters for PSA.

3. HOMOGENEOUS GROUPING OF COMPONENTS

A pilot study was performed [6], driven by previously performed work [2], with the purpose to evaluate the conditions for adoption of the grouping criteria used within German ZEDB into the TUD framework (i.e. the TUD database, T-Book, as well as underlying classification principles and routines). One underlying concern here was that application of current T-Book distributions is restrained because they are derived component-wise and then weighed together plant-wise. Such parameters are not well suited for PSA since neither parameter sampling nor event sampling will be fully applicable. Event sampling might be the more reasonable approach; however it may yield non-conservative results, dependent on the structure of the model. Moreover, there might be parameters from other sources in the model that shouldn't be event sampled (wherefore parameter sampling is the default approach in e.g. RiskSpectrum®).

The main objective of the work described in [6] was to evaluate the possibilities of adopting the ZEDB grouping strategy into the TUD framework and to benchmark current groups and attributes, as well as structures and purposes of the databases used. On the basis of this benchmark, the scope and objectives to fully apply the ZEDB grouping strategy to pumps in the T-Book was outlined. The benchmark was carried out in close dialogue with the TUD office and RiSA (developer of the ZEDB software).

ZEDB is an MS-Access based software used to store and process failure data from 21 nuclear power plants: 19 German, 1 Dutch and 1 Swiss. Integrated with the database are modules for validity, consistency, and homogeneity checks, as well as for computation of reliability parameters. As in T-Code (the software used to derive the reliability parameters in the T-Book), a two-stage Bayesian model is used (this is sometimes referred to as a "super-population approach") [9]. A large part of the information in [6] is related to describing the ZEDB and the T-Book classification frameworks and it is pointed out that the following general conditions have to be fulfilled when putting data together in the ZEDB framework:

- Components in the same group have to have a similar function.
- There has to be a sufficient amount of operational data for the components.

Up to 2004, sets of components in the ZEDB framework were defined using the technical descriptors of the components. Pump sets were then established using "pump type", "pump drive" and "fluid" and then clustered by nominal discharge head and nominal mass flow rate. As from the ZEDB Analysis of 2004, the sets have been split into smaller groups according to their function, thereby strengthening the motives for assuming homogeneity. This enhanced grouping strategy was first applied to pumps and has been applied to other component types in the subsequent issues of the ZEDB Analysis report from 2006 [9]. However, since data is more or less sparse, it is not always possible to derive function oriented groups. Thus, for some pump types, group populations are still used. Group populations and function oriented populations are thus coexisting although the latter is the ideal. Group populations thus represent the old scheme and are used as a secondary choice if no function oriented population is available for the component at issue. Only functional oriented populations are mutual exclusive.

The TUD database contains information on approximately 600,000 components whereof 23,000 of these are represented in the T-Book with reliability parameters for use in PSA. As in ZEDB, a component is defined as a functional part of the plant and a variety of attributes can be registered for each object. However, none of these attributes are used in queries to define the component groups. Component groups only exist in T-Book, each corresponding to a certain T-Book table which is directly connected to the relevant object ID codes via the event records. These groups have been established beforehand in an expert eliciting process which was originally carried out at the prospect of the first issue of T-Book in 1982. At the prospect of the fifth T-Book in 2000, the pump groups were revised and adjusted to the ICDE CCF database project. The revision resulted in new categories with respect to flow and pump head. Furthermore, BWR and PWR components are not separated but may be referred to the same group dependent on the attributes. Since the attributes are not used in queries, they are not mandatory in the database, but allowed to have a varying degree of completeness. Hence, if components are to be reclassified according to a new set of attributes the TUD database is probably not sufficiently detailed, whereas plant databases should be.

As previously mentioned, the work described in [6] also included a benchmark of the function oriented pump groups in the ZEDB Analysis of 2006 [9] and the T-Book version 6 [10]. The main purpose with this benchmark study was to see if the ZEDB grouping criteria could be used also in the T-Book for justification of homogeneity assumption. Thus the important questions were:

- Do the groups match?
- What is making the difference?
- Is it possible to overcome discrepancies?

As a rule, the groups match quite well, although with three main sources of mismatch which need to be dealt with. The differences identified relate to either differences in design or population mismatches where some systems are not separated in the T-Book as they are in ZEDB or a system is divided into subgroups in ZEDB but not in the T-Book. Furthermore it was concluded that the T-Book need to separate strictly between BWR and PWR components in order to be able to utilize the ZEDB grouping criteria. The differences identified was however judged to be manageable and it is noted that the aim here was not to copy the ZEDB scheme but to obtain credible grouping criteria for the T-Book and to be able to verify these against ZEDB groups.

It is also noted in [6] that a re-grouping of T-Book components implies a reduction in the T-Book populations and there is therefore a risk that data is lost. It is at the same time noted that this is exactly the reason why function oriented groups have not always been possible to establish in ZEDB. However, the T-Book comprises more historical data which may compensate for this along with the fact that the T-Book comprises fewer plants compared to ZEDB.

Based upon the conclusions from the pilot study [6] the work has continued with studying the pump populations in the T-Book [11]. The purpose of this work has been to develop function oriented grouping criteria that can be used for regrouping of the pump populations used in the T-Book and to perform statistical tests that will verify that the new groups are homogeneous. The basic assumption that was applied was that the new groups shall be considered to be homogeneous if the opposite cannot be demonstrated by statistical methods.

In order to achieve the function oriented groups the five criteria listed below were used where the third criteria (medium) was added to the already existing grouping criteria in the T-Book, since water chemistry is judged to be of importance when it comes to component reliability (e.g. due to corrosion and wear). Other modifications have also been made to the existing grouping criteria.

- Type of plant (BWR or PWR)
- Type of pump
- Medium distributed by the pump (de-ionized water, contaminated water, sea water, borated water, oil)
- Operational mode (in operation, in stand-by or intermittent operation)
- Mass flow
- Pump head

The test method that was applied measures if the component failures (e.g. spurious stop) are evenly distributed over the group population over time. The hypothesis to be evaluated in this case was that all components within the group could be represented by the same failure intensity. Based on empirical data a test variable (Q) was defined and compared against a critical value. If the test variable is greater than a given critical value then the hypothesis about homogeneity is rejected and the group is considered as being inhomogeneous. The test variable Q, often referred to as *goodness-of fit*, is defined according to Equation (1).

$$Q = \sum_i \frac{(Q_i - E_i)^2}{E_i} \tag{1}$$

Where:

    Q_i: Number of observed failures for pump i.
    E_i: Number of expected failures for pump i.

Assume that a group consists of i pumps with the same constant failure intensity for a given failure mode. Further assume that the variable Q_i can be described by a Poisson distribution, see Equation (2), and that this would render in a total number of N failures during time period T. If those assumptions are valid, the number of occurred failures would be distributed according to Equation (3) (Multinomial distribution).

$$\langle Q_i \rangle = E_i = \lambda t_i \tag{2}$$

$$P(O_1 = n_1, O_2 = n_2, \dots, O_N = n_N) = \frac{N!}{\prod_i n_i} \prod_i \pi_i^{n_i} \tag{3}$$

Where:

$$\pi_i = \frac{n_i}{N} = \frac{t_i}{T} \tag{4}$$

$$\sum_i n_i = N, \sum_i t_i = T \tag{5}$$

Then the expected number of failures E_i is given by (6):

$$\pi_i N = \frac{N}{T} t_i \tag{6}$$

A Monte Carlo technique can then be used in order to simulate the expected distribution of Q given N and the cumulative distribution of Q (CDF(Q)) can be generated. The observed number of failures (q_s) is significant, i.e. the group is inhomogeneous if the probability of getting a value that is higher than q_s is small, which has been defined as less than 5% as shown in equation (7). The method is exemplified below.

$$P(Q > q_s) = 1 - CDF(Q = q_s) < 0.05 \tag{7}$$

Assume that we have two recorded failures in a group of four components according to table below.

Table 2: Example of operational experience

Component	No. of failures	Time in operation
A	2	100
B	0	100
C	0	100
D	0	200

The simulated distribution and the Chi2 distribution are illustrated in Figure 1. The result based on the multinomial distribution is that the observations for the four components are homogeneous with a confidence of >95% (1-CDF($Q = q_s$) = 0.119). On the other hand, the result based on the Chi2 distribution indicates that the components are inhomogeneous (1- χ_3^2 ($Q = q_s$) = 0.046). This discrepancy can be explained by the fact that Q cannot be approximated with a Chi2 distribution based on such sparse data. If the number of observed failures would be six, instead of only two, then the result would be as illustrated in Figure 2 in which the Chi2 distribution is a better approximation of the multinomial distribution.

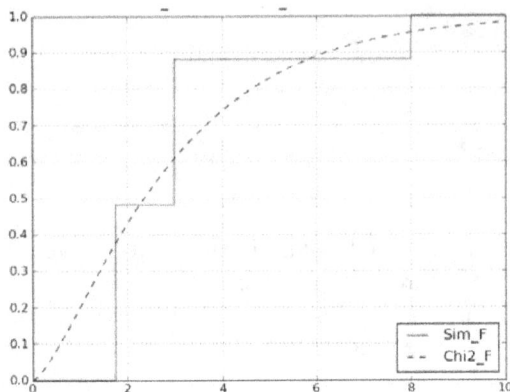

Figure 1 – Example where red line equals 1-CDF(Q) and black line equals $\chi_3^2(Q)$. No. of simulations are 100.000.

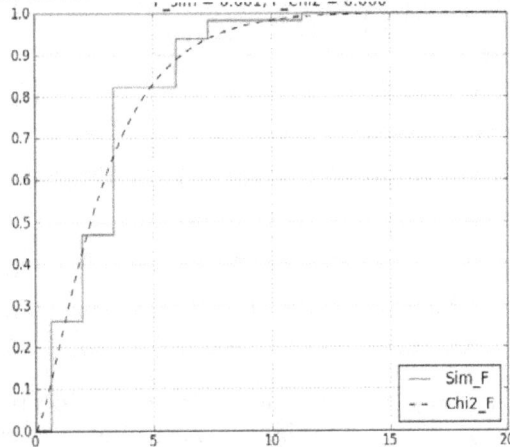

Figure 2 – Result of the example with six observed failures instead of two. Red line equals 1-CDF(Q) and black line equals $\chi_3^2(Q) \chi_3^2(Q$.

Based upon what has been presented above, new pump component groups have been defined compared to current version of the T-Book (version 7, [1]) and these are suggested to replace the existing component groups since all but 1,5% of the groups have been demonstrated to be homogeneous, i.e. the variation between the likelihood of pump failure is small within each group. It is also recommended that data within each group is pooled, i.e. super components should be created for the groups. This is judged to have a positive effect on the PSAs that use the failure data in the form of more precise estimations, decreased conservatism and more consistent treatment of uncertainties, since parameter sampling can be used for uncertainty analysis in the PSA model. There are however some exceptions where it is recommended that the current strategy used in the T-Book is kept, i.e. without pooling of data.

In a similar way as for the pumps work is currently ongoing in defining new groups for batteries, diesel generators and valves.

4. MULTI-PARAMETRIC MODEL FOR STANDBY COMPONENTS

Since the 3rd version of the T-Book (1992), the so called $q_0+\lambda t$ model has been used to estimate failure probability for stand-by components. In this model, q_0 is derived from failures occurring at the time of demand while λ_s characterizes failure mechanisms that are active during the standby time. In reference [2], as has been described earlier, this methodology has been evaluated and the conclusions as presented in [2] are that the model does not work in a satisfactory manner for several of the component groups. The main reason for this conclusion is due to lack of support in the data, i.e. sparse data. It was also demonstrated that the model gives about the same results as the simpler λt model when the amount of data is sufficient. For some component groups, a constant failure probability, q_0, can be assumed to represent the dominating contribution to the component unavailability estimation, and it was concluded in [2] that it is important to identify these component groups.

As a result of the above presented conclusions from [2], one part of the work in producing version 8 of the T-Book is to develop evaluation criteria that will support deciding which component groups should be evaluated with the $q_0+\lambda t$ model and which should be evaluated using either λt or q_0. The purpose of the work presented in [12] is to evaluate and present selection criteria to facilitate this choice and to plan how the selection will be carried out during the implementation phase. The method should be able to demonstrate to what degree the data from experience supports the use of the multi-parametric model $q_0+\lambda t$.

The approach that was used in [3] has been used in order to evaluate and verify the use of $q_0+\lambda t$. An important aspect to consider is that the sought after criterion is based on a technical, physical or functional basis in order to be unbiased with regard to specific data. In [3] it was found that component test interval is of importance in order to demonstrate the validity of using $q_0+\lambda t$ and this parameter can therefore be used as an independent variable during the search for the criteria in question.

All stand-by components were analyzed and compared using the methods presented in [3] (version 6 of the T-Book), the data however was updated to reflect the most recent version of the T-Book [1]. The hypothesis was that the more data the more obvious it would be to identify common patterns. Therefore, based upon the conclusions from [3] the process to verify the use of $q_0+\lambda t$ involved the following steps:

1. Based on the given data; calculate λ_s' using the $\lambda_s t$ model and q and λ_s'' using the $q+\lambda_s t$ models.
2. Determine the difference $y = \Delta Q$ between the unavailability calculated using the different data settings

$$\Delta Q = \frac{Q_\lambda - Q_{q+\lambda}}{Q_{q+\lambda}} \tag{8}$$

Where:

$$Q_m = 1 - \frac{1}{\lambda_s \cdot TI}(1-q)\left(1-e^{-\lambda_s \cdot TI}\right) \tag{9}$$

for $Q_m = f(\lambda_s't)$, then $y = 0$

TI is represented by the groups average test interval since this is when the models are considered to be comparable, as illustrated by the red circle in Figure 3.

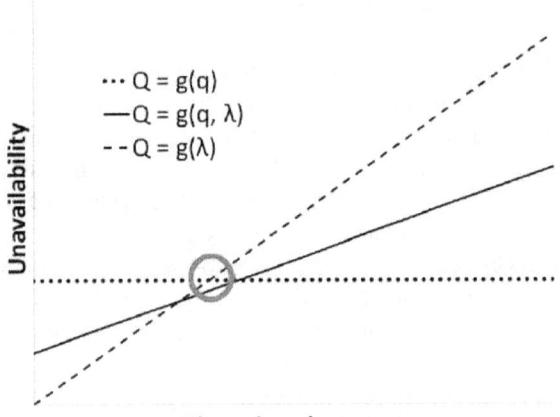

Figure 3 – Intersection between the curves described by each of the models q, $\lambda_s t$ and $q+\lambda_s t$.

The next steps are to:

3. Calculate $y = q/Q$ for the components. This function describes the contribution from q to the total unavailability.
4. Plotting of $y = \Delta Q$ and $y = q/Q$ respectively against the deviation in TI^* where TI^* is the true "actual" test interval, i.e. the quote between number of activations (planned and "spontaneous") and the time in stand-by for each component in the group. This test interval is what is being used in the calculation algorithm T-Code which is used in the T-Book.

Several different measurement of deviation can be used during interpretation of the plot. Standard deviation (average deviation from the mean value) for a groups TI^* can be used but since standard deviation is sensitive to extreme values ("outliers") another type of measurement should also be used, e.g. the median deviation MAD (Median Absolute Deviation).

$$MAD = median_i \left(\left| X_i - median_j \left(X_j \right) \right| \right) \tag{10}$$

In a comparison when different measurements of deviation are used, deviations in data can be explained, e.g. a component group seems to have a large deviation but is nevertheless not analyzed satisfactorily using $q+\lambda_s t$. The explanation for this can be that one or some components in the group significantly differ from the other components and those components give rise to an increased mean value at the same time as the other components in the group are allowed to have a very similar test interval. In these cases the measurement of variation is not valid since it measures a behavior that is reflected by data not supported by the used model. The work presented in [12] used both standard deviation and MAD for this purpose.

Figure 4 shows plotting of $y = \Delta Q$ against standard deviation in TI^* and therefore demonstrates the difference in percentage between Q_λ (unavailability using $\lambda_s t$ model) and $Q_{q+\lambda}$ (unavailability using $q+\lambda_s t$ model) as a function of TI^*.

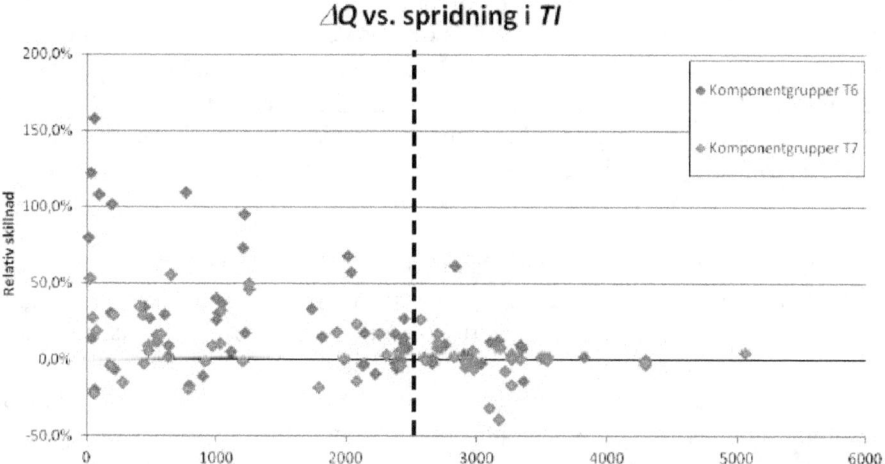

Figure 4 —T-Book tables evaluated by ΔQ as a function of deviation in TI*. ΔQ expresses the difference between Q calculated with the q+λst model and the λst model, respectively. Thus, a value of 0 % implies that the two models yield identical results, a positive value that the q+λst model yields a larger Q than the λst model, and a negative value that λst yields the higher Q.

The result pictured in Figure 4 implies that the smaller the spread in TI^*, the larger ΔQ will be, i.e. with a minor deviation, $q+\lambda st$ is overestimating the unavailability. The most probable explanation for the picture is that with less information the hyper-prior used in the Bayesian update has greater impact on results. This hyper-prior is to be updated with data, i.e. if there are few data points, e.g. small deviation in TI^*, the hyper-prior will remain affecting the result to a considerable extent. However, Q is not overestimated for all tables. Some of the tables in the left region have a ΔQ near 0%, and sometimes it is even negative. This might be revealing an inconsistency in the updating of the hyper-prior, i.e. in the numerical algorithm that is pacing around the "rough estimate of central point" to find data supply. This conclusion is supported by the convergence of $Q_{q+\lambda s}$ and $Q_{\lambda s}$. Moreover, the convergence of $Q_{q+\lambda s}$ and $Q_{\lambda s}$ might be due to a successively decreasing contribution from q with increasing amount of information available. This was analyzed through plotting the ratio of q to $Q_{q+\lambda s}$ against the deviation in TI as presented in Figure 5.

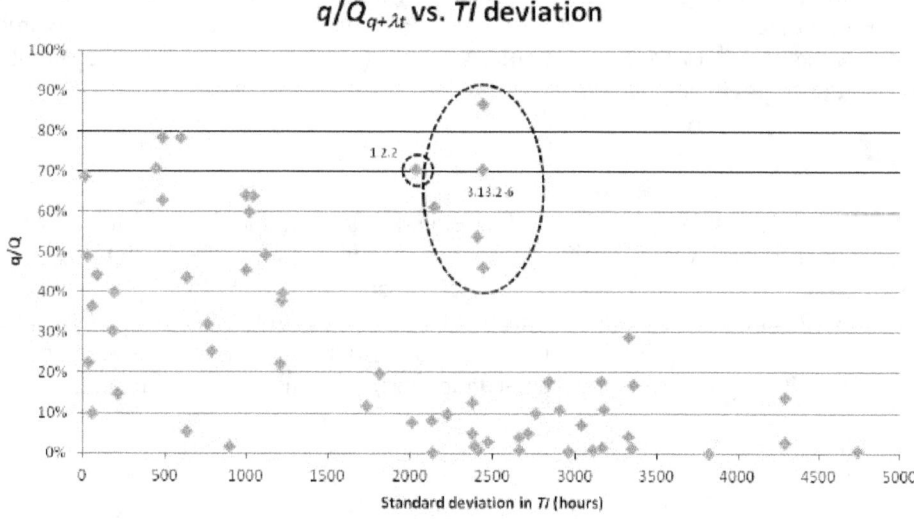

Figure 5 – q/Q as a function standard deviation in respective component groups test interval where each dot represents a table in the T-Book, i.e. a failure in a stand-by component)

The plot of $y = q/Q$ in Figure 5 shows a general tendency of large q for small deviations in TI^* and that the importance of q gets less the larger the deviation is. This can also explain why the more supporting data there is, the smaller ΔQ gets (q a-priori is large, which overestimates $Q_{q+\lambda}$ but the more data there is available the more precise the estimation will get). However, the fluctuations in ΔQ for small variations in TI^* cannot be explained by a large q. These fluctuations are more likely due to numerical problems (instability) when data is sparse. It can also be noted that in some cases q gives a large contribution even for components where the amount of supporting data is larger.

The procedure is then repeated using MAD instead of standard deviation and the resulting plots looks quite similar. The results imply that the $q+\lambda_s t$ model is working when the deviation in TI^* is large, which is logical. It is however not apparent when it starts to work satisfactorily. The dashed line in Figure 4 represents an attempt to make such a distinction. It shall also be noted that when the deviation in TI^* is large ΔQ gets smaller, i.e. it is not important which model is used. The simplest solution may therefore be to use the simplest model, $\lambda_s t$, for all components. However, the importance of q does not vanish completely and for some groups it contributes significantly. Therefore the conclusion is:

> **The $q+\lambda_s t$ model shall be used in the range where it can be proven to work and for component groups where the importance of q is significant.**

6. CONCLUSIONS

The work performed in this paper is significant progress in terms of the quality of the data presented in the T-Book. However, even though it might seem to be rather straightforward there are, and will always be, considerations that need to be taken into account, such as:

- Is it possible to establish a clear and definitive criterion that supports the choice of which method to use for deriving failure probability q and intensity λ?
- Can it be demonstrated that the $q+\lambda_s t$ model is suitable to use even though this will mean that it will not be possible to pool data? For component groups with sparse data this means that the benefit of pooling can be greater than the benefit of using $q+\lambda_s t$ model.
- Can it be demonstrated that the reliability data derived is not optimistic for all components?

For the time being it seems that a systematic process where different aspects are being tested from case to case, and from T-Book to T-Book (new versions), is the most suitable way to progress. Work is currently ongoing regarding redefining the component groups in coming version of the T-Book and to apply the different tests outlined in this paper in order to decide what model to use. In the coming version 8 of the T-Book the plan is to have two parallel versions (old and new methodology), but in future versions the old approach will be phased out.

Acknowledgements

Many individuals and organizations have contributed to the work related to the T-Book and we would like to acknowledge the Nordic PSA Group (NPSAG) for its continuous support, the TUD office, and RiSA and VGB in Germany. We also want to send a special thanks to Kurt Pörn who passed away some years ago. Another person that deserves special thanks is Vidar Hedtjärn Swaling who started working with the T-Book already during his master thesis, and has during all his time as an employee with Lloyd's Register Consulting (formerly known as Scandpower) continued to work in the area.

References

[1] TUD Office (2010), *T-Book, Reliability Data of Components in Nordic Nuclear Power Plants*, 7th Edition, ISBN 978-91-633-6144-9

[2] Swaling, V.H., Gabrielsson, A. (2010), *Principles for data analysis and parameter assessment, part 1*, NPSAG Report 21-001:01 (Available from www.npsag.org, paper presented at PSAM11)

[3] Swaling, V. H (2011). *Principles for Data Analysis and Parameter Assessment. Part 2 – Evaluation of the q+λst model*. NPSAG-report 21-001:02 (Available from www.npsag.org, paper presented at PSAM11)

[4] Swaling, V. H.,(2006). *Statistiska metoder för härledning av indata till säkerhetsanalyser inom kärnkraftsområdet/Statistical methods for deriving input to nuclear power safety assessments*. ISSN: 1650-8319, UPTEC STS06 014 (http://www.utn.uu.se/sts//images/exjobb/0605_Swaling.pdf)

[5] Höge, E., (2009), *Test and Analysis of Homogeneity Regarding Failure Intensity of Components in Nuclear Power Plants*, Department of Mathematics Uppsala University, U.U.D.M. Project Report 2009:2

[6] V. Hedtjärn Swaling, NPSAG Report 28-004:01, *Function Oriented Classification of Components – Pilots Study: Comparison of the German and Nordic Classification Systems*, June 2011.

[7] Lanner, L., (2009), *Homogeneity assumptions regarding reliability parameters and their impact on Probabilistic Safety Assessment*, Department of Mathematics Uppsala University, U.U.D.M. Project Report 2009:21

[8] Sjöström, J (2011), *Evaluation of Standby Component Reliability Modeling in Nordic Nuclear Power Plants*, KTH School of Industrial Engineering and Management, Division of Energy Technology, EGI-2011-008MSC

[9] VGB Powertech (2006), *Centralized Reliability and Events Database (ZEDB) – Reliability Data for Nuclear Power plant Components*, December 2006, Essen, ISSN 1439-7498.

[10] TUD Office (2005), *T-Book – Reliability Data of Components in Nordic Nuclear Power plants*, ISBN 978-91-633-6144-9.

[11] Sunde, E.P., Scandpower Report, 210528/R1, *Funktionsorienterad klassificering av komponenter - tillämpning på T-bokens pumpar (Function oriented classification och components – applied on T-Book pumps)*, March 2012

[12] Swaling V.H., Sunde E.P., Lloyd's Register Consulting Technical Note 211163-TN-003, *Förslag till hantering av $q_0+\lambda_t$ /Suggestion for treatment of $q_0+\lambda_t$ model model*, December 2013

Trend analysis of input data to Nordic PSA

Ostrovskii Dimitri[a], Lindahl Pär[b]

[a] ÅF consulting, Gothemburg, Sweden
[b] OKG AB, Oskarshamn, Sweden

Abstract: In Swedish and Finnish NPPs the "T-book" is one common source for reliability parameters used in PSA. These parameters are calculated based on the assumption that component reliability does not change with time. This assumption is e.g. violated if components degrade, due to ageing effects, or improve, due to improvements in maintenance strategies. It is thus relevant to ask how PSA results may be affected by the time independence assumption. To approach an answer, a non parametric test method, the Wilcoxon rank sum test, was used to analyze how observations of malfunctions deviate from a Poisson-distributed set that represents the case when malfunctions are independent of each other and occur with a constant frequency. It was found that deviations from the Poisson distribution, trends, can be detected in the gathered data, and that corrections for reliability trends may affect the PSA results significantly.

Keywords: PSA, Reliability Parameters, Trend Analysis, Ageing

1. INTRODUCTION

1.1 Background

Reliability data used to analyze the safety of Swedish and Finnish nuclear power plants, through PSA, is presented as reliability parameters for component groups. These reliability parameters, published in what is called the "T-book", describe a frequency or probability of a specific failure mode for the component group. Reliability parameters are calculated from empiric information under the constraints of a set of assumptions and postulates. One postulate states "The reliability parameters are trendless" - in other words, the true values of the components' reliability parameters are unchanged with time. While physical components age and at some point become unusable, they are assumed to be replaced by new components or returned to "mint" condition, resulting in a constant average failure frequency, or failure probability.

Parameters are estimated by analyzing observations of failure rates of components. For each parameter, as the number of observations increases, the probability distribution converges around a value close to the true reliability parameter value.

This study is an investigation of the validity of the postulate on absence of trends. In case this postulate is invalid for some or all of the T-book components, this may have an impact on the validity of the PSA results. If this postulate *is* valid, this opens a field for discussion on how to best analyze new component failure data. Given that the rate of convergence for the parameter estimates is slow, due to an already large pool of input data, one may consider changing focus to methods for early discovery of immerging trends due to e.g. changing ageing characteristics within the component group.

1.2 Reliability parameters in the T-book

The T-book updated regularly by the TUD-council (TUD-kansliet), which has a system of reporting failures that is connected to most of the Nordic NPPs. Every power plant is responsible for reporting its failures to the TUD-database. This process is described in the T-book [4].

The T-book presents components in groups of identical/similar components. For example one component group is "Centrifuge pump, MC-pump" [4] (table 1.4.1). There exist a total of 66 components in this particular group, summed over all the covered nuclear power plants. The failures are assumed to be independent of each other, i.e. the components failure intensity is assumed to be unchanged through time. This leads to a Poisson-distribution of failures.

$$P(X = x|\lambda) = e^{-\lambda t} \frac{(\lambda t)^x}{x!}$$

(1)

Where $P(X = x|\lambda)$ is the probability of x events occurring given a rate parameter λ during the time t.

Further the T-book distinguishes between different component classes; stand-by and continuously running components. In the case of continuously running components the rate of failure is symbolized by λ_d [h^{-1}]. For stand-by components it is relevant to consider the time since the last activation as well as a probability of failure just after this activation. This is modeled with a so called "$q_0 + \lambda t$"-model. Here q_0 [use^{-1}] symbolizes the constant, i.e. time independent, base line probability of the component to fail on start and λ_s [h^{-1}] describes the increasing probability with respect to the time since the component was last tested. Each of the reliability parameters is presented individually.

Please note that failure modes attributed to stand-by components are in some cases presented as those for continuously running. An example of this case is failure mode "spurious stop" for a stand-by component, for this failure mode a measure of the number of uses is irrelevant as the component is in operation by definition. From here on out components defined as continuously running are components with failure modes characterized by one parameter, either λ_d or λ_s, because these types of components failures are dependent only on total running or stand-by time.

The reliability parameters are derived using a two-step Bayesian method [3]. In the first step the prior is updated using observations made for an entire component group, and in the second step plant-specific observations are accounted for. This method produces two "levels" of reliability parameters. Generic parameters adhere to a component group for all participating NPPs, while plant specific parameters describe the component group for each specific plant.

2. TREND ANALYSIS

As a first step in this study a method of data analysis was chosen. Four statistical test methods were compared en terms of efficiency and sensitivity on simulated random data. For an in-depths description of this analysis see [1]. It was found that the Wilcoxon Rank-Sum test (rank sum test) was best suited for the purposes of this study; this finding is in itself interesting. As stated in the discussion on appropriate test methods in [1]:

"During this stage a greater attention was paid to the choice of test methods rather than the input data. Probably the most interesting result was that the test method discussed in literature, the Z-test, was poor in comparison to the rank sum test. An explanation for this might be that the Z-test is in fact superior in efficiency when dealing with a single exact distribution but is sensitive to small variations in the input data and thus not compatible with the methodology of this thesis. This would give rise to an interesting question of exactly how sensitive this test method is, as a pure Poisson process is only the ideal case while real failure distributions surely deviate from it."

2.1 Strategy

In order to conclude whether PSA results need to be questioned, with respect to trend in reliability data, a detectable trend needs to be significant for the PSA results themselves. For reliability data where trends cannot be detected, this can e.g. be tested by postulating a barely detectable linear trend in reliability data and extrapolating reliability parameters in time to account for a change due to the time dependent trend. But in order to postulate a barely detectable trend in data that appears to be trendless one needs to find the magnitude of trends that can be detected.

To reach a conclusion on the validity of the postulate on absence of trends the study was divided into two consecutive steps. First the magnitude of theoretical trends that can be detected by the chosen test method was determined. Then actual data was tested for such trends.

2.2 Choice of studied components

Input data to the Rank-Sum test, the test method used in this project, must consist of two data sets. The test determines if the data sets originate from the same distribution. For the purpose of this study a data set, for example the observed distribution of failures, was compared to data sets simulated using a known rate parameter – containing a trend or trendless. If observed data compared to simulated trendless data yields a low probability of matching distributions, a trend is found in the observed data.

The data used in this study consisted of vectors containing 0's and 1's. Each position in the vector corresponded to a unit of time, 1 hour of operation for continuously running components or 1 use for stand-by components. Every non-zero element in a certain data set was interpreted as a failure – these were Poisson-distributed with a given rate parameter (λ in eq. 1). For distributions generated with an increasing rate parameter with time, in practice the rate parameter was increased for each entry in the data set.

In the study of actual component groups two stand-by and two continuously running components were chosen. Table 1 shows the chosen component groups and failure modes. The set of component groups were selected based on best available prior knowledge, and was verified - post analysis - to be a sufficient sample for the purpose of the study.

Table 1. Chosen component groups and failure modes.

Component group	Failure mode	Component type
Heat exchangers	Inadequate cooling capacity	Continuously running
Centrifuge pumps	Spurious stop	Continuously running
Diesel generators	Failure to start	Stand-by
Closing valves	Failure to maneuver	Stand-by

2.3 Detectability

To study the magnitude of hypothetical trends that can be detected by the rank sum test method, a set of distributions $\bar{X}_i$ were generated. These distributions represented each of the chosen components, matching the number of observations for each respective component with the length of the distribution. Further these distributions were generated using the constant reliability parameter given by the T-book.

A second set of distributions $\bar{Y}_{\alpha_i}$ were generated. These distributions also matched each component in number of observations but were generated using an increasing, trended, reliability parameter:

$$\lambda = \lambda_0 + \alpha \left(\frac{i}{I} - \frac{1}{2} \right) \qquad (2)$$

Where λ is the trended reliability parameter, λ_0 the trendless parameter, α the magnitude of the trend, i the position of the vector element and I the total length of the vector.

This specific trend model was chosen so that the trended parameter has a mean same mean value as the trendless parameter. The parameter α was increased until a trend could clearly be detected. "Clearly detected" in this case means that the p-value of the rank sum test is less than 0.05[*].

---

[*] The p-value is defined as the probability of observing the given data given that the null hypothesis is true. In this case translated into the probability of the two data sets originating from the same distribution.

Note that stand-by component groups are represented by two reliability parameters in the T-book. These two parameters were weighed into one, based on documented times between activations, as described in [1], chapter 3.2.

2.3.1 Results

Sets of simulations were made for each trend magnitude α. The results are presented in figures 1 and 2, as a mean p-value and variance for each tested magnitude. A summary of the results is presented in table 2.

Table 2. Summary of result of the detectability study.

Component group (failure mode)	Detectable trend magnitude
Heat exchangers (inadequate cooling capacity)	$1.9 \cdot 10^{-5}$ h^{-1}
Centrifuge pumps (spurious stop)	$3.5 \cdot 10^{-5}$ h^{-1}
Diesel generators (failure to start)	$3.9 \cdot 10^{-3}$ use^{-1}
Closing valves (failure to maneuver)	$5.9 \cdot 10^{-3}$ use^{-1}

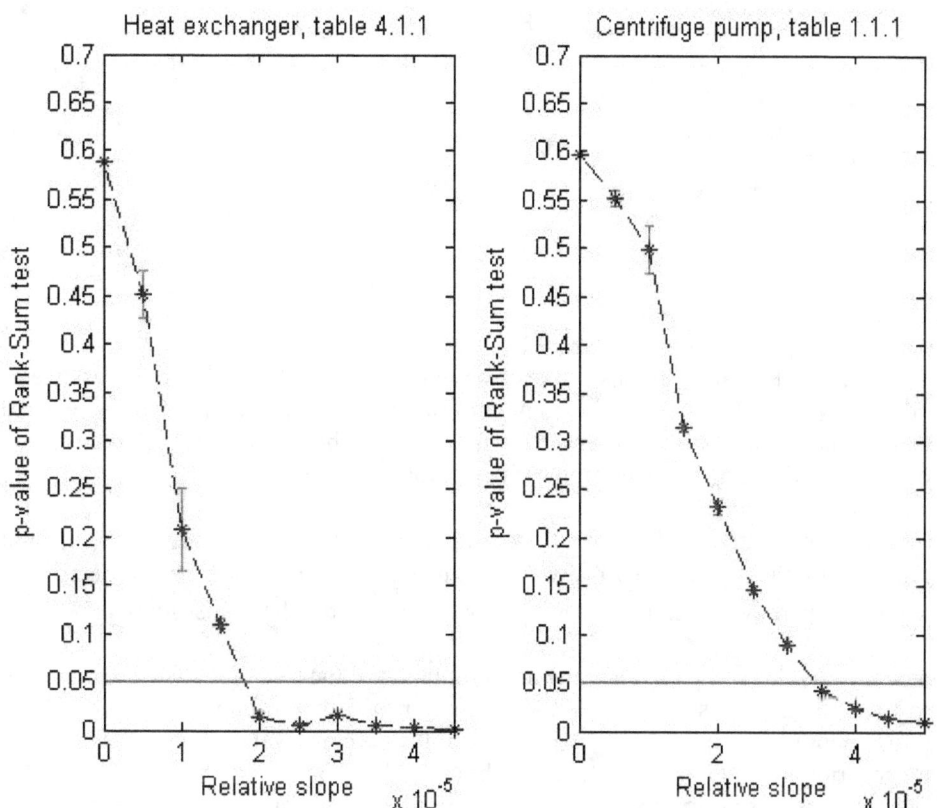

Figure 1: Simulations run on components presented in tables 1.1.1 (right) and 3.13.1 (left). Results are averaged from 10000 and 2000 simulations respectively. Bars indicate standard deviation 2σ.

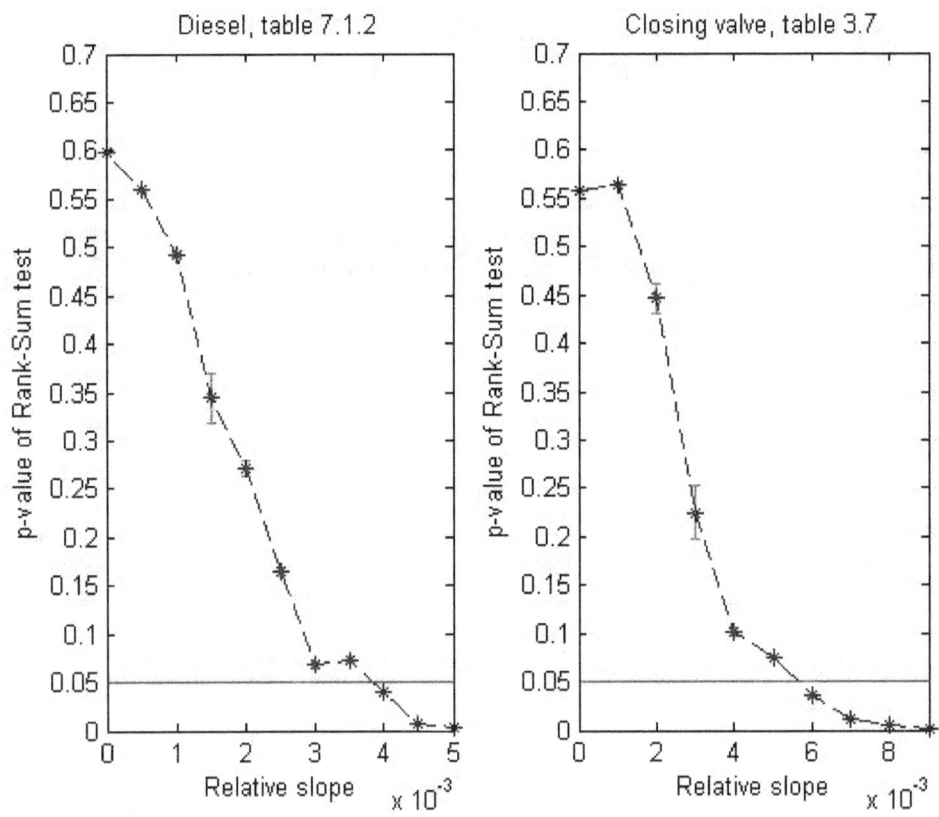

Figure 2: Simulations run on components presented in tables 3.7 (right) and 7.1.2 (left). Results are averaged from 30000 and 20000 simulations respectively. Bars indicate standard deviation 2σ.

2.4 Trend analysis

Because of the lack of actual distributions of the observed failures, the "real distributions" were approximated. This was done by placing the number of observed failures randomly in between the time interval when these failures were observed. Data on the number of observed failures at set time points was obtained from the different editions of the T-book [4], [5], [6] and [7].

When studying real distributions of failures the same approach was implemented as described in chapter 2.3. In this case the distributions $\bar{Y}_{\alpha_i}$ were exchanged to the actually observed distributions, and compared using the rank sum test to the randomly generated, trendless, distributions $\bar{X}_i$.

Using the same methodology as previously the approximation of the true time distributions were analysed. Input data to the rank sum test were: a distribution with the difference in number of observed failures between T-book version randomly distributed in the time intervals between the publishing's of T-books, and a distribution with the same total number of failures placed randomly in the total time interval. This procedure was performed several times to obtain an average p-value for each component versus the random distribution. Results of these calculations are presented in table 3. Pairs of sample distributions for each component group are shown in figures 3 and 4 to illustrate a possible difference in the distributions.

2.4.1 Results

Table 3: Summary of results of tests on actual data.

Component group	Average p-value	Trend suspicion
Centrifuge pump	0.6163	Trend highly unlikely
Closing valve	0.4965	Trend highly unlikely
Heat exchanger	ca. 0	Trend likely
Diesel generators	0.0128	Trend likely

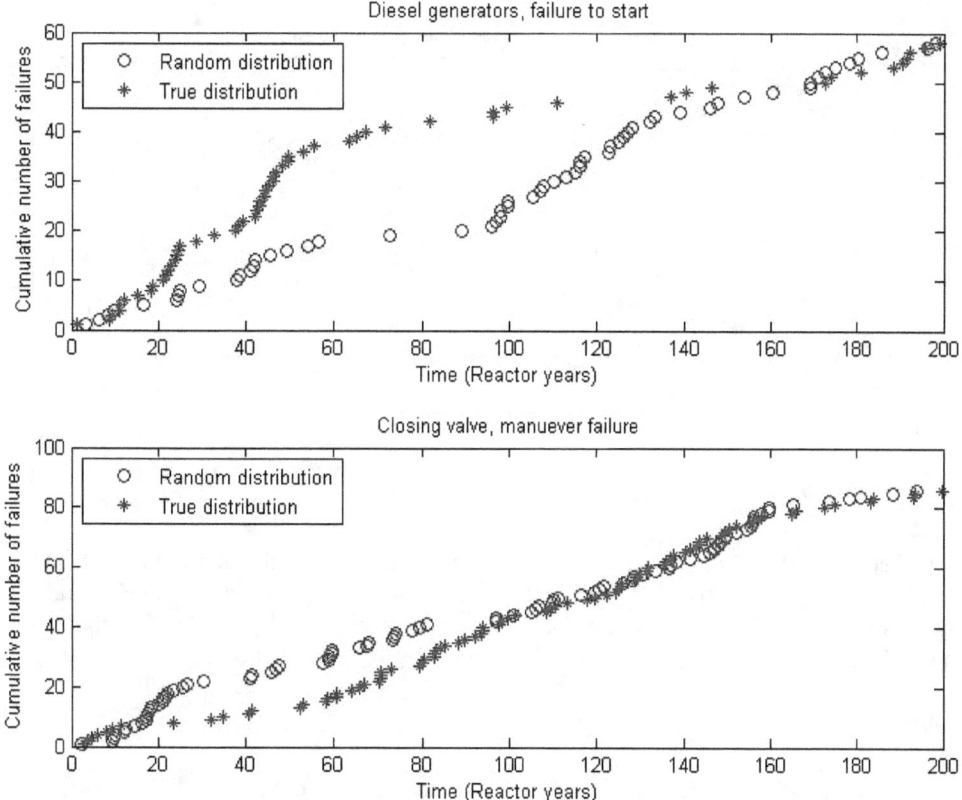

Figure 3: The approximations of the true and random distributions for the diesel generators and closing valve. The failures are presented cumulatively against time in reactor years.

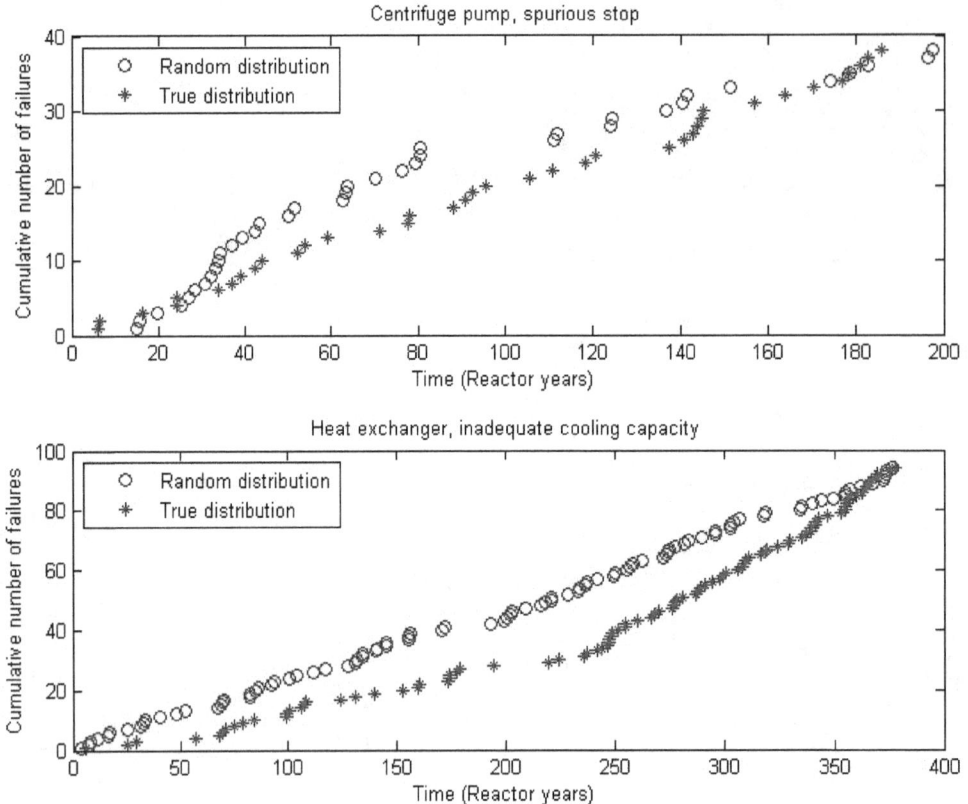

Figure 4: The approximations of the true and random distributions for the centrifuge pump and heat exchanger. The failures are presented cumulatively against time in reactor years.

As shown in Table 3 and Figures 3 and 4 trends were detected for diesel generators and heat exchangers. Further it is clear from the figures that one of each kind of trend, increasing and decreasing, can be observed by the test method. This result is interesting, as it implies that the analysis method is valid and can be used as a basis for the study of trends in reliability data.

2.5 Summary

This project proposes a methodology of screening observed component failures, in order to assess the validity of the postulate on absence of trends in reliability parameters based on these observed failures. The methodology is based on a statistical test method that determines whether two data sets are drawn from the same distribution. Assuming complete absence of trends in reliability data, component failures will ideally follow the Poisson-distribution. With a high number of observations, the average trendless parameter represents the rate parameter of a Poisson-distribution.

Thus, proving that a Poisson-distributed data set, generated with a certain rate parameter, is drawn from a different distribution than the observed failures attributed to the same reliability parameter, disproves trendlessness.

The rank sum test is in this paper shown to be able to detect trends in failure data.

3. DISCUSSION

3.1 Uncertainties

Some approximations were made in this project. Most of these originate in the fact that an exact time distribution of the observed failures is not available and approximate distributions were used. There is no way to estimate how precise these distribution are, therefore this is considered the largest source of uncertainty in the results. The way in which the time distribution was produced was however considered the best and only reasonable option given the available data.

For a more detailed discussion on this subject see [1].

3.2 Conclusion based on results

Even though many approximations were made, clear conclusions can be drawn from the results. It is apparent that trends can be observed in reliability data today. For the two component groups in which trends were detected the average p-values were 0.0128 (Diesel generators) and 0 (Heat exchangers).

P-value results should be interpreted as the probability of the two data sets being drawn from the same distribution. This means, as an example, that there is only a 1.28% probability that the observed failures of the diesel generators are Poisson-distributed, thus making the postulate on absence of trends invalid. The end user of this methodology is urged to decide before implementation the statistical limits on when the observed data is deemed to no longer be randomly distributed.

Applying the methodology proposed in this paper on all component groups presented in the T-book, or any other collection of operational experience, can show which components' or component groups' reliability parameters are subject to time dependency. Such knowledge may potentially have an impact on the PSA results for a nuclear power plant.

Figures 3 and 4 summarize the result visually. It is shown that trends can be both positive (increasing reliability) and negative (decreasing reliability), which is an unexpected result.. Based on the presented results the postulate on absence of trends can be rationalized, as two of the studied components show very good overlap with random distribution. Nevertheless the discovery of time dependent trends in some component in combination with a lack of trends in others validates the analysis method and motivated further study of reliability data.

The presence of positive trends opens the possibility of the postulate on absence of trends being conservative. This is possibly due to improved testing methods that lead to higher reliability of components. A new possible implication of the postulate is thus discovered, a possibility of over-conservatism that ads a new dimension to future studies.

Concluding this project, the trendlessness postulate may still be a sufficiently good approximation in general. However, it is also clear that there exist components for which the validity of this postulate may be questioned; this should be confirmed by a more detailed study. A preliminary investigation of the impact on PSA results - comparing nominal parameter values according to the T-book with values assuming barely detectable trends - points at the diesel generators as components where trends may have a significant impact. Based on this result, combined with a falsification of the trendlessness hypothesis in this case, it is reasonable to assume that PSA results need to be reevaluated. However, since the reliability of the diesel generators seem to be improving with time according to figure 3, the nominal PSA results are likely to be conservative. A more realistic reliability parameter estimate will in this case reduce the safety importance of the diesel generators, and reduce the overall core damage frequency for the PSA model.

The project recommends further study of time dependent trends, preferably in reverse order – starting with a parameter sensitivity study, followed by trend analysis of the identified parameters significant to PSA. If negative trends cannot be detected in reliability parameters most significant to PSA of a plant, the postulate on absence of trends is acceptable. The end user is also urged to account for positive trends, making an evaluation of the desired level of conservatism in future PSA.

3.3 Acknowledgements

Special thanks are extended to: Magnus Gudmundsson, Vattenfall research and development, for providing input data to and help with T-code; Dan Kristensson, TUD-representative at OKG, for taking time to answer questions and provide background information on failure reporting and recurrence during an interview.

References

[1]   Ostrovskii, D. (2013) *"Trend analysis in input data for PSA"*, Master's thesis, Göteborg : Chalmers University of Technology (June 2013).

[2]   F. T. W. Lee J Bain, Max Engelhardt, *"Tests for an increasing trend in the intensity of a poisson process: A power study"*, Journal of American Statistical Association, Volume 80, Issue 390, June 1985, pages 419-422.

[3]   K. Pörn. *"The two-stage Bayesian method used for the T-Book application"*, Reliability Engineering & System Safety, Volume 51, Issue 2, February 1996, Pages 169-179.

[4]   TUD-kansliet. *"T-boken, tillförlitlighetsdata för komponenter i Nordiska kraftreaktorer"*, 7th edition, ISBN: 978-91-633-6143-2, Strömberg Distribution, 2010.

[5]   TUD-kansliet. *"T-boken, tillförlitlighetsdata för komponenter i Nordiska kraftreaktorer"*, 6th edition, ISBN: 91-631-7231-3, Distro Pack, 2005.

[6]   TUD-kansliet. *"T-boken, tillförlitlighetsdata för komponenter i Nordiska kraftreaktorer"*, 5th edition, ISBN: 91-630-9862-8, Vattenfall support grafiska, 2000.

[7]   TUD-kansliet. *"T-boken, tillförlitlighetsdata för komponenter i Nordiska kraftreaktorer"*, 4th edition, ISBN: 91-7186-303-6, Vattenfall support grafiska, 1994.

Preparation of Implementation Standard Concerning Severe Accident Management in Nuclear Power Plants

Shinya Kamata[a*], Koji Okamaoto [b], and Tomoyuki Sugiyama [c]

[a] Japan Nuclear Safety Institute, Minato-ku, Tokyo, Japan
[b] The University of Tokyo, Tokai-mura, Naka-gun, Ibaraki, Japan
[c] Japan Atomic Energy Agency, Tokai-mura, Naka-gun, Ibaraki, Japan

Abstract: The Great East Japan Earthquake with a magnitude of 9.0 (The 2011 off the Pacific coast of Tohoku Earthquake) occurred on March 11, 2011, and the beyond design-basis tsunami descended on the Fukushima Daiichi Nuclear Power Plant by the earthquake. Eventually, the core cooling systems of the units 1, 2 and 3 could not operate stably, they all suffered severe accident, and hydrogen explosions were triggered in the reactor buildings of units 1, 3 and 4.

In the light of these circumstances, Atomic Energy Society of Japan (AESJ) decided to establish a standard that consolidates the concept of maintaining and improving severe accident management. The standard also provides technical requirements for renovation and addition of the equipment, the formulation of procedures, and strategies. All these items enable the minimization of risks so as to prevent severe accidents, or otherwise enable the mitigation of impacts of severe accidents once occurred.

Keywords: Earthquake, Tsunami, Severe Accident, Accident Management, PRA, Risk

1. INTRODUCTION

The Severe Accidents Management (SAM) Standard has been under discussion since December 2011, while Severe Accidents Management Subcommittee was set up under the System Safety Technical Committee (STC) for the Standards Committee (SC) of AESJ.

The standard provides technical requirements and satisfactory methods concerning maintenance and improvement of severe accidents management in existing nuclear power plants.

With the primary purpose of maintaining and improving the accident management capability, the standard demands strategies, in both hardware and software aspects, which include the effective use of risk information based on PRA as a tool and the upgrading of hardware, as well as the placement of highly-skilled resident staff at power plants, plus the on-going assessment of skills required, including education, and enhanced training and development of skills to respond accordingly to various scenarios of severe accidents, including the low-frequency, high-impact events.

The standard may have applicability to other nuclear facilities such as spent fuel reprocessing facilities. However, if the standard will be applicable to other facilities, the feature (specification and replacement of components, systems, building, and structures) of facilities in safety design should be considered.

The paper describes the basic concept of the SAM Standard and exemplification which will be formally issued soon.

2. DETAIL OF THE SEVERE ACCIDENT MANAGEMENT STANDARD

2.1. Structure of Severe Accident Management Standard

The standard is composed of the 12 chapters by reference to the requirements of NS-G 2.15[1], and the specification items are clearly described in the text and appendixes of each chapter. In addition, in the appendices and interpretations, the applicable examples are provided to help users understand the specification items of the standard as necessary.

The 12 chapters are as follows.

---

* kamata.shinya@genanshin.jp

1. Scope and application
2. Citation standard
3. Glossary
4. Principal requirements of the SAM standard
5. Extraction of nuclear power plant vulnerabilities
6. Identification of nuclear power plant capability
7. Development of accident management strategy
8. Accident management guidelines
9. Establishment of emergency response organization
10. Verification and validation
11. Education and training
12. Maintenance and update of accident management strategies

2.2. Requirements for Evaluation Step

2.2.1. Principal Requirements of the SAM Standard

The objectives of accident management for prevention and mitigation in this standard are as follows:

a) Preventing core damage; b) Terminating severe accident progression; c) Maintaining containment integrity; d) Minimizing radioactive material release; e) Achieving long term stable state.

In order to accomplish these objectives, accident management should be established and improved according PDCA (Plan-Do-Check-Act) cycle. As shown in Figure 1, this standard is composed of the 12 chapters describing requirements associated with appendices, and each steps of PDCA cycle are approximately as follows.

- Identification of vulnerabilities and capabilities of the plant ;
- Development of accident management in order to reduce those vulnerabilities following a structured approach, and confirmation of the feasibility ;
- Classification of each element of developed accident management (hardware and software) based on the importance from the view point of risk reduction;
- Verification and validation of the effectiveness of the developed accident management, considering risk reduction by integration of hardware and software;
  Reflecting findings obtained through education and training on the improvement of the accident management so that it leads to capability enhancement;
- Maintaining the up-to-date knowledge and insights, and incorporating them into accident management as appropriate.

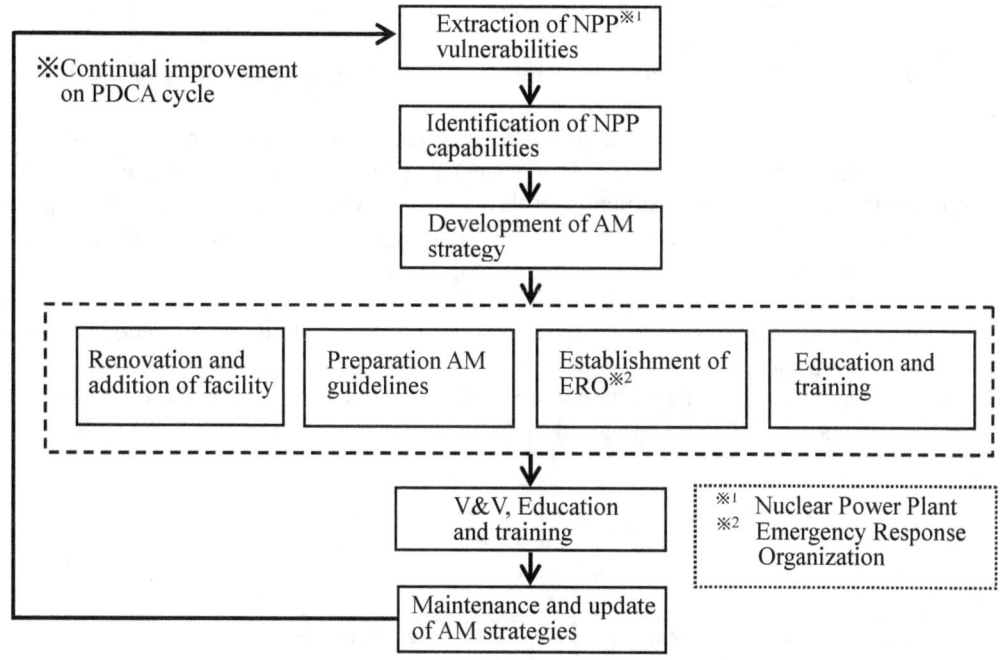

Figure 1   PDCA cycle for Accident Management

2.2.2. Extraction of Nuclear Power Plant Vulnerabilities

Plant vulnerabilities should be extracted by identifying the important accident sequences which may lead to a severe accident beyond the design basis according to the flow in Figure 2.

(1) Assumption of events

Events should be assumed according to the following items a) and b) with an aim to assure comprehensiveness as much as possible:

a) Following events should be considered as the initiating events leading to the loss of safety function at the target plant:

1) single events (internal events and external events), 2) combined events, 3) events leading to the significant loss of safety functions

b) Following events should be considered with a view to support, replacement or recovery of safety function:

1) loss of social infrastructure,   2) damage to multiple plants

(2) Extraction of important events to be assumed

From the single events and combined events among those assumed in the above item a) of item (1) excluding events leading to the significant loss of safety functions, important events should be extracted by performing the "preliminary qualitative screening" and then "quantitative screening based on a bounding or demonstrably conservative analysis" in a stepwise manner.

(3) Identification of important accident sequences

Important accident sequences should be identified for each of the important events to be assumed which were extracted in item (4) by performing a probabilistic risk analysis (PRA) (Level 1 and Level 2), deterministic assessment or engineering judgment, or a combination of these methods.

PRA is used to identify important sequences at least for internal events. The use of PRA is desirable to identify important sequences for external events, too. However, for an event for which an applicable evaluation method has not been fully developed, a deterministic assessment or engineering judgment should be used to identify important sequences.

(4) Identification of important sequences through PRA

When PRA is applied, important sequences should be identified by defining the individual groups, extracting important sequence groups according to the degree of frequency of occurrence, and identifying important sequences accordingly.

(5) Identification of important sequences through deterministic assessment and engineering judgment

In applying the deterministic assessment and engineering judgment to an external event, accident sequence groups are set referring to the accident sequence groups set for internal events, and a sequence that is judged to occur most frequently in the group is identified as an important sequence. Deterministic assessments and assessments using engineering judgment for external events include the FIVE method for internal fire events and seismic margin assessment method.

(6) Extraction of plant-specific vulnerabilities

Plant-specific vulnerabilities should be extracted according to the systems and components which may cause the important accident sequences identified in item (4), (5). Regarding the events which may lead to the significant loss of safety functions, plant-specific vulnerabilities should be extracted using the deterministic assessment and engineering judgment.

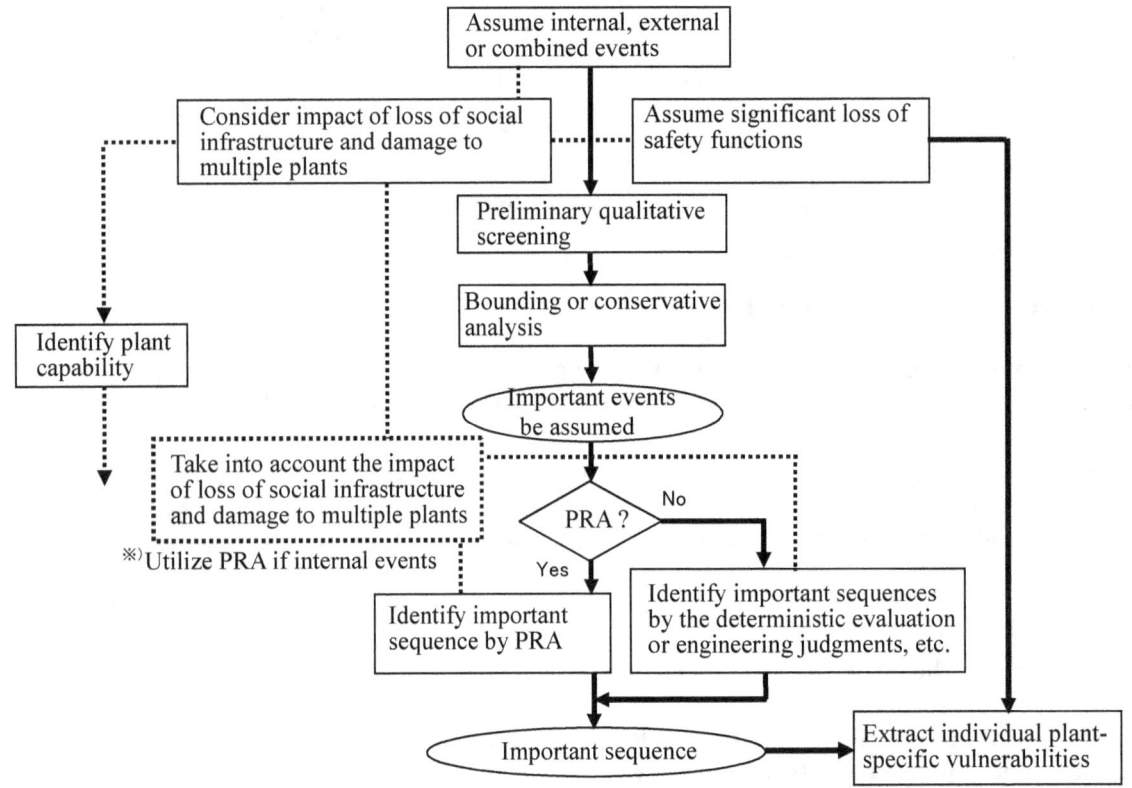

Figure 2   Flow chart of extraction of NPP vulnerabilities

2.2.3.   Identification of Nuclear Power Plant Capability

In parallel with the identification of dominant accident sequence and the plant vulnerability described in the previous section, all the plant capability which is effective for prevention and mitigation of severe accident should be identified as the basis of accident management development.

(1) Available system and procedure

All the potential use of system capabilities such as non-dedicated function, beyond-design basis performance, recovery of damaged equipment and temporary lineup should be investigated. It should be also examined whether those systems have adequate performance to prevent core damage or containment failure, or at least, delay the timing of failure in case they do not have sufficient capability.

(2) Organization, personnel competence, and working environment

The plant organization should be functional with high confidence for emergency response even during night-time or days off. Under the degraded environmental condition caused by the postulated hazard, the adequacy of personnel competence and training experience for the accident management should be confirmed.

(3) On-site and off-site support

If any support relies on the adjacent plants within the site, the availability should be verified considering external hazard which may affect multi units or damage propagation by the shared system among plants.

2.2.4. Development of Accident Management Strategies

(1) Formulating an accident management strategies

Feasible accident management strategies should be formulated based on the results of Section 2.2.2, "Extraction of nuclear power plant vulnerabilities," and Section 2.2.3, "Identification of nuclear power plant capabilities."

Typical considerations for formulating an accident management strategy are as follows.
- Formulate actions that prevent severe accidents, suppress the progress of accidents, maintaining the integrity of the containment as long as possible, minimizing releases of radioactive material and achieving a long term stable state;
- Allow for independence from those facilities that are likely to have lost function in a target sequence, which may or may not cover all the capabilities of the plant;
- Consider available equipments with the restriction on access root to get to the field, operation time and severe accident environmental condition and etc.;
- Ensure the accident management strategy which is properly formulated to respond to external events as they occur, by choosing where to install and store to retain the effectiveness of accident management.

(2) Systematic assessment of accident management priorities

The actions should be selected to take by considering their importance to meet the progress of a target sequence and also assess the priorities of their implementation.

(3) Effectiveness of accident management

In formulating an accident management strategy, analyses and assessments should be conducted to prevent the occurrence of severe accidents or verify the effective functioning of accident management from the perspective of mitigating the impacts of severe accidents. Typical considerations for validating accident management are as follows.
- Use widely verified analysis codes when conducting thermal-hydraulic analyses;
- Adhere to optimal analysis assessments in principle, interpreting the analysis results with the limits and uncertainties of the models taken into account; and
- Assess the positive and negative effects that the formulated risk management strategy may have upon mitigating the risks of uncovered vulnerabilities quantitatively or qualitatively via a probabilistic risk assessment or alternative method.

(4) Establishment of Management Classes

a) Defining Management Classes

In order to ensure reliability of accident management, application of management classes are required as follows.
- Define management classes based on awareness of weightings, risks and other relevant factors, to make intensive and positive use of hardware and software key for safety;
- Establish criteria for assigning management classes to reflect individual plant risk assessments and severe accident environmental conditions; and
- Review the management classes as accident management modifications to reflect the application of the latest available knowledge.

b) Applying management classes

In implementing management classes in accident management, the following concepts should be adhered:

- Ensure that accident management measures conform to basic safety requirements according to the management class level, including facility independence, quake resistance and positional separation.
- Assign management classes to reflect their definitions.

Management classes should be assigned to reflect the required considerations for plant risk assessments and accident management. Figure 3 shows a typical example of management class assignment flow.

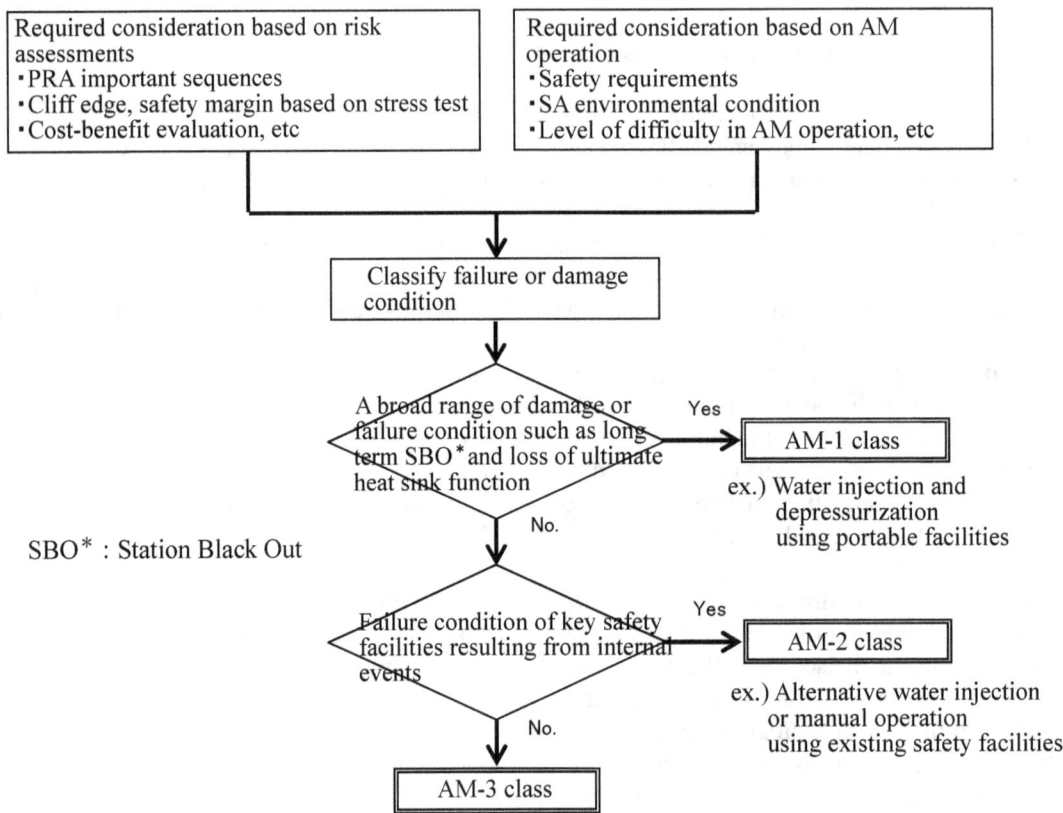

Figure 3   Typical example of management class assignment flow

2.2.5. Accident Management Guidelines

The procedures and guidelines for prevention and mitigation are prepared based on the information about identified vulnerability, plant capability, and developed accident management as a result of process described in the previous sections. Specifically in the mitigative domain, procedures and guidelines should be flexible (rather than prescriptive) so that possible consequences and associated uncertainties following candidate actions can be examined.

(1) Identification of plant condition

The procedures and guidelines should contain the description of necessary plant parameters, criteria for candidate actions, and their priorities. Where the direct readout of measurements is not available, alternate means for parameter estimation (ex. simple formula, pre-calculated diagram) should be provided.

(2) Prioritization and implementation

The procedures or guidelines should describe anticipated (both positive and negative) impact by candidate action, and prioritization of candidates should be revisited according to the progression of accident. When the candidate measure is not available, procedures or guidelines should provide alternative means or mitigative options. Transition conditions among procedures or guidelines for various accident stages should be clearly defined. They should also include attention to required actions for achieving long-term stable state.

(3) Consideration of environmental condition

Habitability of main control room or technical support center, or environmental condition (radiation, cooling, ventilation, lighting, structural damage) under severe accident which may affect the accessibility to other locations should be taken into account. Power supply for instrumentation, computer, or communication device should be secured as well. In case of multi-unit site, procedures and guidelines should anticipate simultaneous occurrence of severe accident.

2.2.6. Establishment of Emergency Response Organization

In order to promote the effective application to the prevention and mitigation, for severe accident, the emergency response organization should be preliminary defined within the documentation of the accident management program.

The roles of personnel involved in severe accident management should be considered in three categories as follows.

The emergency director recognizes plant states including impact to off–site and the impact by decision making, and is responsible for results of response with decision making.

The technical support center staff evaluates accident management and its validity and proposes it to the emergency director. He exchanges information on phenomenon progress, presence or absence of available facility and the degree of impact with control room staff based on knowledge of severe accident and procedure.

The control room staff provides input to evaluations of the technical support center on the basis of their knowledge of the capabilities of plant equipment and instrumentation. He implements a part of accident management which emergency director decided.

Concerning transfer of responsibilities and decision making authority, the transfer should be made at the time that the response can be continued. For example, the transfer should not be made if new decision maker can not prepare first direction of the decision making.

2.2.7. Verification and Validation

(1) Verification of accident management

The department in charge of developing accident management measures should verify accident management.

Developed accident management measures which secure nuclear safety based on defense in depth approach[2] should be checked for plant operation and shutdown states. In verification, it is recommended that risk information based on the probabilistic method as probabilistic risk assessment is referred.

(2) Validation of procedures and guidelines

Validation should be carried out to confirm by staff not involved in developing accident management that the actions specified in the procedures and guidelines can be followed by staffs to manage emergency events. Possible methods for validation of the procedures and guidelines are the use of a full scope simulator or plant analyzer code (if available).

(3) Validation of accident management by independent third party

Independent third party review should be performed in order to check the validation of accident management as follows.

a) A review team should be consisted of members with expertise, independence and fairness.

b) A review should be performed at least once every ten years. The second review should be considered if accident management is changed significantly even fewer than ten years after the previous review.

c) Several items of review might be skipped if the result of independent review performed at the same nuclear power plant site exists. etc.,

2.2.8. Education and Training

The education and training are primary elements to maintain competence and enhance capability for accident management.

(1) Role and required competence

Individuals of the personnel (from top management to engineer including member from contractor supporting plant operation) at site should have required competence for the accident management according their assigned role, and continue to enhance their capability. The education and training should be commensurate with the tasks and responsibilities of members during accident management, and the program (objectives, basis, implementation and feedback) should be systematically developed.

(2) Practicality and flexibility

In order to promote practicality, the training condition (plant behavior, working environment and human factors) should simulate actual situation to the extent reasonable, and the drill of communication and corporation with external organization should be also planned during appropriate period in the program.

The extremely low-frequency, high-impact events are usually excluded from vulnerability identification and accident management development. However, they should not be disregarded in the course of education and training without pursuing the promotion of flexible ability for mitigation. Those extremely low-frequency, high-impact events should be revisited as exercises to explore novel management.

(3) Update and feedback

In order to maintain the latest knowledge and insights, and incorporate them into accident management where needed, planning of education should be programmatic. The latest knowledge is not limited to new findings, but also includes design change of plant systems. The education should also be performed without delay on the occasion of new assignment. The lessons learned and good practice obtained through education and training should be shared among utilities.

2.2.9. Maintenance and Update of Accident Management Strategies

(1) Search and surveillance

About the following items that may affect the development, the availability, and the validity of accident management strategies, they should be supervised continuously.

a) Change of equipment of plant, manuals, and implementation structure

b) Change of the environmental conditions (peripheral people's state, an external hazard factor, etc.) around plant

c) Revision of the reference engineering documentation drawn up and used when deciding upon accident management strategies

d) Research on the phenomenon of a severe accident, the latest knowledge of analysis tools in and outside the country

e) New knowledge based on an accident and a trouble example in and outside the country

(2) Check of effects

The followings will be checked when it is judged as a result of search and survey that it has significant effect on the development, the availability, and the validity of accident management strategies.

a) Identification of dominant sequences and plant vulnerability performed in section 2.2.2 "Extraction of nuclear power plant vulnerability"

b) Effectiveness assessment performed in section 2.2.4 "Development of accident management strategies"

c) Adequacy validation performed in section 2.2.7 "Verification and validation"

(3) Reexamination

In the following cases, it reexamines about maintenance of accident management strategies.

a) When it is judged that there is a problem by the adequacy validation performed in section 2.2.7 "Verification and validation"

b) When it is judged that reexamination is required among what was judged that an improvement is required in section 2.2.8 "Education and training"

c) When significant effect is identified in item (2)

d) Within at least ten years after the last examination

3. CONCLUSION

The SAM standard provides risk assessment based hardware measures as well as software measures for staff education/training and updating procedures in order to enhance the staff's capability for dealing with safety issues. The combination of hardware and software measures based on the risk assessment enables a scientific and rational approach to apply to scenarios of various severe accidents including low-frequency, high-impact events, and assures safety with functionality and flexibility. Furthermore, the standard requires the accident managements to be continuously improved by implementation of the PDCA cycle.

The AESJ is asking for public comment on the draft version of the SAM standard as of the end of February, 2014. After establishment and publication of the SAM standard, with regard to effectiveness assessment for accident management and V&V of severe accident analysis code, the detailed guideline will be prepared as appendices of the standard.

Acknowledgements

The authors would like to acknowledgement the contributions of the Severe Accident Management Subcommittee, the System Safety Technical Committee and the Standard Committee of AESJ to this work.

References

[1] IAEA, " *Severe Accident Management Programs for Nuclear power Plants* "No. NS-G-2.15 (2009)

[2] IAEA, " *Safety of Nuclear Power Plants: Design* " Specific Safety Requirements No. SSR-2/1 (2012)

EPRI Fukushima Technical Evaluation—Evaluation of Flammable Gas Leakage from Fukushima Daiichi Containments using the MAAP5 Computer Code

David L. Luxat[a], Donald A. Dube[a], Andrew S. Dercher[a], Richard Wachowiak[b], Rosa Yang[b] and Jeff R. Gabor[a]

[a] ERIN Engineering and Research, Inc., West Chester, PA, 19380, United States of America
[b] Electric Power Research Institute, Palo Alto, CA, United States of America

Abstract: This paper presents initial results from the investigations of flammable gas transport from the Units 1, 2 and 3 containments into their respective reactor buildings. This study is being conducted as part of the Phase 2 effort of the EPRI Fukushima Technical Evaluation, which is an extension of Phase 1 evaluation (Reference [3]). It builds upon the existing event evaluations conducted by TEPCO (References [1] and [2]) and Sandia (Reference [4]). The analyses are conducted using EPRI's Modular Accident Analysis Program (MAAP), version 5.01. The analyses identify the potential for high temperature conditions in the drywell head region of Units 2 and 3 to contribute to the onset of leakage from each drywell—at drywell pressures below twice design. It is not likely that high temperatures in the drywell head region developed at Unit 1 prior to the onset of leakage from the drywell head flange (at about twice design pressure). The leakage at all units through the drywell head flange has been found to enhance the build-up of flammable gases on the refuel floor. Unit 1 may have experienced flammable conditions on its refuel floor for 10 hours prior to the combustion event. Unit 2 likely did not develop flammable conditions on its refuel floor due to the open blowout panel. At Unit 3, leakage from the hard pipe vent into the Standby Gas Treatment System soft ducting may have allowed hydrogen to build-up at lower elevations—this could have contributed to more damage to the reactor building structure.

Keywords: PRA, thermal-hydraulics, MAAP, Fukushima Daiichi.

1. INTRODUCTION

The combustion of hydrogen and carbon monoxide in the Fukushima Daiichi Units 1, 3 and 4 reactor buildings has become a key signature of the event—both from the perspective of public awareness and the overall course of accident management and remediation. Initial work to enhance safety and accident management procedures/guidelines following the event has identified a number of key actions to be performed. However, the detailed technical basis for specific implementations of actions, such as ventilating a reactor building, is still under development. Insights related to the conditions that gave rise to the different combustion events at Fukushima Daiichi thus serve an important role in establishing the technical basis for on-going and future safety enhancements.

The thermal-hydraulic conditions underlying the transport of hydrogen from the Reactor Pressure Vessel, to the wetwell and drywell, and ultimately to the reactor building and environment, are key to the full understanding of the course of the accident at Fukushima Daiichi. The MAAP5 computer code, used for evaluation of integral plant response, is used in this study to:

    a) Analyze the containment response at Units 1, 2 and 3;
    b) Identify the thermal-hydraulic conditions and types of challenges to containment integrity which may have played a role in the onset of flammable gas leakage into the reactor building; and
    c) The transport of flammable gases in the respective reactor buildings.

2. FLAMMABLE GAS LEAKAGE PATHWAYS FROM MARK I CONTAINMENTS

This section describes the assessment of potential hydrogen leakage pathways from containment to the reactor building and environment that was performed as part of this overall effort. The assessment identified the containment failure location, failure mode, and applicable nuclear units at Fukushima Daiichi. The likelihood of the failure mode is categorized in terms of HIGH, MEDIUM AND LOW based on whether the failure mode and associated phenomena are consistent with observed accident behavior at Fukushima Daiichi, and supported by most analyses and separate effects tests. Likewise, the consequences of the failure mode are categorized based on the potential magnitude of the release pathway (e.g., design leakage (LOW consequence) up to 100 volume % per day or greater (HIGH)).

Drywell head flange leakage is the failure location given most attention in the scientific literature. While overpressure failure via straining of the flange bolts is a predominant failure mode as discussed in a number of studies, this failure mechanism alone cannot explain all of the phenomena. A combination of high pressure, high drywell atmospheric humidity and high temperature conditions leading to elastomer seal degradation best explains the leakage phenomena.

References [7] and [8] indicate that the capacity of head flange against leakage for the Mark I containment is above 0.9 MPa (gauge) (130 psig) at normal temperature, well above observed peak drywell pressures at the three units. Either the pre-load of the bolts was particularly low at Fukushima Daiichi, or some other failure mechanism was occurring. Hence the assignment of LOW to MEDIUM likelihood is made for overpressure alone as the containment failure mechanism. Potential leakage rates via the head flange above 100 volume %/day are possible (HIGH consequence).

The NISA report [8] states that a combination of high pressure and high temperature in the drywell may be the cause of leakage via the head flange. A positive feedback mechanism might explain the behavior. The flange bolts elastically stretch upon high pressure and temperature. Tests [8] indicate that some amount of high temperature gas flow in excess of 350°C (660°F) results in degradation of the elastomer gaskets. This in turn causes greater distortion and opening, resulting in still higher flow rates. The containment system is then in a self-relieving mode, sustaining just enough leakage at a given pressure to equal the rate of gas generation (water vapor, non-condensable gases) and pressure rise with increasing drywell gas temperature. This mechanism may well explain phenomena at several of the units where temperatures in the upper head region in the 600°F range are calculated (see Section 4 below).

Flange distortion caused by a combination of high temperature and pressure that causes a permanent opening in the head flange could be another failure mechanism unlike the self-relieving mode discussed above. Several tenths of a mm opening around the circumference of the flange due to deformation would equate to 10^{-3} to as much as 10^{-2} m^2 opening, the latter equating to thousands of volume %/day leakage (HIGH consequence). This could partially explain the sudden containment depressurization around 90 hours at Unit 2, although it is not the only possible explanation. Therefore, the likelihood that this was the failure mode at Unit 2 is at best a MEDIUM.

Failure of the equipment hatch is next considered. Failure due to overpressure alone is unlikely based on analysis in Reference [7] indicating the capacity to be well above 1 MPa (gauge) (150 psig). A more likely failure mode would be failure of the gaskets around the hatch due to high temperatures. NUREG/CR-4944 [9] and NUREG/CR-5096 [10] indicate from tests that gaskets start to fail as low as 238°C for neoprene, 299°C for ethylene propylene, and 370°C for silicone rubber. Some or all of these temperatures are believed to have been exceeded at each of the units. However, the actual material used at Fukushima Daiichi is unknown at this point, so more definitive assessment is not possible. The likelihood that this leakage pathway existed at one or more units is MEDIUM to HIGH. Based on the potential leakage area, the consequence is MEDIUM to HIGH. However, the release location would be into the lower floors of the drywell, resulting in different hydrogen concentration profiles than head flange leakage.

The personnel airlock is next considered. No design information for Fukushima Daiichi is available as of this writing. NUREG/CR-5118 [11] and SAND90-0119 [12] describe testing which indicate no significant leakage below 427°C with pressure up to 2.07 MPa (300 psig). Above 454°C the inner gasket was degraded and the inner door effectively bypassed. Because of the thermal inertia, the outer door seal remained intact. Thus the leakage path likelihood is LOW to MEDIUM. Based on the potential leakage openings given degraded seals, the consequences are LOW to MEDIUM as well.

For electrical penetration assemblies, NUREG/CR-5334 [13] tests indicate no failures up to pressures of 0.51 to 1.07 MPa (75 to 155 psia) and temperatures from 180 to 370°C (360°F to 700°F) for up to 10 days. The outer seal did not experience harsh conditions during the tests. Because the temperatures at some of the Daiichi units likely reached or exceeded the maximum test temperature of 600°F, it is difficult to extrapolate the test results. The likelihood that leakage via this pathway was experienced is judged as LOW to MEDIUM. From test data [8] and given the possibility of multiple EPA failures, the consequence in terms of leakage area and rate is MEDIUM.

Much attention has also been given to the vent bellows connecting the wetwell to drywell. With regard to failure by overpressure, the NUREG/CR-6154 [14] series of tests showed that bellows can be stretched to two to three times their undeformed length without leaking. Reference [7] indicates low probability of leakage below 1.2 MPa (gauge) (175 psig). Hence, this failure mode is assigned LOW likelihood. With regard to consequence, Reference [7] describes experiments which confirm that bellows steel is extremely tough and resistant to unstable crack propagation. A bellows tear results only in a "leak", and that "rupture" will not occur in the bellows. A LOW to MEDIUM consequence is assessed. However, the tests cited in NUREG/CR-6154 did not consider the effects of severe accident elevated temperatures substantially beyond the test condition of about 218 °C (425 °F). This temperature was clearly exceeded at the three Daiichi units. Thus, the potential for a combination of high temperature and high pressure to have caused deformation of the bellows is assessed as MEDIUM.

Two failure pathways that could have caused containment bypass are melt-through and ejection of molten debris at the bottom of the RPV into the transverse instrument probe tube(s), and melt-through in the reactor cavity sump into the recirculation cooling water (RCW) piping. Based on high dose rates in areas of the reactor building of Unit 1 with RCW components, a HIGH likelihood is assessed for Unit 1 and LOW for Units 2 and 3. While some amount of hydrogen could have been ejected along with the corium debris, based on the restricted openings and low elevation for release in the drywell, LOW to MEDIUM consequence has been assessed.

Finally, based on a series of structural analyses documented in Reference [7], the likelihood of failure by overpressure has been assessed as LOW for the following locations:
- Wetwell access hatch
- Global drywell region
- Global wetwell region
- Other penetrations including steam lines, High Pressure Coolant Injection test line, and containment spray.

The capacities of these potential leakage pathways are typically above 1 MPa (gauge) (150 psig). Depending on the size of the breach, the consequences span the entire range from LOW to HIGH.

For Unit 3, there is a MEDIUM to HIGH likelihood that leakage of flammable gases from the hard pipe vent into the Standby Gas Treatment System (SGTS) soft ducting may have occurred, and is supported by elevated dose rates on the SGTS filters. The consequences are also assessed as MEDIUM to HIGH.

3. MAAP5 REPRESENTATION OF FUKUSHIMA DAIICHI CONTAINMENTS AND REACTOR BUILDINGS

This section describes the MAAP5 models developed to represent the Fukushima Daiichi containments and reactor buildings. The nodalization scheme adopted is the same for Units 1, 2 and 3. Identical containment and reactor building models are used for Units 2 and 3—these units are essentially the same from the perspective of a lumped volume approximation for the containment and reactor building. However, the Unit 1 model is distinct from the Units 2 and 3 models. This is due to the differences in volume between these units.

3.1. Enhanced Representation of Mark I Containment

The model of the containment and reactor building follows a common structure for all three units modelled. These models are enhancements to the current representation of a Mark I containment provided as part of sample user guidance (e.g., the sample model for a plant similar to Peach Bottom). The standard model of a Mark I containment which forms the basis for the user guidance represents the containment in terms of four distinct volumes:
- Pedestal representing the region underneath the RPV
- Drywell representing the volume of containment excluding the wetwell downcomers and the pedestal
- Wetwell downcomers
- Wetwell.

As part of the Phase 2 analysis, enhancements to the containment model have been made to represent additional aspects of plant response in order to assess their impact on the event progression. These enhancements facilitate assessment of the variation of two safety-significant parameters:
- Temperature over the drywell
- Distribution of hydrogen gas within the drywell.

This consists of a refined nodalization of the drywell to explicitly represent the sub-volumes:
- Pedestal representing the cavity region underneath the RPV
- Lower drywell for the volume inside the drywell sphere region
- Drywell cylinder volume in which the atmosphere between the drywell wall and RPV cylinder is assumed to be well-mixed due to circulation flows around the bioshield
- Drywell head volume separated from the drywell cylinder by the refuel seal (flows between the two volumes are through a limited number of openings).

Sub-volumes representing the wetwell downcomers and the torus are the same as in the containment model used in the Phase 1 analysis.

The drywell temperature profile is necessary to identify the potential for thermal challenges to containment penetration elastomeric sealing materials. Depending on the stage of core melt progression, it is possible to have different magnitudes of thermal loading of the drywell sphere, cylinder and head regions. Since the natural circulation flow paths through the drywell do not promote strong heat transport flows, it is possible to develop relatively high temperatures in the drywell head region. The elastomeric drywell head flange seal is thus at potential risk of experiencing high temperatures and undergoing thermally induced degradation.

Understanding the thermal challenge to the integrity of containment penetrations is of interest to assess potential reactor building combustion profiles. The location of leakage from containment can have an effect on the distribution of flammable gases inside the reactor building. This in turn has an important effect on the regions of the reactor building in which sufficient hydrogen can build-up to combust.

In addition to the point of leakage, the distribution of hydrogen within the drywell could affect the magnitude of hydrogen available to leak through points of containment impairment. One possibility addressed by this enhanced drywell model is the potential for hydrogen to stratify in drywell head region. A higher concentration of hydrogen in this region can have the following effects on plant response:

- Enhancement of material degradation at the top of the drywell head
- It is not likely that significant hydrogen embrittlement of metal structures could occur over the time frames of interest
  - However, the presence of hydrogen can enhance the degradation of the elastomeric drywell head flange seal when exposed to a high humidity environment
  - An increase in the rate at which hydrogen leaks out through an impaired drywell head flange seal on to the refuel floor.

3.2. Nodalization of Mark I Reactor Building

The standard model of a Mark I reactor building is simpler, consisting of two volumes:
- Torus room
- Remainder of the reactor building.

Some plant models incorporate additional volumes to represent the Standby Gas Treatment System and a stack. These additional volumes were in the Unit 1, 2 and 3 MAAP5 plant models used in the Phase 1 study [3].

The enhanced reactor building model incorporates the following layout:
- Torus room
- Remainder of reactor building basement
- First floor
- Second floor
- Third floor
- Fourth floor
- Lower half of refuel floor
- Upper half of refuel floor
- Refuel cavity (the region above the drywell head but separated from the refuel floor by concrete shield plugs)
- Standby Gas Treatment System
- Stack.

Of primary interest from the perspective of flammable gas combustion inside the reactor building is the distribution of hydrogen through the height of the refuel floor. For all three affected units, there appears to be a high likelihood of enhanced containment leakage developing through the drywell head flange. This leakage path would have displaced hydrogen (and carbon monoxide) directly on to the refuel floor. Thus, the concentration of flammable gases throughout the refuel floor is a critical accident parameter.

However, there is some indication that bypass flows could have developed at Unit 3 directing hydrogen (and carbon monoxide) from the hard pipe vent into the soft ducting of the building ventilation system. This could have displaced flammable gases on to the refuel floor. It also would have displaced flammable gases on to the fourth floor of the reactor building. This is depicted in Figure 1. This enhanced layout of the reactor building is used below to assess the nature of combustion events for different enhanced leakage scenarios.

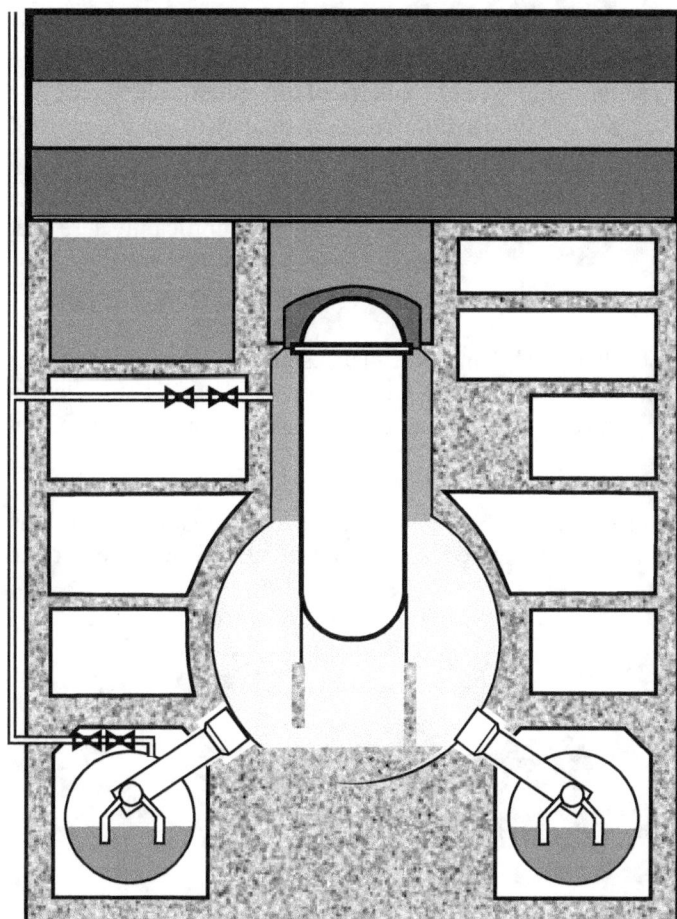

Figure 1: Enhancements to MAAP5 Model of Fukushima Daiichi Containment and Reactor Building

4. MAAP5 SIMULATION OF CONTAINMENT RESPONSE

The release of flammable gases into the Units 1, 2 and 3 reactor buildings depends on the:
- Core melt progression and its influence on total amount of flammable gases generated as well as the rate of generation
- Containment response following the onset of core melting.

This section describes the MAAP5 analyses of the core melt progression and containment response. The analyses are used to identify an estimate of the transient discharge of flammable gases into the Unit 1, 2 and 3 reactor buildings.

4.1. Unit 1 Containment Response

Reference [3] has investigated the alternate Unit 1 accident progression scenarios and compared each against the observed:
- RPV pressure
- Drywell pressure
- Site boundary dose rates.

The following accident progression characteristics are potentially more representative of the Unit 1 event:

- No core cooling with installed systems following the Isolation Condenser being isolated just prior to the arrival of the tsunami
- A steam leak from the RPV into the drywell after T+5 hours, sufficient to depressurize the RPV
- Drywell head impairment around T+12 to T+13 hours, inducing a small leak in the drywell (greater than about 2 square inches)
- RPV lower head breach around T+12 to T+13 hours
- Low water injection rates into the RPV beginning at about T+15 hours (significantly less than 10 gpm)
- Wetwell venting around T+23.8 hours for about 30 minutes.

This type of accident progression is similar to PRA core damage sequences progressing from a Station Blackout (SBO). However, the operation of the Unit 1 Isolation Condenser for about 1 hour after the earthquake delayed the progression of the event (due to the lower decay heat at the time core cooling was lost).

The overall containment pressurization is well-represented based on these event scenario assumptions. The simulation of the drywell pressure transient, compared against observed drywell pressure, is shown in Figure 2.

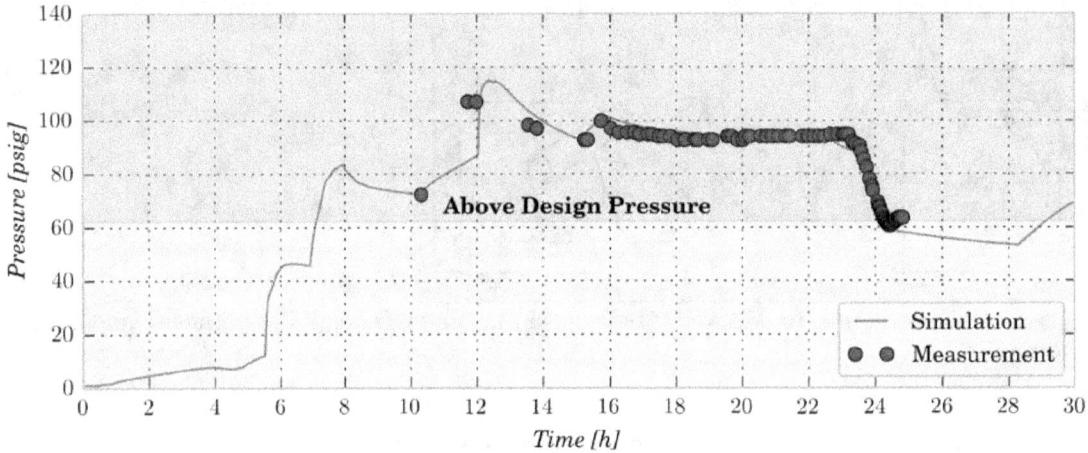

Figure 2: Simulated Drywell Pressure for Unit 1

The impairment of the drywell head is assumed to be due to the rapid rise in containment pressure to nearly 120 psig (around twice design pressure). The drywell head would have lifted against the restraint of the flange bolts. The drywell head seal likely would have degraded over a finite time following drywell head lifting, due to exposure to a steam environment. There may have been a period of nearly an hour before leakage from the drywell head commenced due to the difference between the first observation of pressures around twice design (between T+11 and T+12 hours) and the rise in site boundary dose rates (around T+13 hours).

The temperature in the drywell head region would likely have been relatively low at this time of significant drywell head lifting. Figure 3 shows the simulated distribution of temperatures in the Unit 1 drywell. The relatively short time at which RPV lower head breach is simulated to occur is the primary reason the temperatures in the drywell head region do not exceed 500°F in the simulation. The relocation of core debris into the pedestal around T+12 hours transfers the majority of the heat source to the bottom of the drywell, preventing continued heat losses from the RPV into the drywell cylinder and head regions. Thus, thermal degradation of the drywell head seal would likely not have occurred prior to significant lifting of the drywell head. Leakage from the Unit 1 drywell head flange was thus most likely due to the significant drywell overpressure experienced.

The core melt progression identified through MAAP5 simulations (see, for example, Reference [3]) would have resulted in early generation of hydrogen. Hydrogen generation could have begun around 20 hours prior to the occurrence of the energetic combustion event. Figure 4 shows the in-vessel hydrogen transient simulated using MAAP5.

The magnitude of in-vessel hydrogen generation is quite large in this simulation. Nearly 750 lbm of hydrogen is generated during the in-vessel phase of the core melt progression, based on the simulation. Slow leakage of this amount of hydrogen on to the refuel floor may have resulted in flammable conditions on the refuel floor. However, leakage out of the refuel floor to the environment would likely have prevented flammable conditions being maintained until T+24 hours (i.e., around the time of energetic combustion in the Unit 1 refuel floor).

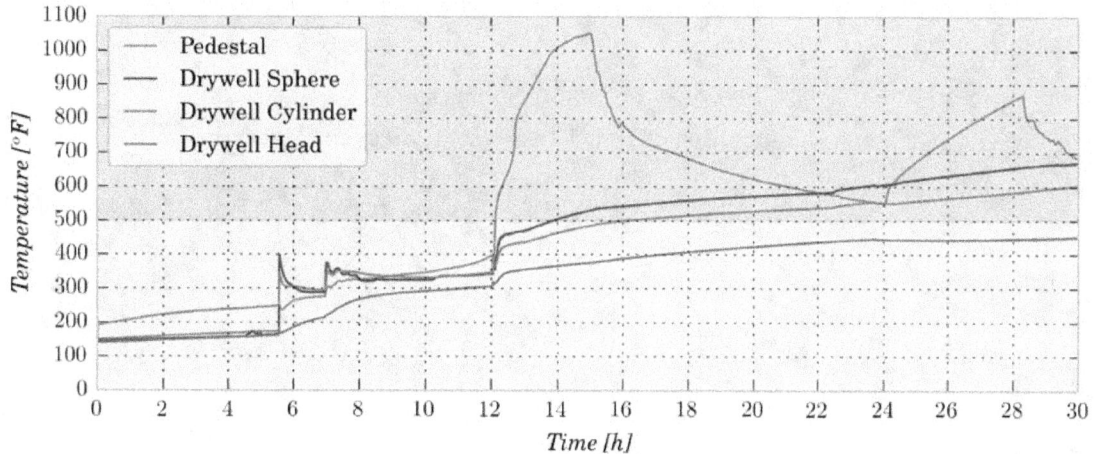

Figure 3: Simulated Drywell Temperature Distribution for Unit 1

The breach of the RPV at about T+12 hours in the MAAP5 simulation initiates core-concrete attack. This process results in the generation of a large amount of hydrogen (and some carbon monoxide[*]) as indicated in Figure 5. The rate of generation is relatively slow. During the early phase of core-concrete interaction (CCI), hydrogen is typically generated at a rate of 1 to 2 kg/s (due to the oxidation of the remaining Zr in the core debris) [6]. In the late phase of CCI, hydrogen is generated by Fe oxidation in the core-concrete debris—this typically occurs at a rate of about 4 g/s [6].

This prolonged generation of hydrogen would have maintained a relatively constant leakage, in the long term, of hydrogen on to the refuel floor following drywell head lifting. Higher rates of water injection to the RPV may have been able to quench the core debris on the concrete floor, and terminate CCI. This could have arrested the long-term leakage of hydrogen on to the refuel floor.

---

[*] Basaltic concrete is assumed for these simulations based on the type of concrete used for the Fukushima Daiichi units. This concrete type generates significantly lower amounts of carbon monoxide (and carbon dioxide) relative to limestone/common sand concrete.

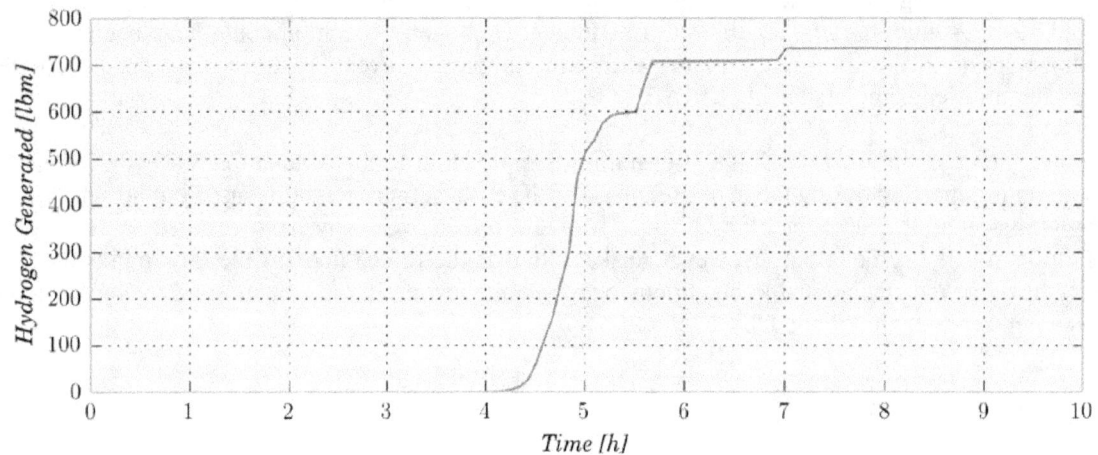

Figure 4: Simulation of Unit 1 In-Vessel Hydrogen Generation

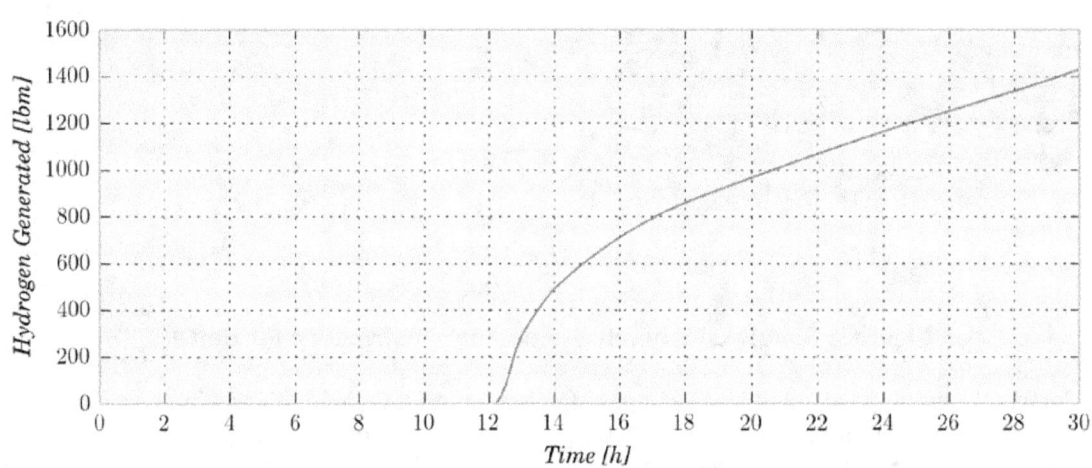

Figure 5: Simulation of Unit 1 Ex-Vessel Hydrogen Generation (Degraded RPV Water Injection Scenario)

4.2. Unit 2 Containment Response

Unit 2 accident progression was investigated in Reference [3] following the same approach adopted for Unit 1. Additional analyses of the Unit 2 accident progression have been performed as part of this Phase 2 effort. These analyses have focused on understanding in greater detail the accident progression after RCIC water injection failed.

The characteristics of the event are as follows:
- Core cooling was likely maintained until about T+67 hours due to operation of RCIC in an unattended mode
- After the loss of RCIC at this time, core cooling was not restored until after T+75 hours and following depressurization of the RPV by deliberate opening of an SRV
- The RPV partially re-pressurized 3 times between T+75 hours and T+85 hours most likely due to SRVs re-closing (see Figure **6**)
- Between T+80 and T+81 hours, the drywell pressure rose by nearly 40 psig due to a combination of enhanced hydrogen generation and steam generation
- Core melt progression was relatively stable until about T+94 hours when core melt relocation either into the lower plenum or reactor pedestal may have occurred[†]

[†] The MAAP5 simulation for Unit 2 finds core melt relocation to the lower plenum is a possible explanation of the rapid rise in containment pressure.

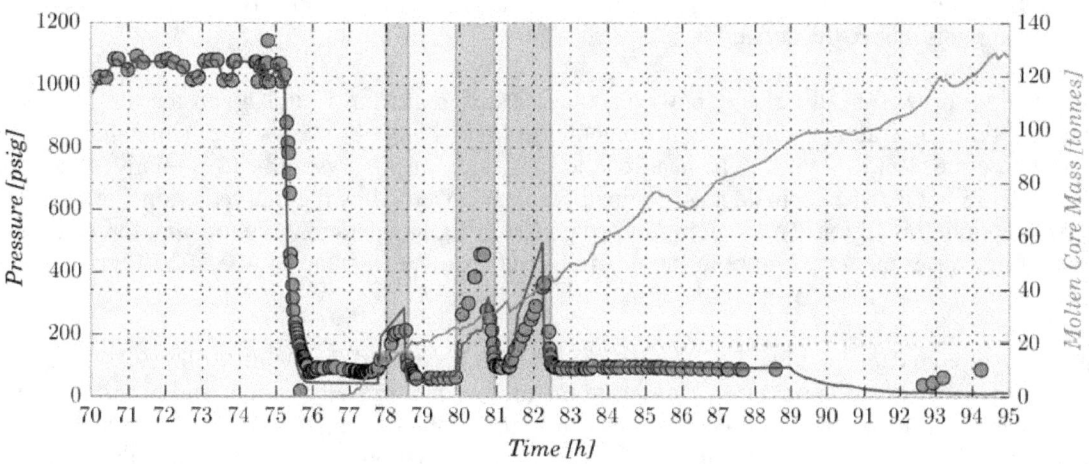

Figure 6: Measured Unit 2 RPV Pressure Transient from T+75 hours to T+85 hours

Cessation of RPV water injection occurred during the periods of RPV partial re-pressurization, shown in Figure **6**. This allowed core heatup to continue, as shown by the continuing melting of core materials. When SRV opening reduced RPV pressure again, RPV water injection restoration would have been to an overheated core. These conditions made hydrogen generation possible. The MAAP5 simulations have indicated sufficient core heatup could have occurred by T+80 hours to promote substantial generation of hydrogen.

The simulated drywell pressure transient is shown in Figure **7**. Superimposed with the drywell pressure transient is the simulated in-vessel hydrogen generation transient. The rapid pressure rise in containment starting around T+80 hours is due to the generation of steam and hydrogen in the RPV. The quenching of overheated core material causes this.

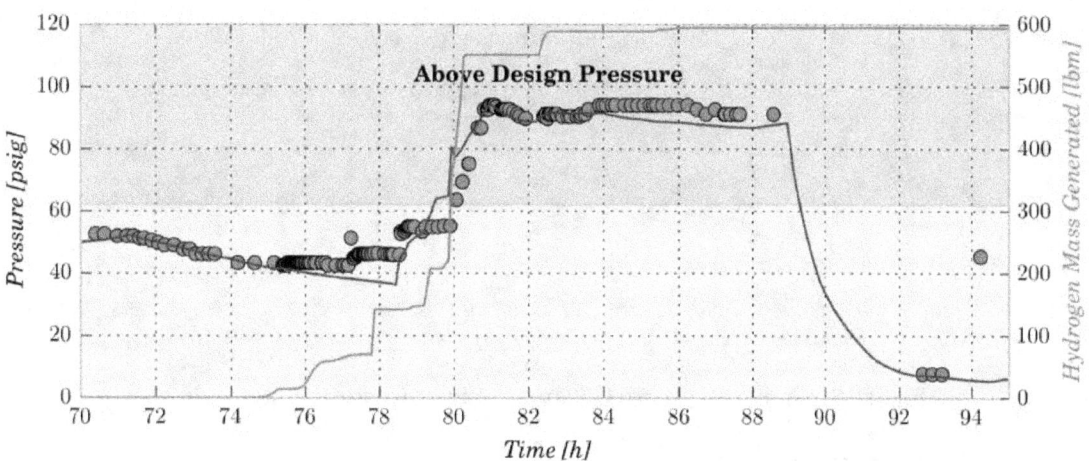

Figure 7: Simulated Drywell Pressure Response for Unit 2

The rise in containment pressure around T+80 hours, however, also indicates the potential for a direct steam leak developing from the RPV into the containment. The rise in RPV pressure around this time (see Figure **6**) coincided with SRV closing. To capture both a rise in RPV pressure and a very sharp jump in containment pressure, MAAP5 simulations have indicated the potential for a steam leak through, for example, a failed in-core instrument tube.

From Figure **7**, the drywell pressure remains relatively constant between T+80 hours and T+89 hours. This is similar to the trend observed between T+15 hours and T+24 hours at Unit 1 (see Figure **2**), which was likely governed by leakage through the drywell head flange. Drywell head leakage at Unit

2 is highly probably based on the observation of high dose rates—of about 100 rem/h—on the refuel cavity seal plugs (above the drywell head).

The cause of drywell head leakage, however, may be different from Unit 1 given the long period of Unit 2 RCIC operation without containment cooling. Figure **8** shows the simulated drywell temperature transient for Unit 2. The heat load to the region of the drywell head is relatively constant prior to T+75 hours, when core damage commences. It is governed by thermal radiation from the RPV upper head into this region. Thus, the prolonged period of RCIC operation, without any forced drywell circulation, causes the temperature in the drywell head region to increase toward 500°F prior to core damage.

The onset of core damage, after T+75 hours, results in a more rapid rate of drywell head temperature increase. This heat load is primarily focused in the area of the drywell cylinder and head.[‡] However, the impact of this enhanced heat load is distinct from that simulated for Unit 1, due to the operation of RCIC until T+67 hours without any forced drywell circulation. By T+78 hours, the simulated temperature in the drywell head region is above 500°F, sustained at these levels for nearly 10 hours. The potential for thermal degradation of the drywell head flange seal is thus possible—by this point drywell head lifting would have started due to containment overpressure (see Figure **7**) and exposed the seal material to a steam environment.

However, the simulation of core melt progression indicates that hydrogen generation from the damaged core may not have occurred over a prolonged period of time. A large amount of hydrogen may have been generated around T+80 hours (about 600 lbm). In the absence of RPV lower head breach, and the occurrence of CCI, current models of in-vessel core melt progression indicate limited hydrogen generation beyond the initial core melting/candling phase.

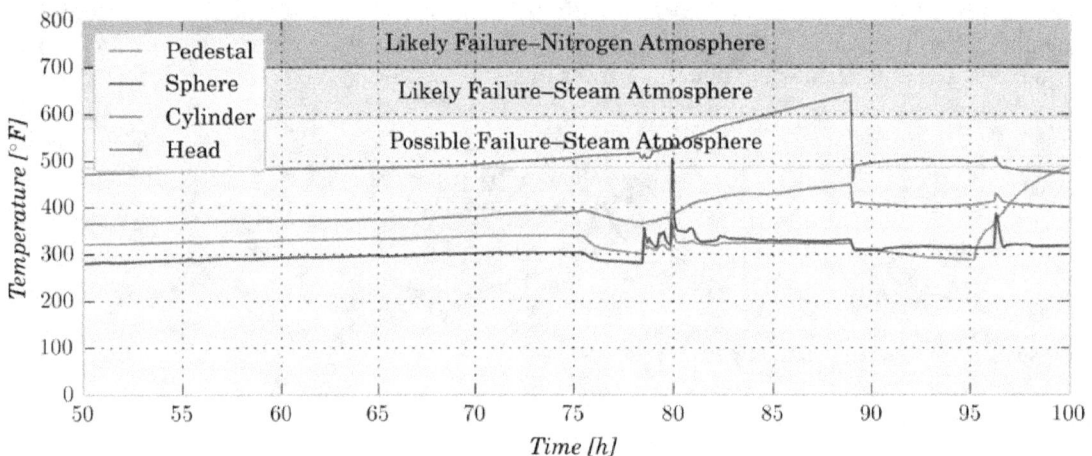

Figure 8: Simulated Drywell Temperature Distribution for Unit 2

4.3. Unit 3 Containment Response

Unit 3 accident progression was investigated in Reference [3] following the same approach adopted for Units 1 and 2. The Phase 2 efforts have focused on evaluating characteristics of accident progression with respect to their potential for leading to a flammable atmosphere in the Unit 3 reactor building.

The key characteristics of the event are as follows:
- Core cooling was maintained after the tsunami until about T+20 hours with the RCIC system

---

[‡]   The temperatures in the lower portion of the drywell are somewhat ameliorated by the assumption of heat dissipation from torus water into the torus.

- From T+21 hours until T+36 hours, operators used the HPCI system in an attempt to maintain water level
- Operation of the HPCI system at low RPV pressure after about T+28 hours likely resulted in a reduction in water injection rate and an inability to maintain water level—by T+36 hours, it is likely that RPV water level had reached TAF
- From T+36 hours, RPV pressure was controlled by SRV cycling and no water was injected due to the high pressure
- After T+42 hours, water injection to the RPV was possible due to the unintentional depressurization of the RPV (it is possible that RPV depressurization occurred due to ADS being triggered [1], [2])
- Fire engine injection was degraded due to flow bypass through the condensate transfer pump (see, for example, Reference [2])—assumed to be at a rate around that required to remove decay heat via boiling of the injection flow
- There is a potential that leakage occurred through the drywell head flange seal starting around T+60 hours—this leakage could have been complemented by some leakage of vent gases into the soft ductwork and discharge on to the fourth and fifth floors of the reactor building.

The drywell pressure response is shown in Figure **9**. The venting of containment after T+42 hours aided in maintaining drywell pressure around or below design. However, between T+60 and T+67 hours (the time of energetic combustion in the reactor building), the drywell pressure escalated and held at design.

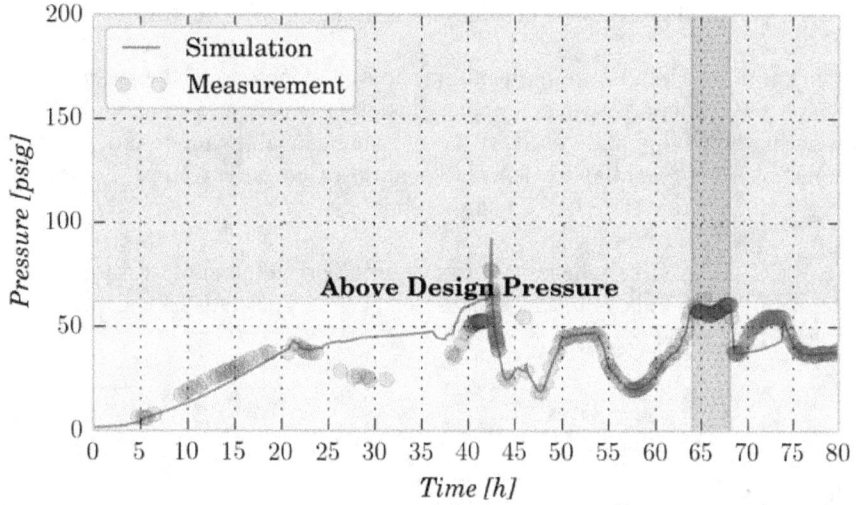

Figure 9: Simulated Drywell Pressure Response for Unit 3

Similar to Unit 2, the heat dissipation from the RPV into the drywell would have been prolonged. Wetwell sprays were used between T+21 hours and T+36 hours to control drywell pressure. For a period of a few hours prior to T+36 hours, drywell sprays were used as well. However, drywell sprays primarily affect the drywell sphere, having minimal effect on temperatures in the cylinder and head regions.

Figure **10** shows the simulated drywell temperature transient for Unit 3. These results for the multi-node MAAP5 drywell have been benchmarked against measured Unit 3 temperatures in the first 20 hours following the earthquake. The simulation beyond T+42 hours is thus judged to provide a reasonable representation of the drywell thermal response. The temperature in the drywell head region exceeds 500°F beyond about T+42 hours, driven higher due to the onset of core damage after T+36 hours. By T+60 hours, the drywell head atmospheric temperature is around 600°F.

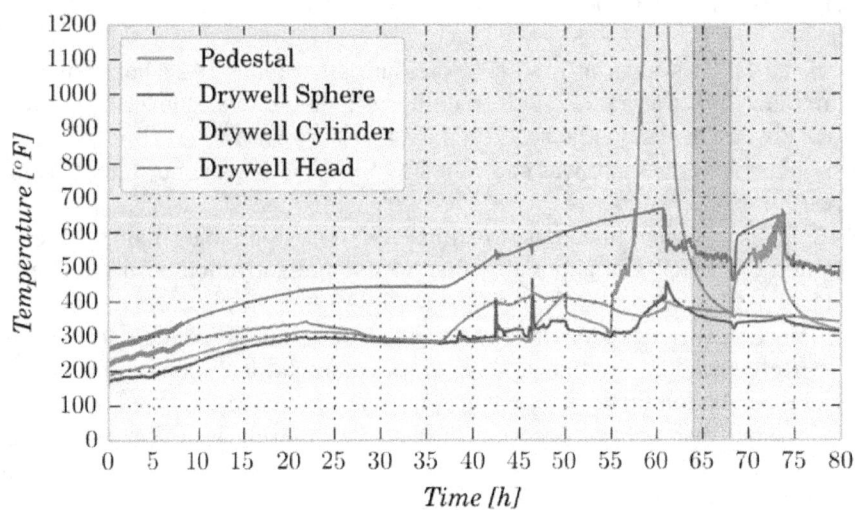

Figure 10: Simulated Drywell Temperature Distribution for Unit 3

There is high likelihood that thermal degradation of the drywell head seal could have occurred by T+60 hours. Leakage from containment is found with MAAP5 simulations to be necessary to explain the drywell pressure holding constant between T+60 and T+67 hours. Thus, leakage from the Unit 3 containment via the drywell head flange could have commenced by T+60 hours.

The MAAP5 simulation of Unit 3 core melt progression has highlighted the potential for RPV lower head breach around T+60 hours. However, this conclusion is sensitive to small variations in the rate of water injection during HPCI operation at low RPV pressure as well as the rate of RPV water injection by fire engine pumps. The potential for some relocation of core debris into containment at Unit 3 appears to be highly likely from the MAAP5 simulations.

The occurrence of CCI is also likely necessary to explain the development of flammable conditions in the Unit 3 reactor building. This is discussed further below. The hydrogen generation transient for Unit 3 is presented in Figure **11**.

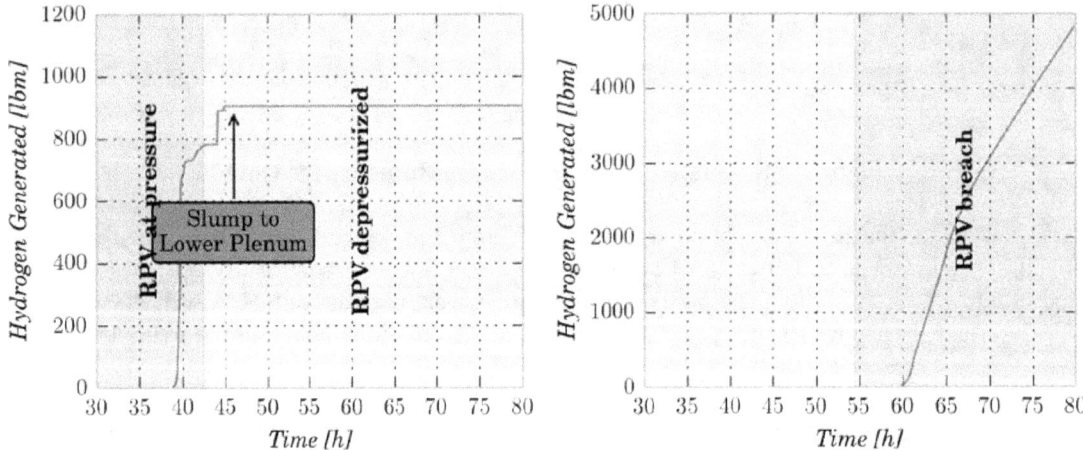

Figure 11: Simulated Unit 3 Hydrogen Generation

5.   MAAP5 SIMULATION OF REACTOR BUILDING FLAMMABLE GAS DISTRIBUTION

5.1. Simulation of Unit 1 Flammable Gas Distribution

Figure **12** shows the simulated distribution of hydrogen in the Unit 1 reactor building. The accumulation of hydrogen on the refuel floor is relatively slow. After the onset of drywell head lifting and leakage, approximately 2 hours elapses before the concentration of hydrogen exceeds that sufficient to support an energetic combustion event. By the time of the energetic combustion event at Unit 1 (T+24.8 hours), the hydrogen concentration on the refuel floor is about 17%.[§]

An energetic combustion event could have occurred at any point between T+15 and T+24.8 hours. The energetic combustion event, however, may not have occurred until T+24.8 hours due to the lack of an ignition source. It may not have been until this point in time that efforts to restore power to the plant generated the necessary spark to ignite the refuel floor atmosphere.

There is a slow build-up of hydrogen at lower elevations. This is due to natural convection flow through the building, facilitated by large area openings between each floor and the stairwells. By approximately T+24.8 hours, the concentration of hydrogen at lower elevations is not found to exceed the lower limit for flammability (about 4% in dry air). This distribution of hydrogen is consistent with the type of damage that occurred to the Unit 1 reactor building—the damage to the structure was localized to the refuel floor.[**]

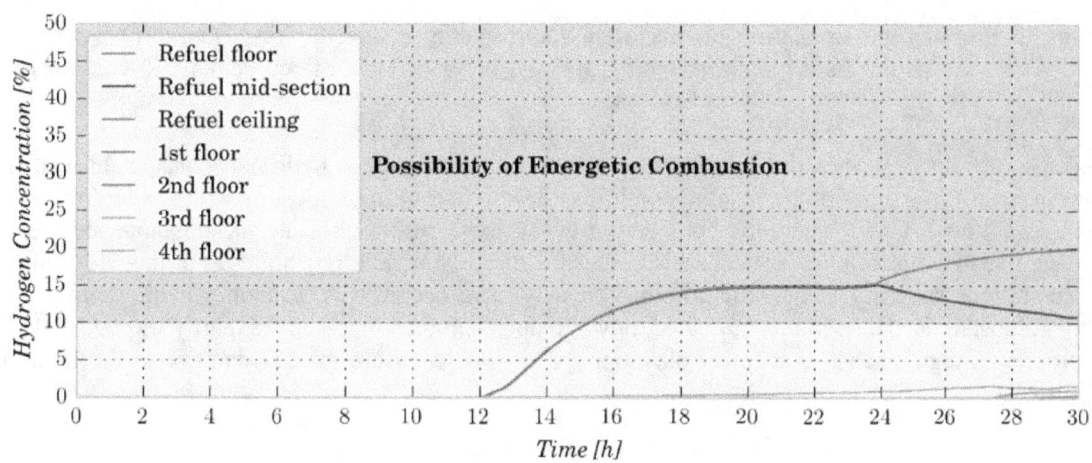

Figure 12: Simulation of Reactor Building Flammable Gas Distribution for Unit 1 (Degraded Water Injection)

5.2. Simulation of Unit 2 Flammable Gas Distribution

The build-up of hydrogen on the refuel floor at Unit 2 would have been significantly different relative to Units 1 and 3. After the energetic combustion of hydrogen occurred on the Unit 1 refuel floor, it is likely that the Unit 2 refuel floor blowout panel was dislodged. The rarefaction phase of the

---

[§]   The combustion of hydrogen is artificially suppressed in these simulations to mimic the absence of an ignition source. The concentration of hydrogen beyond T+24.8 hours is an artifact of the simulation and not reflective of the hydrogen concentration in an open refuel floor after this time.

[**]   It should be noted that the nature of damage to the Unit 1 structure would also have been influenced by the limited resistance to pressure loading provided by the sheet metal siding at the elevation of the refuel floor. This would have resulted in the combustion pressure wave being vented to atmosphere more readily. By contrast, the Units 3 and 4 refuel floors were part of the concrete super-structure. The refuel floor wall panels for these units would have experienced more over-pressure before yielding and venting the combustion pressure wave to the atmosphere.

compressional wave generated by the combustion event would have resulted in a negative pressure outside the Unit 2 reactor building. The resulting positive differential pressure between the inside and outside surfaces of the Unit 2 blowout panel would have been sufficient to open the blowout panel.

MAAP5 simulations find that the open blowout panel is sufficient to limit the hydrogen concentration to below 4% on the refuel floor. This is despite a relatively large area leakage assumed from the drywell head in order to depressurize containment to atmosphere.

5.3. Simulation of Unit 3 Flammable Gas Distribution

The build-up of flammable gases in the Unit 3 reactor building is influenced by a number of factors not relevant to Unit 1 and 2. These reflect the overall uncertainty in the accident progression at Unit 3. Uncertainties arise because of the over 2 day period after T+42 hours during which the core melt would not have been stabilized:

- The core status (i.e., the potential for RPV breach) can have a significant impact on the magnitude of hydrogen generated—MAAP5 simulations tend to indicate that continued hydrogen generation is limited once core melting compacts the debris (and reduces the exposed surface area to participate in oxidation)
- Leakage points from the Unit 3 containment—there is a high likelihood that drywell head leakage occurred, though the potential for leakage of flammable gases from the hard pipe vent into the SGTS soft ducting is less well understood but supported by elevated dose rates on the SGTS filters.

Figure **13** presents the simulation of reactor building hydrogen distribution. This corresponds to a simulation with partial hard pipe vent bypass (on the order of 10% of the vent flow) commencing around T+55 hours.

Leakage of vent flow into the soft ducting is found to complement hydrogen leakage through the drywell head. This supports flammable conditions developing in the Unit 3 reactor building by T+67 hours. Unlike the Unit 1 energetic combustion event, these results indicate the potential for greater damage to lower elevations, particularly the fourth floor. Flammable gas build-up at this elevation to concentrations supporting energetic combustion may have occurred. The nature of the combustion, with a substantial debris plume directed upward upon combustion, can be partly explained by a significant energy discharge at lower elevations. This would ensure greater damage to the building superstructure.

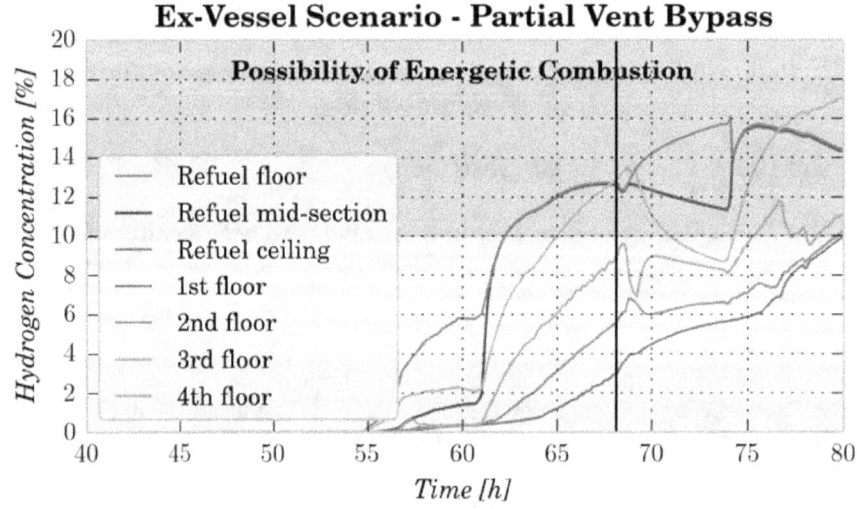

Figure 13: Simulation of Reactor Building Flammable Gas Distribution for Unit 3

6. CONCLUSION

This paper presented initial results from the investigations of flammable gas transport from the Unit 1, 2 and 3 containments into their respective reactor buildings. This study is being conducted as part of the Phase 2 effort of the EPRI Fukushima Technical Evaluation, which is an extension of Phase 1 evaluation (Reference [3]). It builds upon the existing event evaluations conducted by TEPCO (References [1] and [2]) and Sandia National Laboratories (Reference [4]).

The analyses reported in this paper have identified the potential for high temperature conditions in the drywell head region of Units 2 and 3 to contribute to the onset of leakage from each drywell. The drywell pressure at which this leakage would have commenced is around 1.3x to 1.0x design pressure for Units 2 and 3, respectively.

This indicates the potential for a drywell head flange failure mode influenced by degradation of the elastomeric seal at high temperatures. The lifting of the drywell head flange due to high drywell pressures (around design pressure or higher) would have exposed the seal to a high humidity atmosphere. High humidity enhances the degradation of elastomeric seals at elevated temperatures.

The high temperatures developed due to the long time that existed between the loss of active containment cooling and forced circulation and the onset of core damage. Drywell head temperatures at Unit 1, by contrast, were likely much lower when leakage from the drywell head flange commenced. The leakage at Unit 1 is identified through simulations to be caused by very high overpressures developing in the drywell—approximately twice design pressure. These pressures developed very early in the accident (about T+12 hours). At this time, the temperature in the drywell head region was well below temperatures at which elastomeric seal degradation could occur.

Thus, the effect of drywell head temperature may have contributed to the onset of leakage from Units 2 and 3 around, or just above, design pressure.

The leakage at all units through the drywell head flange has been found to enhance the build-up of flammable gases on the refuel floor. Unit 1 may have experienced a flammable refuel floor for a period of 10 hours prior to the actual combustion event. The persistence of flammable conditions likely resulted because of the lack of power, and an ignition source on the refuel floor. Unit 2 most likely did not see flammable conditions developing on the refuel floor because of the effect of the open reactor building refuel floor blowout panel. At Unit 3, flammable conditions developed relatively quickly after leakage from the drywell head may have commenced (at T+60 hours). MAAP5 simulations indicate that leakage from the hard pipe vent to soft ducting may have contributed to the build-up of flammable gases at lower elevations. This would have caused more damage to the Unit 3 reactor building, relative to that observed at Unit 1.

References
[1] "Fukushima Nuclear Accident Analysis Report," TEPCO, Tokyo, Japan, (2012).
[2] "Evaluation of the situation of cores and containment vessels of Fukushima Daiichi Nuclear Power Station Units-1 to 3 and examination into unsolved issues in the accident progression, Progress Report No. 1," TEPCO, Tokyo, Japan, (2013).
[3] "Fukushima Technical Evaluation: MAAP5 Phase 1 Analysis," EPRI, Palo Alto, CA, (2013).
[4] R. Gauntt et al., "Fukushima Daiichi Accident Study (Status as of April 2012)," SAND2012-6173, Sandia National Laboratories, Albuquerque, NM, (2012).
[5] "Severe Accident Management Guidance Technical Basis Report, Volume 1: Candidate High-Level Actions and Their Effects," EPRI, Palo Alto, CA, 1025295, (2012).
[6] B.R. Sehgal, ed., "Nuclear Safety in Light Water Reactors: Severe Accident Phenomenology," Academic Press, New York, New York, USA, (2012).
[7] B.W. Spencer et al., "Risk-Informed Assessment of Degraded Containment Vessels," NUREG/CR-6920, SAND2006-3772P, Sandia National Laboratories, Albuquerque, NM, (2006).

[8] "*Technical Knowledge of the Accident at Fukushima Dai-ichi Nuclear Power Station of Tokyo Electric Power Co., Inc.,*" (Provisional Translation), Nuclear and Industrial Safety Agency, (2012).

[9] T. L. Bridges, "*Containment Penetration Elastomer Seal Leak Rate Tests,*" NUREG/CR-4944, SAND87-7118, Sandia National Laboratories, Albuquerque, NM, (1987).

[10] D.A. Brinson and G.A. Graves, "*Evaluation of Seals for Mechanical Penetrations of Containment Buildings,*" NUREG/CR-5096, SAND88-7016, Sandia National Laboratories, Albuquerque, NM, (1988).

[11] J.T. Julien and S.W. Peters, "*Leak and Structural Test of a Personnel Air Lock for LWR Containments Subjected to Pressure and Temperature Beyond Design Limits,*" NUREG/CR-5118, SAND88-7155, Sandia National Laboratories, Albuquerque, NM, (1989).

[12] D.S. Horschel et al., "*Insights into the Behavior of Nuclear Power Plant Containments during Severe Accidents,*" SAND90-0119, Sandia National Laboratories, Albuquerque, NM, (1993).

[13] D.B. Clauss, "*Severe Accident Testing of Electrical Penetration Assemblies,*" NUREG/CR-5334, SAND89-0327, Sandia National Laboratories, Albuquerque, NM, (1989).

[14] L.D. Lambert and M.B. Parks, "*Experimental Results from Containment Piping Bellows Subjected to Severe Accident Conditions, Volume 2: Results from Bellows Tested in Corroded Conditions,*" NUREG/CR-6154, SAND94-1711, Sandia National Laboratories, Albuquerque, NM, (1995).

Prediction of Complex Thermal-Hydraulic Phenomena Supplemented by Uncertainty Analysis with Advanced Multiscale Approaches for the TALL - 3D T01 Experiment

Angel Papukchiev[a], Marti Jeltsov[b], Clotaire Geffray[c], Kaspar Kööp[b], Pavel Kudinov[b], Rafael-Juan Macián[c] and Georg Lerchl[a]

[a]Gesellschaft fuer Anlagen- und Reaktorsicherheit (GRS) mbH, Garching n. Munich, Germany
[b]KTH Royal Institute of Technology, Stockholm, Sweden
[c]Technische Universitaet Muenchen (TUM), Garching n. Munich, Germany

Abstract: The thermal-hydraulic (TH) system code ATHLET was coupled with the commercial 3D computational fluid dynamics (CFD) software package ANSYS CFX to improve ATHLET simulation capabilities for flows with pronounced 3D phenomena such as flow mixing and thermal stratification. Within the FP7 European project THINS (Thermal Hydraulics of Innovative Nuclear Systems), validation activities for coupled thermal-hydraulic codes are being carried out. The TALL-3D experimental facility, operated by KTH Royal Institute of Technology in Stockholm, is designed for thermal-hydraulic experiments with lead-bismuth eutectic (LBE) coolant at natural and forced circulation conditions. No tests have been performed up to now. GRS carried out pre-test simulations with ATHLET – ANSYS CFX for the TALL-3D experiment T01, while KTH scientists perform these analyses with the coupled code RELAP5/STAR CCM+. In the experiment T01 the main circulation pump is stopped, which leads to interesting thermal-hydraulic transient with local 3D phenomena. In this paper, the TALL-3D behavior during T01 is analyzed and the results of the coupled pre-test calculations, performed by GRS (ATHLET-ANSYS CFX) and KTH (RELAP5/STAR CCM+) are directly compared. Moreover, this work is supplemented by uncertainty and sensitivity analysis for the T01 experiment, carried out at the Technische Universitaet Muenchen.

Keywords: Gen IV, Liquid Metal Flow, Phenomena Modelling, Uncertainty and Sensitivity Analysis

1. INTRODUCTION

In specific nuclear reactor transients and accidents like boron dilution, main steam line break or pressurized thermal shock, three-dimensional coolant mixing and stratification phenomena play an important role and might have a remarkable impact on transient results. Unfortunately, such phenomena cannot be predicted correctly by system codes, based on 1D lumped parameter approach. To overcome this limitation, these numerical tools are coupled with modern CFD programs, which are capable to predict 3D flow behavior in complex geometries and can provide detailed distributions of the physical quantities in space and time. In a coupled 1D-3D numerical analysis, the CFD code simulates only the regions with complex 3D effects, while the rest of the facility is calculated with the 1D system code [1,2].

Within the European project THINS, research activities for the development and validation of advanced multi-scale simulation tools for Gen IV reactors are being carried out. Different approaches have been developed and implemented by GRS and KTH for the coupling of the system codes ATHLET and RELAP5 with the 3D CFD programs ANSYS CFX and STAR CCM+, respectively. In order to validate the newly developed coupled 1D-3D tools, dedicated thermal-hydraulic experiments are performed in Sweden. The TALL-3D thermal hydraulic loop is an integral, two-circuit, well instrumented, 7 m high experimental facility, operated by KTH (Fig. 1, left). The primary circuit consists of three vertical legs, filled with LBE. One heater is installed in the leftmost, main heater (MH) leg, and another one in the middle, 3D test section leg. A heat exchanger (HX) is placed in the upper part of the rightmost leg. The 3D test section, which is domain of complex 3D effects and source for challenging thermal-hydraulic feedback to the rest of the loop is shown in Fig. 2 (right) [3]. Inside the test section pool, there is a metal plate which prevents the liquid metal leaving the test section pool without extensive mixing with the heated fluid inside. This LBE mixing is well pronounced at forced circulation conditions. At natural circulation conditions, stratification occurs in the test section pool, which is enhanced by the heated upper part of the pool wall. In one of the specified experiments, T01, the main circulation pump is stopped. This leads to interesting thermal-hydraulic transient with local 3D phenomena like LBE mixing and stratification affecting the overall loop behavior.

All parameters, which are used for modeling of complex experiments including initial and boundary conditions as well as physical models, might be source of uncertainty. Such uncertainty can significantly impair the quality of code validation process. Both a sensitivity analysis (SA) for the identification of the most influential input uncertain parameters and a uncertainty analysis (UA) for the determination of the range of variability of output variables due to the uncertainty of input parameters are important for a good understanding of model properties and eventually for validation of the codes. Based on the methodology developed by GRS [4,5], TUM performed UA and SA of the TALL-3D experiment T01.

In this paper, the TALL-3D behavior during T01 is analyzed and the results of the coupled pre-test calculations performed at GRS (ATHLET-ANSYS CFX) and KTH (RELAP5/STAR CCM+) are directly compared. Emphasis is given to the understanding of the thermal-hydraulic behavior of the TALL-3D facility during test T01. In a next step, the results from the supplementing uncertainty and sensitivity analyses are discussed.

2. SHORT DESCRIPTION OF THE TALL-3D FACILITY MODELING WITH ATHLET-ANSYS CFX AND RELAP/STAR-CCM+

2.1. Modeling of TALL-3D with ATHLET-ANSYS CFX

Since the TALL-3D test section pool has a rotational symmetry, it was decided to generate a 2D CAD model of the test section [6]. This model served as an input for the ICEM CFD software, which has been used to prepare a 2D hexahedral grid. Then, the 2D grid was revolved to 1° rotational symmetry sector of the test section. In this way, a real 3D CFD mesh was generated and used for the simulations. Grid sensitivity studies were performed, resulting in a final mesh with 48.000 elements (Fig. 2). The mesh quality is very good with a minimum orthogonally angle of 88°, an expansion factor of 4 and a maximum aspect ratio of 465.

For the correct modeling of the buoyant LBE flows in the TALL-3D test section, buoyancy terms in the momentum equation and in the production terms of the turbulence model equations have been included. In the calculations, the SST turbulence model has been used [7]. In the transient CFD and coupled simulations, Second Order Backward Euler transient scheme was selected, while for the numerical transport of the quantities (velocity, temperature, etc.) through the solution domain, a High Resolution advection scheme was used [8]. An adaptive time stepping scheme (time step sizes between 0.1 and 0.5 s) was used in all coupled simulations.

Four priority chains (flow paths) were used for the simulation of the whole experimental facility with ATHLET. These describe the primary and secondary circuit of the facility. Figure 3 shows the coupled ATHLET-ANSYS CFX model (here shown with a symmetry plane of a full model of the 3D test section).

2.2. Modeling of TALL-3D with RELAP/STAR-CCM+

Both STAR CCM+ and ANSYS CFX domains comprise of pool-type section with part of inlet and outlet pipes [3]. The 2D axisymmetric mesh consists of 74.661 polyhedral cells (see Fig. 4, left). The outlet and inlet pipes are modeled with 25 prism layers and the rest of the domain is modeled with 15 layers on the wall. Inlet mass flow rate and temperature were defined at the inlet boundary and pressure was defined at the outlet. In transient simulations, unsteady implicit time integration scheme is used with second order temporal discretization. Segregated solver with second order upwind convection scheme is used for the flow [9]. Density as a polynomial function of temperature is used as the equation of state. Mixed convection turbulence inside the 3D test section is predicted using a Realizable K-Epsilon turbulence model with Buoyancy Driven Two Layer formulation developed by Xu et al. [10].

The corresponding RELAP5 nodalization of TALL-3D geometry can be seen in Figure 4 (right picture). The model consists of pipe structures connected together by single junctions and time-dependent junctions. Time-dependent volumes are used for expansion tank and secondary side inlet and outlet. The main heater (MH) is simulated as a pipe with flow area comparable to the flow area in the annulus at the facility.

Fig. 1: TALL-3D facility

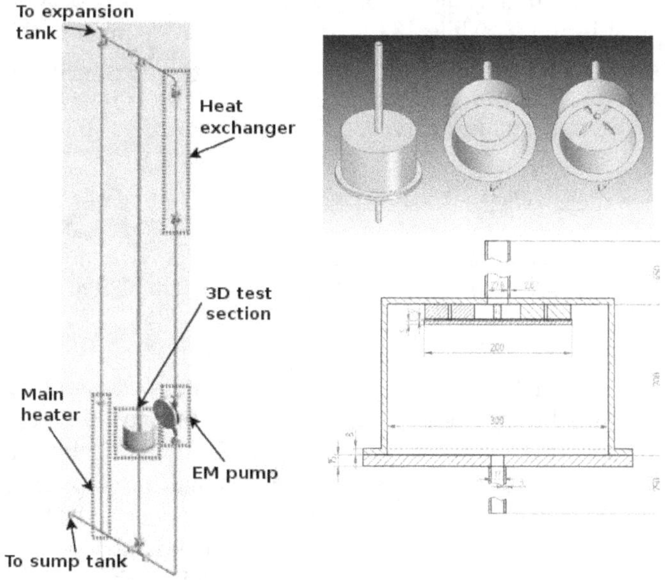

Fig. 2: Numerical mesh

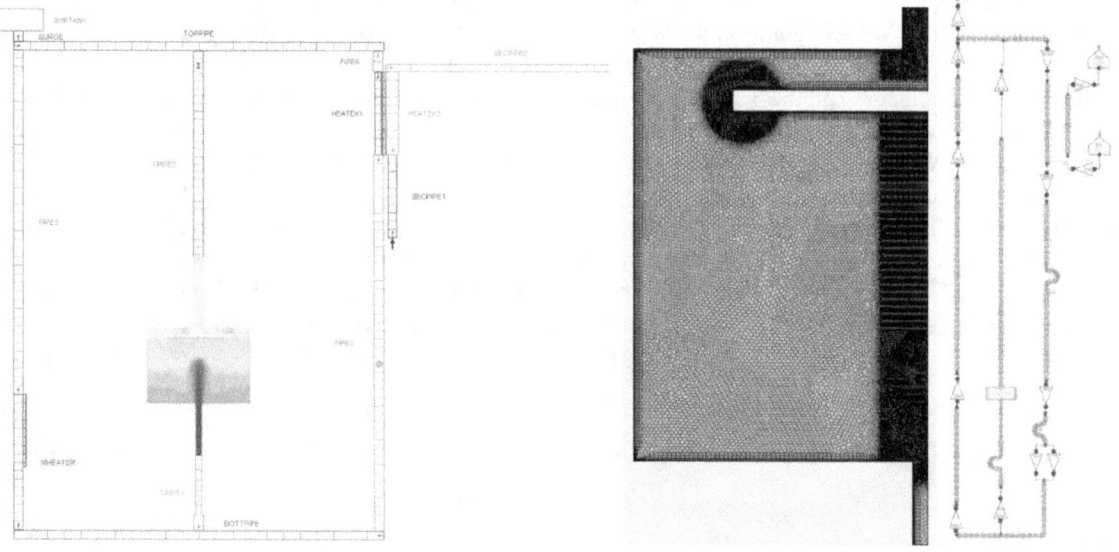

Fig. 3: ATHLET-ANSYS CFX model   **Fig. 4: RELAP5/STAR CCM+ model**

2.3. Transient Test T01 and Boundary Conditions

Since no measured data are available, only specified data and boundary conditions have been used for the developed simulation models. Later, experimental data from the steady state TALL-3D commissioning tests will be used for fine calibration of the coupled models. In the test case selection process, emphasis was put on transients with strong 3D effects in the 3D test section. It is expected, that these will have an impact on the whole TALL-3D facility. Their simulation requires the application of advanced coupled codes like ATHLET-ANSYS CFX and RELAP5/STAR-CCM+. Pressure and temperature will be measured at different locations in the facility. This data will be then used for the validation of both programs.

The planned transient test case "Test T01 – Forced to natural circulation with both heaters always switched on" was selected for comparison of the two simulations. This test case represents a loss of flow transient in the facility. It starts from a steady state forced convection with a specified mass flow rate of 4.77 kg/s in the HX leg. This mass flow rate is not distributed evenly in the rest of the primary circuit – 2.84 kg/s are expected to flow through the MH leg, and 1.93 kg/s through the 3D test section leg. The temperature at the 3D test section inlet is 563 K, while 488 K are specified at the inlet of the secondary circuit (secondary mass flow rate: 4.0 kg/s). During the whole transient, both, the main and the 3D test section heaters are in operation at 5 kW power each. At time 120 s, the electromagnetic pump is switched off. The simulation time of this transient test was specified to be approx. 1500 s.

2.4. Analysis and Comparison of the Simulation Results

With the specified boundary conditions, three calculations have been performed – two at GRS (ATHLET stand alone and coupled simulation with ATHLET-ANSYS CFX) and one coupled at KTH (RELAP5/STAR-CCM+).

The electromagnetic pump was tripped at 120 s simulation time with a run down time of one second, which leads to a rapid mass flow rate decrease in all three primary legs. Within ten seconds, the LBE mass flow rate in the HX leg drops from 4.77 kg/s to 0.5 kg/s and then stabilizes at about 0.75 kg/s in all three calculations. Due to the LBE density distribution in the primary circuit and the difference between the geodetic heights of both heaters (lower part of the facility) and the HX (upper part of the facility), natural LBE circulation occurs in the primary circuit of the TALL-3D loop immediately after the pump trip. In this early phase of the transient, a mass flow increase in the MH leg is observed (Fig. 5), while the mass flow rate in the 3D test section leg decreases further (Fig. 6).

Because of the intensive natural circulation in the MH leg, less LBE flows in the 3D test section leg. In the stand alone ATHLET calculation, the LBE flow even reverses for approx. 90 s, while in the RELAP5/STAR CCM+ calculation the flow reversal occurs with a delay of 45 s and lasts 45 s. In the coupled ATHLET-ANSYS CFX simulation, the LBE flow decreases to 0.06 kg/s, but does not reverse. This imbalance of the LBE flow distribution in the primary circuit occurs, although both heaters are kept operated at 5 kW power each. The reason for this imbalance is related to the different volumes of the MH pipe and the 3D test section pool. Since the volume of the MH pipe (4.92E-4 m^3) is significantly smaller than the one of the 3D test section pool (141.37E-4 m^3), LBE is heated more rapidly in the MH (up to 640-650 K, see first temperature peaks around 160 s in Fig. 7), and its outlet temperature increases faster than the temperature at the outlet of the test section outlet pipe (Fig. 8). The heated LBE is lighter (lower density) and rises faster in the MH pipe. This enhances the natural circulation in this leg, while at the same time it impedes the LBE circulation in the 3D leg.

Fig. 5: Mass flow rate at the MH inlet

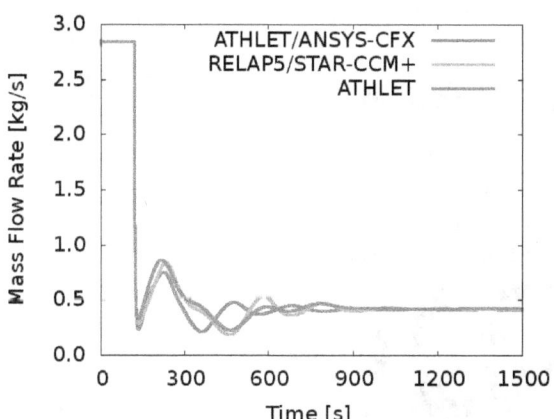

Fig. 6: Mass flow rate at the 3D section inlet

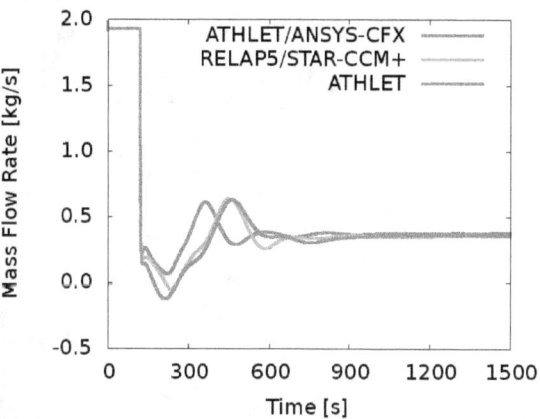

Fig. 7: Temperature at the MH outlet

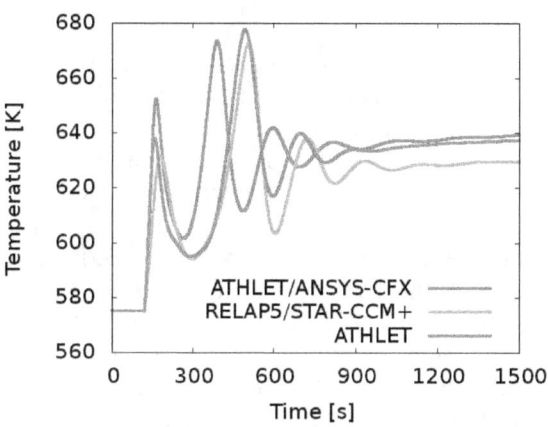

Fig. 8: Temperature at the test section outlet

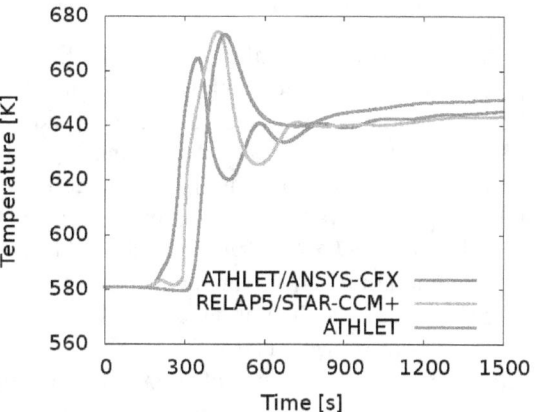

The temperature at the test section outlet (Fig. 8) in the ATHLET-ANSYS CFX simulation starts to increase 40 s after pump trip and, in the ATHLET simulation, after 90 s. The reason for the faster temperature increase in the coupled simulation is the positive LBE flow through the 3D test section leg. It is eventually enhanced by more intensive heating in this calculation, since wall structures are not present in the current ANSYS CFX model. In ATHLET, the heater power of 5 kW is applied to the test section wall, which is 6.75 mm thick, while in the coupled simulations a heat flux corresponding to 5 kW power is applied directly to the fluid. In this way, the specific heat capacity of the test section stainless steel wall is neglected in the coupled simulation. This leads to a constant heat flux during the whole ATHLET-ANSYS CFX simulation and to higher temperatures. This phenomenon enhances the natural circulation in the 3D test section leg and hinders flow reversal there.

Figure 9 shows the LBE temperature distribution in the 3D test section at forced circulation just before the initiation of the pump trip, calculated with ATHLET-ANSYS CFX (left) and ATHLET (right). The test section pool (0.2 m high) is represented in ATHLET as a 1D pipe with 20 nodes. Since the heater is installed around the upper half of the test section pool wall, only the nodes in this part are actually heated, while the ones below the heater are still filled with cold LBE ($T_{LBE}=T_{inlet}$). In this way, for the steady state of TALL-3D (pump is running, heaters at nominal power), a 1D system code like ATHLET or RELAP5 will predict stratification in the test section (see Fig. 9, right). In the coupled calculation, a large swirl develops in the CFD domain. The cold LBE first hits the plate and then moves to the side pool wall, taking the heat away from it and transporting it to the center and the bottom part of the pool. As a result, a mixed flow pattern in the 3D test section pool is observed. The mean LBE temperature in the vicinity of the bottom plate is approx. 578 K (Fig. 9, left), while the 1D approach of ATHLET predicts 563 K cold and stratified LBE at the same location (Fig. 9, right).

Fig. 9: Temperature distribution in the 3D pool

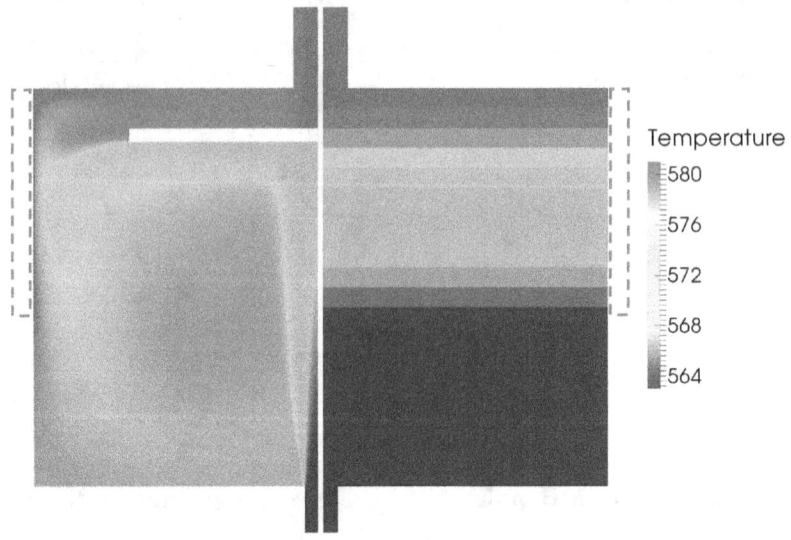

Figure 10 shows the evolution of the LBE temperature at the inlet of the test section inlet pipe. After pump trip, the LBE flows slowly through the primary HX tube at natural circulation conditions. It is better cooled down by the cold secondary side, which is kept at constant mass flow rate of 4.0 kg/s and 488 K LBE temperature. Approximately 110 s after pump trip, the LBE temperature in the RELAP5/STAR-CCM+ simulation increases suddenly, because of the flow reversal in Fig. 6. Since the 3D pool is well mixed, warm LBE from the bottom part of the test section pool (temperature distribution in STAR-CCM+ very similar to the one in Fig. 9, left) flows back to the test section inlet pipe. Although ATHLET predicts also a flow reversal (Fig. 6), this temperature increase is not present. The temperature at the bottom of the pool is the same as the one in the inlet pipe, due to the "artificial" stratification. In the ATHLET-ANSYS CFX simulation no temperature increase is observed, because LBE flow does not reverse at all. Approximately 110 s after pump trip, the cold LBE reaches the inlet of the test section, and the temperature calculated by ATHLET-ANSYS CFX starts to decrease.

Figures 11 and 12 show the temperature distribution in the 3D test section before the initiation of the transient and at its end, approx. 1500 s after pump trip. At high LBE velocities the test section is mixed, while at low velocities and heaters switched on, buoyancy effects play an important role. At natural circulation conditions, thermal stratification establishes in the CFD domain. Cold and heavy LBE stratifies at the bottom of the test section, while lighter, hot LBE can be found in the upper half of the test section pool.

Fig. 10: Temperature at the test section inlet

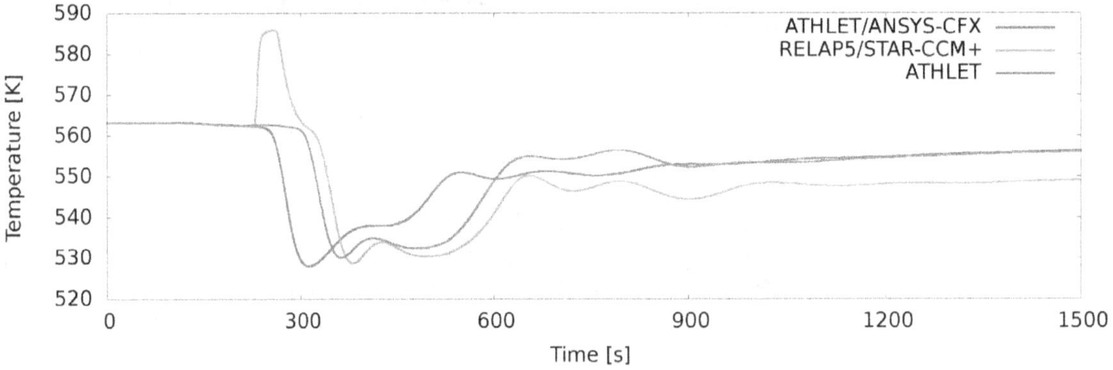

Fig. 11: T_{LBE} distribution at forced convection   Fig. 12: T_{LBE} distribution at natural convection

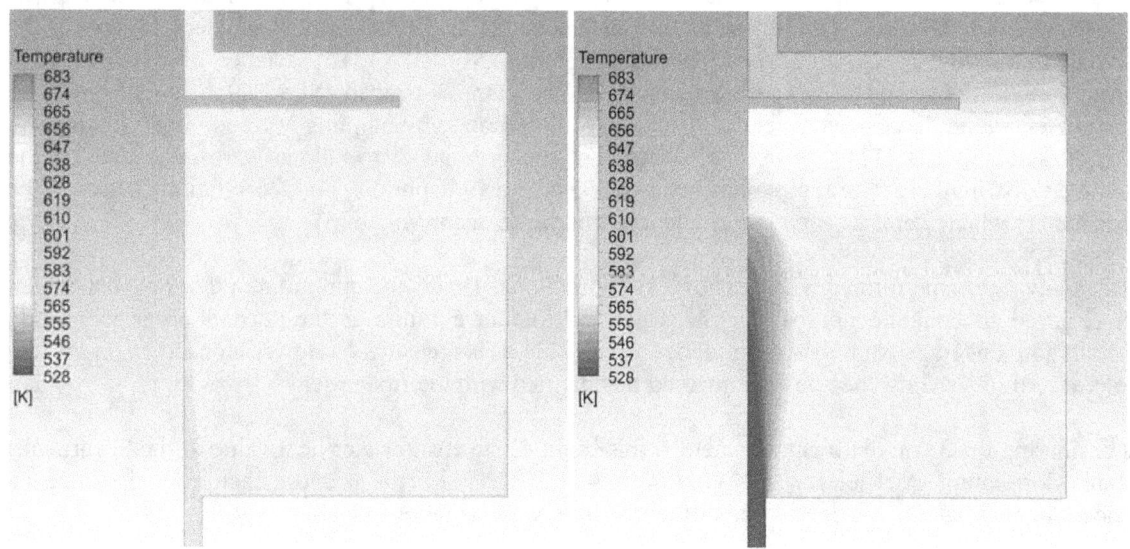

A good overall agreement between the ATHLET stand-alone and the coupled RELAP5/STAR CCM+ calculation results can be observed. The occurrence of the main thermal-hydraulic phenomena in time is almost identical. The small differences in the temperatures might be explained with the different heat transfer correlations for LBE implemented in ATHLET and RELAP5. Moreover, two different simulation approaches are used in these calculations - the 1D (ATHLET) and the 1D-3D approach (RELAP5/STAR CCM+). It is surprising to observe such good agreement between the 1D and the more sophisticated 1D-3D RELAP5/STAR CCM+ simulation approach. ATHLET-ANSYS CFX results differ from the results of the other two calculations, but are still in agreement with them. The experimental data, which is expected to be available soon, will allow better analysis of the strengths and weaknesses of the different simulation approaches.

3. UNCERTAINTY AND SENSITIVITY ANALYSIS OF THE TALL 3D FACILITY USING ATHLET – ANSYS CFX

Once a Best-Estimate (BE) model has been designed, the influence of model input uncertainty needs to be taken into account. Uncertainty and Sensitivity Analysis is a powerful technique to help assess the sensitivity of the model to several modeling assumptions. This method provides information on the variability induced in the model output by the model input uncertainty and helps to identify the key input parameters. This information might be used to improve model accuracy and eventually to validate a model against experimental data.

3.1. Model Adaptation for Uncertainty Analysis

The variation of the input parameters applied in the scope of an Uncertainty and Sensitivity Analysis induces perturbation in the model which makes it deviate from the Best-Estimate run. These deviations can result in a system that is not stable by the start of the transient. To ensure that the state in which the system can be found at the start of the transient is actually a stable steady-state, two simple control systems (proportional-integral-derivative) have been implemented in the ATHLET TALL-3D model. The first controls the electromagnetic pump, thus ensuring that the total mass flow rate in the primary side (HX leg) is kept constant during steady-state. The second one controls the secondary LBE mass flow rate, and thus ensures power balance between primary and secondary side.

3.2. Computer Experiment

The screening analysis performed here considered 33 uncertain parameters. Due to the lack of experimental data, only uniform distributions have been applied. For all the parameters, except for the turbulent Prandtl number (No 32) and the LBE properties (No. 10 to 13), a variation of ± 10% or ± 5% around the best-estimate value has been specified. The pump is tripped at t = 500 s. The model input parameters values have been generated using the Simple Random Sampling Method. The Kolmogorov Goodness-of-Fit test [11] was used to check that the original distributions were respected by the sampled distributions. The sample has been checked to avoid spurious rank correlations between the parameters which were assumed independent in this work, according to [5].

The analysis was performed with a set of 153 simulations. This is the minimum number of runs which is required to compute non-parametric two-sided tolerance limits at the second order with 95% population coverage and 95% confidence level [11]. This ensures a lower conservatism of the tolerance limits against the one which would be obtained with the first order.

Considering the 33 model input parameters, this sample size ensures a critical value of the Spearman's Rank Correlation Coefficients (SRCC) lower than 20% [11]. This is more than enough, since an uncertain parameter is not considered influential if its value is below 40%.

4. NON-PARAMETRIC UNCERTAINTY ANALYSIS

In the following paragraphs the non-parametric tolerance limits (TL) are presented along with the BE run (all model input parameter values are set to their mean value). These represent the spread of the model output which can be expected due to the model input uncertainty. Over all the possible cases (there is infinity of them) and given the model input uncertainty considered here, we are 95% sure that in 95% of the cases the model output will lie within the tolerance limits.

The main results from the non-parametric uncertainty analysis are listed below:

- The uncertainty over the input parameters induces relatively small variations in terms of mass flow rate, except for the mass flow rate in the 3D leg (Fig. 13) at the early stages of the transient (at approx. 600 s, that is 100 s after the pump trip) (Fig. 14).

- The variations induced on the pressure (not represented here) are very small (less than 1000 Pa)

- Heat transfer phenomena appear to be the most influenced by the model input uncertainty. The temperatures at 3D pool inlet and outlet (Fig. 15) and their difference (Fig.16) show relatively large variations (up to 15 K).

Fig. 13: Mass flow rate 3D leg (Full)   **Fig. 14: Mass flow rate 3D leg (Transient)**

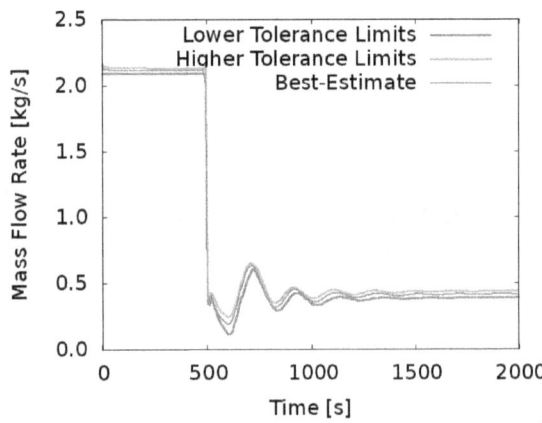
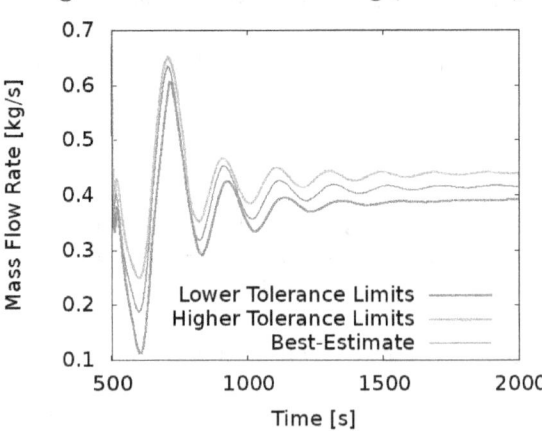

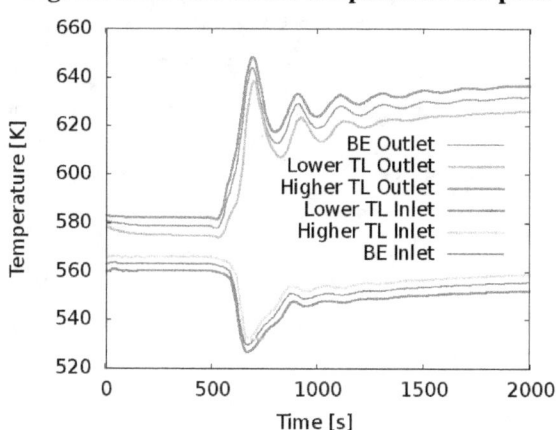

Fig. 15: Inlet and outlet temperature 3D pool

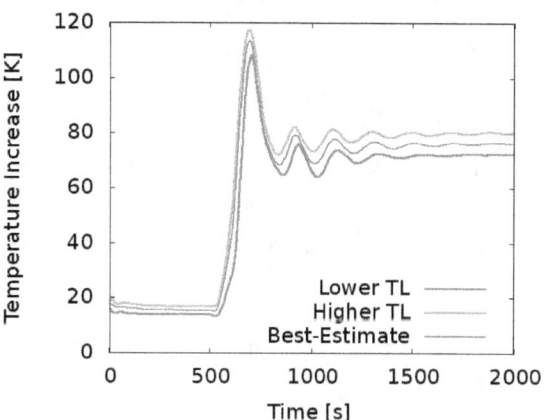

Fig. 16: Temperature increase over the 3D pool

5. SENSITIVITY ANALYSIS

Once the model output variability has been quantified, the model input parameters which account for this variability can be identified. This part of the analysis is called Sensitivity Analysis. The latter is performed here using correlation coefficients as a measure of the strength of the relation between the model input parameter and the monitored output variable. In the following, Spearman's Rank Correlation Coefficients (SRCC) [11] were used to quantify this influence. These coefficients are calculated assuming a monotonic relation between the input and the output, and quantify the strength of this monotonic relationship. To check whether this relation actually explains most of the model output variability, the Multiple Determination Coefficient must be calculated [4] and its value provided along with the SRCC values.

5.1. Scalar Sensitivity

At each time step, ANSYS CFX was configured to provide the maximum local LBE temperature in the whole CFD domain. The maximum of these values over the whole transient can be determined per each run. Sensitivity analysis has been performed on this data.

Following model input parameters are the most influential on the maximum LBE temperature (Fig. 17):

- The initial inlet temperature (31) is positively correlated as a starting point for the transient. It is still very influential since the maximum temperature is reached shortly after the pump trip when the LBE has been virtually blocked in the pool.

- The heat capacity of LBE (13) is negatively correlated. A lower value tends to increase the difference in temperature when a given amount of power is added to the fluid.

- The power of the 3D heater (15) is obviously positively correlated.

- The turbulent Prandtl number (32) tends to reduce turbulent heat transfer and enables higher local temperature elevation.

For the time at which the maximum temperature is reached (Fig. 18), the delaying effect of the LBE heat capacity and of the heat losses through the insulation is highlighted by the positive correlation. The power of the 3D heater is negatively correlated since a high power induces faster increase of LBE temperature and faster initiation of natural circulation.

Fig. 17: SRCC observed maximum temperature in the CFD part

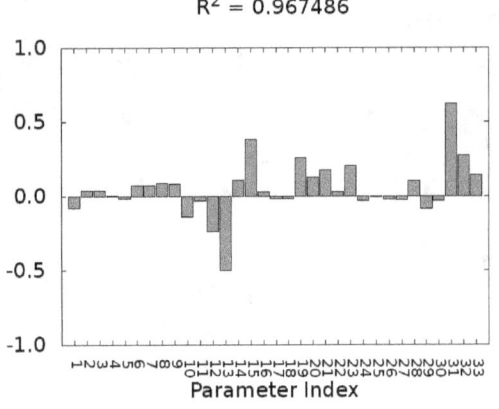

Fig. 18 SRCC time of maximum observance in the CFD part

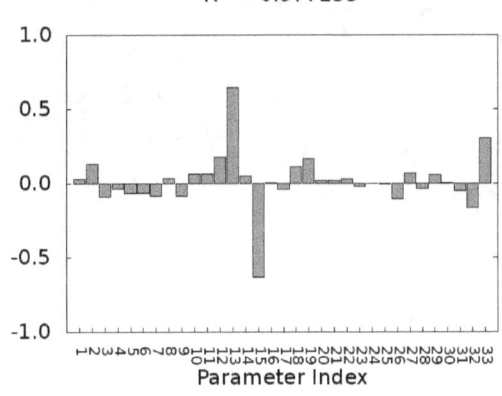

5.2. Time-Dependent Scalar Sensitivity

The multiple determination coefficient values are not represented here, but it remains close to 1 during the whole calculation. The present analysis investigates the influence of the input parameters on the mass flow rate in the 3D leg (Fig. 19).

During the steady-state phase, the pressure form loss (PFL) related terms due to the fuel pin spacers in the MH leg and the Rotamass mass flow meter and the roughness in the 3D pool are the most influential.

During the transient:

- The more power is transferred to the fluid by the 3D pool heater, the higher the density differences driving the flow. This explains the positive correlation during most of the transient.
  Within the time frame 700 – 800 s, the correlation turns negative. This is due to the transition from forced to natural circulation. The 3D pool contains a large amount of LBE. When the pump is tripped, this amount of fluid is blocked in the pool and heated up, and hence energy gets accumulated in the pool. When the mass flow rate starts to increase again after 600 s simulation time, the large volume of the pool and the energy accumulated there induce thermal inertia in the system behavior. The higher the power of the 3D heater, the longer it will take the HX to remove the accumulated energy. Therefore, the outlet temperature of the pool as well as the HX temperature remain higher and the density difference remains smaller which leads to a lower mass flow rate in the 3D leg and to the negative value of the correlation coefficient.

- The value of the heat capacity influences the temperature variations where heat transfer occurs. Globally large temperature variations lead to large density differences which drive the natural circulation: hence the negative correlation during most of the transient. The same phenomenon as previously discussed for the power of the 3D heater applies here for the time frame 700 – 800 s.

- The turbulent Prandtl number is globally positively correlated. If its value is increased, turbulent thermal mixing is reduced which increases the outlet temperature of the 3D pool. The density difference driving the natural circulation is then increased.

- The higher the heat losses in the 3D pool, the lower the temperature increase. This reduces the density differences (inlet and outlet of the 3D pool): hence the negative correlation.

Fig. 19 SRCC mass flow rate 3D leg

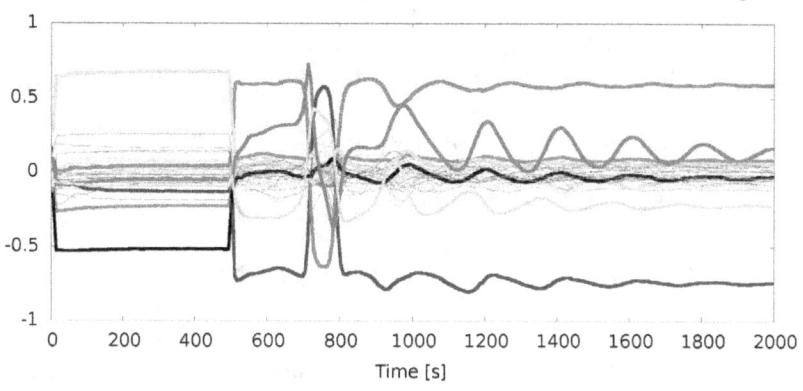
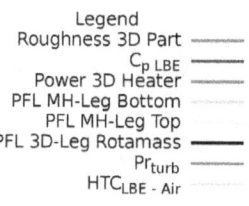

6. CONCLUSION

This paper provides comparison of the results from the first blind simulations of the TALL-3D facility, performed at GRS (ATHLET, ATHLET-ANSYS CFX) and KTH (RELAP5/STAR-CCM+). In the specified by KTH transient test T01, a pump trip is simulated and the natural circulation phenomenon for liquid metal flows is being studied. The performed simulations show, that the selected geometry of the TALL-3D facility together with the chosen boundary conditions result in a very interesting and challenging experiment. This involves complex physics including liquid metal forced and natural circulation, combined with heat transfer and 3D mixing and stratification phenomena. All three simulations show similar behavior of the main thermal-hydraulic parameters. The most important difference between ATHLET-ANSYS CFX and the other two simulations is the fact that the mass flow rate does not reverse in the 3D test section leg.

Although a very good general agreement is observed between the 1D ATHLET and the 1D-3D RELAP5/STAR-CCM+ calculations, it was shown, that the 1D simulation is not always adequate for the simulation of geometries with pronounced 3D flow effects like mixing or stratification. This reveals the need for the development and application of advanced coupled thermal-hydraulic programs such as ATHLET-ANSYS CFX and RELAP5/STAR-CCM+. Moreover, the comparison with the experimental data will allow better analysis of the strengths and weaknesses of the different simulation approaches.

The Uncertainty Analysis leads to the conclusion that most important model output variability is to be found in the heat transfer related processes which also affect the mass flow rate in the 3D leg during the transition phase.

The Sensitivity Analysis helped identifying the most influential parameters on the calculation. There are the LBE fluid properties, the power of the heaters, the turbulent Prandtl number, some of the pressure loss related terms, especially the spacer in the MH leg and the Rotamass flow meter in the 3D leg.

Influential parameters were identified for global variables (Scalar Sensitivity) which can be safety relevant. The time-dependent Sensitivity Analysis helped identifying phenomena and provided detailed information and understanding on the influence of the model input parameters on the simulation results as well as on the hierarchy of the input parameters during the three phases of the transient (forced circulation, transition phase and natural circulation).

Acknowledgements

This work is supported by the EC 7th Framework Project THINS (Thermal-Hydraulics of Innovative Nuclear Systems).

References

[1]   A. Papukchiev and G. Lerchl, *"Development and Implementation of Different Schemes for the Coupling of the System Code ATHLET with the 3D CFD Program ANSYS CFX"*, Proc. of the NUTHOS-8 Conference, Shanghai, China, October 10-14, 2010

[2]   M. Jeltsov et al, *"Development of a Domain Overlapping Coupling Methodology for STH/CFD Analysis of Heavy Liquid Metal Thermal-hydraulics"*, Proc. of the NURETH-15 Conference, Pisa, Italy, May 12-17, 2013

[3]   M. Jeltsov et al, *"Multiscale Calculation Results on TALL-3D"*, THINS Deliverable D5.2.03 (revised), KTH, 2013.

[4]   H. Glaeser, B. Krzykacz-Hausmann, W.Luther, S.Schwarz and T.Skorek *"Methodenentwicklung und exemplarische Anwendungen zur Bestimmung der Aussagesicherheit von Rechenprogrammergebnissen"*, Technical Report, GRS, 2008.

[5]   E. Hofer, *"Sensitivity analysis in the context of uncertainty analysis for computationally intensive models"*, Computer Physics Communications, Volume 117, Issues 1–2, 1 March 1999, Pages 21-34

[6]   A. Papukchiev and G. Lerchl, *"System-CFD Multiscale Input Decks for the TALL-3D Facility"*, THINS Deliverable D5.2.02 (revised), GRS, 2013.

[7]   F. Menter, *"Two-Equation Eddy-Viscosity Turbulence Models for Engineering Applications"*, AIAA-Journal, Vol. 32, pp. 269 – 289 (1994).

[8]   ANSYS, *"ANSYS CFX 14.5 User's manuals"*, 2012.

[9]   CD-adapco, *"Star-CCM+ User Guide version 7.04.011"*, 2012.

[10]   W. Xu et al, *"A New Turbulence Model for Near-Wall Natural Convection"*, Int. J. Heat Mass Transfer, 41, 1998, pp. 3161-3176.

[11]   W. Conover, *"Practical Nonparametric Statistics"*, Wiley Series in Probability and Statistics, Wiley, 1999

Cost-Effectiveness of Vehicle Barriers and Setback Distance for Protecting Buildings from Vehicle Bomb Attack

Nathaniel Heatwole[a]

[a] University of Southern California, Los Angeles, USA, heatwole@usc.edu

Abstract: Decision-making regarding implementing measures to protect buildings from vehicle bomb attack is often undertaken using highly judgment-based risk processes. This paper presents a quantitative risk-cost model for using vehicle barriers to create setback distance around a new office building. The model explicitly considers both the attack probability, and the damages in the event of an attack (both target building and collateral), as well as how both of these might change as mitigation measures are implemented. The attack damages are assessed using a new empirical blast model, which adapts the estimation methods used by the U.S. Geological Survey for earthquake damages, and is based on data from three well-studied vehicle bomb attacks. Monte Carlo simulation is used to carry the uncertainty in the inputs through to the final results. The model outputs are the mitigation costs, the attack damages, the "breakeven" attack probability (at which the benefits of the mitigation justify its costs), and the cost per statistical life saved (assuming an attack). The results suggest that this mitigation option is cost-effective only when the attack probability (for the case without the mitigation measures present) is rather high.

Keywords: terrorism risk, vehicle bombs, Monte Carlo simulation, value of a statistical life

1. BACKGROUND

Vehicle-borne improvised explosive devices (VBIEDs) are a favored terrorist weapon. Davis [1] calls them "stealth weapons of surprising power and destructive efficiency" – the "poor man's air force" – and notes that over a person of 25 years, VBIED attacks have occurred in at least 58 countries. However, decision-making regarding blast protection for buildings is often undertaken using highly judgment-based risk processes. First, a design basis threat (that is, size of bomb) is specified, and a portfolio of mitigation measures is selected. The damages with the mitigation are then assessed and, if deemed to be *reasonable*, the cost is examined. If either the damages or the mitigation cost are deemed to be *unreasonable*, the portfolio of mitigation measures is reworked. As such, the attack probability tends to be treated as binary, with the benefits and costs of the mitigation examined somewhat separately of one another [2,3,4,5,6,7,8,9,10]. The need for more risk-informed methods for blast protection – including greater consideration of uncertainties – has been widely recognized [6,9,11,12,13,14,15,16,17,18].

2. PREVIOUS WORK

Various works (e.g., [11,13,15,16]) examine protective design using a quantitative risk framework, but rely on highly simplified assumptions regarding the avoided damages and costs (e.g., 90% reduction in risk for a 10% increase in building construction costs). Foo et al. [14] offer a blast risk assessment method for buildings; however, their model does not account for progressive structural collapse, and many aspects of it are not overly transparent.

3. BLAST PRIMER

When a high explosive detonates, a blast (or shock) wave is created that propagates through the air at multiple kilometers per second. Upon reaching a particular point, the pressure at that point rises abruptly to a *peak overpressure*, P – perhaps many times atmospheric (ambient) pressure. The overpressure then falls back to atmospheric quasi-exponentially – ending the *positive phase*. The duration of the positive phase is measured in milliseconds, and the time integral of the overpressure during this period is the *impulse*, I. The overpressure then drops below zero for a time (the *negative phase*), before eventually returning to atmospheric. Soon after detonation, the blast wave engulfs the building – first loading the windows, façade, and exterior columns, and then entering the structure – pushing on the floor slabs and exposing the occupants directly to the airblast – and finally loading the structure globally, with inward pressure on all sides. Common causes of death and injury in VBIED attacks include flying debris (e.g., window glazing), being propelled against a rigid object (e.g., wall), and, in particular, structural collapse [6,8,19]. In analyses of blast phenomena, the cube root scaled distance is often used, defined as

$$Z \equiv \frac{R}{W^{1/3}} \tag{1}$$

where R (m) and W (kg) are the standoff distance from and TNT equivalent mass of the charge (or the mass of TNT needed to produce the same effects), respectively. Theoretical and empirical analyses over a wide range of charge weights show that the detonation of two explosives of identical shape – but unequal masses – yields the same peak overpressure at a given scaled distance [6,19].

The scaled distance (Equation 1) predicts that blast loads will decrease rapidly with distance, and so many authors contend (e.g., [5,6,8,16,20]) that setting a building back from all vehicle-accessible points is one of the most cost-effective blast mitigation options. This paper presents a quantitative risk-cost model for using vehicle barriers to create setback distance around a new office building. Many aspects of blast risk analysis are highly uncertain [6,11,12,13,15,17], so a stochastic approach is taken, with Monte Carlo simulation (100,000 trials) used to propagate the uncertainty in the inputs through to the predictions.

4. MITIGATION COSTS

The costs are assumed to be comprised of expenses for additional land (assuming no land around the building in the no mitigation case) and for the vehicle barriers themselves, and to be repaid over a 30 years, at 8% interest, with a 10% down payment, and to become sunk once incurred. The building footprint (and lot) is rectangular, with length (left to right) and depth (front to back) of L and D (m), respectively. The vehicle barriers span the entire lot perimeter, creating a setback distance of R (m). The land cost is the product of the additional land area to be purchased and the unit land cost, U_{land} ($/m²$), or

$$C_{land} = (U_{land}) \cdot (2 \cdot R) \cdot \left[(2 \cdot R) + L + D \right] \tag{2}$$

Land costs are highly variable. The Federal Bureau of Investigation (FBI) has field offices in 56 cities, each of which also has a federal courthouse. For as many of these as possible (or for a nearby lot, if data for the selected lot could not be located), the lot size and market (not tax assessed) value of the land (not the improvements) were obtained from the website of the local property tax assessor. According to Case [21], this represents the best source of data for local land values. For these properties, U_{land} (or the ratio of the total land value to the area of the lot) is then well-modeled by a lognormal distribution with mean and standard deviation (non-logged) of $1,090/m²$ and $4,100/m²$, respectively ($n=91$; $\chi^2=11.6$; $p=0.31$).[1] The mean error in the fitted cumulative distribution function (CDF) is –1.5% (90% c.i. –20% to +17%).

For the vehicle barrier, we use anti-ram bollards, which are long, thick cylindrical posts, set in concrete foundations – one of the most common blast mitigation measures. The total bollard cost is

---

[1] Unless otherwise noted, all costs are in 2012 US$, with inflation adjustments made using the Producer Price Index, across all commodities (data from http://www.bls.gov/ppi/data.htm). Note that a *high* p-value is sought, as this is the probability that the data are drawn from the distribution.

$$C_{bollards} = \left(\frac{U_{bollard}}{spacing} \right) \cdot \left[(8 \cdot R) + 2 \cdot (L + D) \right] \tag{3}$$

where $U_{bollard}$ is the unit bollard cost ($/bollard), *spacing* is the center-to-center gap (for which we use a uniform distribution over 1–2 m), and the bracketed term is the perimeter of the lot. For 8-inch diameter concrete-filled pipe bollards, the U.S. Veterans Administration [22] lists a unit cost of $1,100 (in 2012$), and to account for the variability in these costs, $U_{bollard}$ is modeled as a lognormal distribution with a coefficient of variation (CoV; or the ratio of the standard deviation to the mean) of 10%.

For the building specifications, we use information from Persily & Gorfain [23], who for 100 randomly chosen U.S. office buildings present distributions for the: number of floors, F; gross floor area, A_{gross}; and occupant density, ρ (mean=0.027/m^2, assuming gross floor area is 50% greater than occupied floor area). To solve for the distributions for L and for D, the area of the building footprint ($L \cdot D$) is approximated as the ratio of A_{gross} to F (with the correlation between the two set equal to one), and the aspect ratio of the building's footprint (L/D) is specified as a triangular distribution with mode of 2.6 (see Table 2, Oklahoma City) and range of 1–4. We use a floor height of 3.9 m (average of the values from Table 2).

5. NEW EMPIRICAL BLAST DAMAGE MODELS

This section presents a series of new empirical models for VBIED attack damage estimation, some of which were formulated by adopting the methods used to estimate damages from worldwide earthquakes.

5.1. PAGER Earthquake Damage Model

The U.S. Geologic Survey's (USGS) Prompt Analysis of Global Earthquakes for Response (PAGER) model estimates fatalities [24] and property damage [25] from earthquakes. PAGER models the loss ratio (LR) – or the expected damage divided by the maximum potential damage – using the CDF of the lognormal distribution (which is bounded between zero and one), as a function of the level of ground shaking (assessed using the Modified Mercalli Index, MMI), with parameters fitted to empirical data.

5.2. Blast Flux

For earthquake hazards (previous), an intensity metric is already available in the form of the MMI. For the case of VBIED attacks, we formulate an analogous intensity metric using the concept of flux, which relates to the flow of a quantity through a surface. The *blast flux* through a building is defined as

$$\beta \equiv \left(\frac{1}{Z} \cdot \frac{1}{\sqrt{L^2 + D^2 + H^2}} \right) = \frac{W^{1/3}}{R \cdot \sqrt{L^2 + D^2 + H^2}} \tag{4}$$

where L, D, and H are the building's length, depth, and height, respectively. In Equation 4, the first term in the parenthesis assesses the blast intensity, using the (inverse) scaled distance (Section 3), and the square root term gives the distance from one (extreme) building corner to another, thereby accounting for the distribution of the occupants in all three dimensions. The occupant vulnerability is then specified as

$$V_{blast} = 10^3 \cdot \beta \cdot \sqrt{V_w \cdot V_s} \tag{5}$$

where V_s and V_w ($0 < V \le 1$) are vulnerability coefficients associated with different building structural and wall types, respectively, which are specified using information from FEMA [7] (Table 1). In Equation 5, the factor of 10^3 is used so as to yield V_{blast} values on the order of 1–100 (see Table 2), and the square root is used on the basis that V_w and V_s each imply some unit of vulnerability per unit of β. In the calculations, a uniform distribution is used over all wall and all structural system coefficients in Table 1.

Table 1: Blast Vulnerability Coefficients for Different Building Wall and Structural Systems [7]

Wall Type	Structural System Type	V_w or V_s
Cast-in-place reinforced concrete	Reinforced concrete shear walls, or bundled tubes	0.10
Curtain wall	Braced exterior frame	0.74
Precast panels, or reinforced masonry	Frame with core, or precast	0.84
Massive unreinforced masonry	Precast tilt-up, or frame with (concrete or masonry) infill, or reinforced masonry, or belt truss	0.93
Light frame or slender unreinforced masonry	Light metal frame, or brick, or timber	1.00

Note: the source provides two sets of V-values, depending on whether the bomb is more or less than 100 ft away from the building. However, when converted into relative terms (as shown above), these discrepancies disappear.

5.3. Empirical Blast Data

The mortality and morbidity models are based on data from the three well-studied VBIED attacks, or:
1) AMIA building (Jewish community center, Buenos Aires, Argentina, 1994);
2) Oklahoma City (Murrah federal building, U.S., 1995); and
3) Khobar Towers (barracks for U.S. Air Force personnel, Dhahran, Saudi Arabia, 1996).

All sources were peer-reviewed journal articles, books, government reports, and some media accounts in major newspapers available through *LexisNexis* [26]. The prime references were: Oklahoma City–[27]; AMIA–[28]; and Khobar Towers–[29]. In cases where multiple estimates of a quantity were available, the average was taken (with identical values considered as one). Three degrees of injury are examined: killed (K), hospitalized injuries (HI), and non-hospitalized injuries (NHI; in general, emergency room treated and released). The building occupancies were estimated on the basis of a density of $0.027/m^2$ (Section 4). For Oklahoma City (where the number of building occupants is known reasonably well), the estimated occupancy is 369, which compares well with the actual occupancy of 361 [27]. Various data related to these bomb attacks is summarized in Table 2. The calculated blast vulnerabilities (V_{blast} – Equation 5) are 8.8 for Khobar Towers, 30.5 for Oklahoma City, and 43.9 for AMIA (all $kg^{1/3}/m^2$).[2]

5.4. Mortality and Morbidity in Target Building

Table 2 indicates that $LR(K)$ is increasing in V_{blast}. This is cogent: as β increases (Equation 4), so should the number of persons in the target building who are killed. So $LR(K)$ is modeled using the CDF of the lognormal distribution (Section 5.1), with mean and standard deviation of μ=3.53 and σ =0.96 (both logarithmic), respectively (selected by minimizing the sum of squares error). The mean error in the predicted number of deaths is –17%.[3] Both $LR(HI)$ and $LR(NHI)$ are U-shaped in V_{blast} – presumably because as V_{blast} increases, persons who would have only been injured at lower V_{blast} values are instead killed. So we model these loss ratios as piecewise exponential functions that asymptotically approach zero as V_{blast} increases without bound, as indicated in Table 3. The number of victims in each injury group (x) is then equal to $N \cdot LR(x)$, where N is the total number of building occupants.

---

[2] The blueprint for the AMIA building (obtained from D. Ambrosini – see Table 2) was also examined. In two cases (Khobar, AMIA), those victims whose locations (target building vs. other) were unknown were allocated in proportion to the number of victims known to be in each location. When a source notes only the total number of victims, their locations are assessed using the portions from either the "prime" source for the attack (Khobar), or the average of the portions from the "prime" sources for the other two bombings (AMIA). Finally, the dimensions of the Khobar Towers building were estimated based on drawings in the sources (presented to scale).

[3] The errors are: –40% for Khobar Towers (K-predicted=11; K-actual=18); +3% for Oklahoma City (K-predicted=167; K-actual=163); and –13% for the AMIA building (K-predicted=71; K-actual=82).

Table 2: Empirical Mortality and Morbidity Data for Vehicle Bomb Attacks ($n=3$)

	Description, Notation (Units, if applicable)		Khobar Towers[a]	Oklahoma City	AMIA Building
Building and Bomb Information	Bomb size, W (kg TNT)		10,270 (6)	1,810 (4)	270 (5)
	Standoff distance to building, R (m)		28 (8)	4.1 (8)	2.7 (3)
	Scaled distance, Z (m/kg$^{1/3}$)		1.3	0.34	0.42
	No. of occupants (*estimated*), N		138	369	118
	No. of floors (*above ground*), F		8 (4)	9 (10)	6 (5)
	Floor-to-floor height (*average*), H_F (m)		3.4 (1)	4.0 (3)	4.2 (1)
	Building length, L (m)		40-*near*, 10-*far*[a] (1)	63 (6)	17 (1)
	Building depth, D (m)		9-*near*, 28-*far*[a] (1)	24 (6)	43 (1)
	Blast flux, β (kg$^{1/3}$/m^2)		0.030	0.039	0.046
	Wall vulnerability coefficient, V_w		0.85[b]	0.74[b]	1.00[b]
	Structural vulnerability coefficient, V_s		0.10[b]	0.85[b]	0.93[c]
	Blast vulnerability, V_{blast} (kg$^{1/3}$/m^2) (Equation 5)		$(4.6 + 4.2) = 8.8$[a]	30.5	43.9
Mortality and Morbidity – Target Building	No. Killed, K		18 (4)	163 (3)	82 (2)
	No. Hospitalized Injuries, HI		16 (2)	50 (1)	9 (1)
	No. Non-Hospitalized Injuries, NHI		19 (4)	111 (3)	21 (2)
	Loss Ratios, $LR(x)$	$LR(K)$	0.13	0.44	0.69
		$LR(HI)$	0.11	0.14	0.075
		$LR(NHI)$	0.14	0.30	0.18
Mortality and Morbidity – Elsewhere	No. Killed, K		1 (4)	5 (3)	3 (2)
	No. Hospitalized Injuries, HI		47 (2)	33 (1)	32 (1)
	No. Non-Hospitalized Injuries, NHI		351 (4)	372 (3)	113 (2)

Note: values in parenthesis indicate the number of sources/estimates underlying the (average) value presented.
[a] Building was T-shaped, so V_{blast} was calculated separately for each segment, and the results were then summed.
[b] Personal communication, E. Hinman, President, Hinman Consulting Engineers, San Francisco, CA, 2014.
[c] Personal communication, D. Ambrosini, Professor of Structural Engineering, National University of Cuyo, Argentina, 2013.

The loss ratios for the target building as a function of V_{blast} are plotted in Figure 1a, along with the empirical data on which the models are based (data points aligned vertically by bombing event). The total number of persons in the target building who are affected (injured or killed) reaches a local maximum at V_{blast}=30.5 kg$^{1/3}$/m^2 (corresponding to Oklahoma City), then decreases somewhat, before asymptotically approaching one. The number of injuries (both hospitalized and non-hospitalized) also peaks at V_{blast}=30.5 kg$^{1/3}$/m^2, and then asymptotically approaches zero as V_{blast} increases without bound.

5.5. Blast Flux Limitations

The blast flux (Section 5.2) has various limitations. For example, some building areas may present greater risk because of the particular things (e.g., window glass) nearby. The damage in one area might also not be independent of the damages elsewhere (for instance, as the blast interacts with and imparts energy to the building, less residual energy is available to damage the remainder of the building). The blast flux is also based on only three attacks, collectively covering a somewhat narrow range of Z-values (0.34–1.30 m/kg$^{1/3}$). Some of these issues are partially addressed through the use of empirical data to fit the model parameters, although it remains unknown how truly applicable the model might be to the "next" VBIED attack. The primary value of the blast flux model is that it:
1) is a simple metric to assess the (aggregate) vulnerability of a building's occupants to blast;

2) can be applied to any building and any size of bomb; and
3) is consistent with the damage estimation methods used for earthquakes (a hazard area in which empirical data are far more abundant than for VBIED attacks).

5.6. Property Damage to Target Building

The damage to the target building itself is modeled using Wilton and Gabrielsen [30], who review a variety of U.S. government tests wherein dwellings (one and two story brick, wood, and concrete block construction) were exposed to high explosive and nuclear detonations. Although based on data for residential dwellings, and not commercial office buildings, their model considers both the damage to individual building components, and the replacement cost of each component, and expressed damage in dollar terms. Their data are well-modeled (adjusted $R^2=0.91$; $n=19$) by a log-linear function of the form

$$\ln(DC) = -3.0 + (3.6 \times 10^{-5}) \cdot (P) + (1.1 \times 10^{-4}) \cdot (I) \tag{6}$$

where the damage cost, DC, is specified as the portion of the building replacement cost ($0 < DC \leq 1$), P and I are in metric units (see below),[4] and all of the coefficients are highly significant ($p<0.001$). R.S. Means [31] lists average construction costs for U.S. office buildings (1–20 floors) of $1,600–$2,000/m^2, so we model the unit replacement cost (U_{RC}) as a lognormal distribution with non-logged mean of $1,800/m^2 and CoV of 20%. Finally, according to Willis & LaTourette [32], from an analysis of the RMS terrorism risk model, the value of the damage to a building's contents is around 60% of the value of the damage to the building itself, and the business interruption losses associated with the event are about 170% of the value of the damage to the building itself, so we inflate U_{RC} accordingly to account for these losses.

The peak overpressure and impulse are evaluated using DoD [33]. For the detonation of hemispherical TNT charges at ground level, the peak overpressure (P; Pascals) is well-modeled (adjusted $R^2>0.99$) by

$$\ln(P) = 14 + 0.28 \cdot (Z) - (3.3 \times 10^{-3}) \cdot (Z^2) - 2.3 \cdot \ln(Z) - 0.29 \cdot \left[\ln(Z)\right]^2 \tag{7}$$

and the impulse (I; Pascal·sec) is well-modeled (adjusted $R^2=0.97$) by

$$\ln\left(\frac{I}{W^{1/3}}\right) = 5.3 - 0.33 \cdot (Z) + (4.4 \times 10^{-3}) \cdot (Z^2) - 0.14 \cdot \ln(Z) + 0.29 \cdot \left[\ln(Z)\right]^2 \tag{8}$$

In each case, $n=200$ points were selected from each curve. Equations 7 and 8 may appear ad hoc, but all coefficients are highly significant ($p<0.01$; $p<0.001$ for all but one). From a review of peak overpressure curves from several sources, Baker et al. [19] assess the variation across them as a factor of +/−2, so we multiply P and I each by a lognormal distribution with (non-logged) mean of 1 and CoV of 50%.

5.7. Collateral Damages

The numbers of collateral deaths and injuries (that is, *outside* the target) are non-monotonic in W (Table 2) – perhaps in part because of the low sample size (Section 5.5). So the numbers of collateral deaths and injuries are modeled as lognormal distributions, using the (non-logged) means and standard deviations of the values in Table 2. The collateral property damage (PD_c; $) is estimated using data from Oklahoma City (Table 2). For that attack ($W=1810$ kg), which damaged a total of 348 buildings, the estimated cost to the state of Oklahoma was about $1.06 B in 2012$ [34] – including rebuilding costs, police officer overtime, and lost tax revenue. For lack of other data, PD_c is assumed to be linear in W, with a slope therefore equal to (5.9×10^5).

---

[4] Note that the source expresses damage costs in percent rather than fractional terms, and also use non-metric units.

Table 3: Empirical Blast Mortality and Morbidity Models for the Target Building ($n=3$)

Damage Quantity		Model Parameter		Notes
Loss ratio–killed, LR(K)		μ	3.5	Section 5.4
		σ	0.96	
Loss ratio–hospitalized injury, LR(HI)	$V_{blast} \leq 30.5$	a	0.10	
		b	(8.7×10^{-3})	
	$V_{blast} \geq 30.5$	a	0.52	
		b	−0.044	$LR(x) = a \cdot \exp(b \cdot V_{blast})$
Loss Ratio–non-hospitalized injury, LR(NHI)	$V_{blast} \leq 30.5$	a	0.10	
		b	0.035	
	$V_{blast} \geq 30.5$	a	0.98	
		b	−0.039	

Note: in all cases, the independent variable, V_{blast}, is specified using Equation 5.

5.8. Threat Scenario and Attack Probability Specifications

The minimum setback distance that a decision-making might consider is assumed to be 5 m, and the standoff distance in the no mitigation case, R_0 (m), is modeled as a uniform distribution from 1–5 m. With the bollards, the bomb is assumed to be detonated right at the line of barriers. The bomb size is specified using data from $n=103$ actual VBIED attacks (see Section 5.3 for search procedure description). With the Khobar Towers bomb removed (which was nearly five times the size of the second largest bomb size), W (kg) is modeled as an exponential distribution with mean (and standard deviation) of 340 kg ($\chi^2=14.8$; p=0.14). The mean error in the fitted CDF is −6.1% (90% c.i. −26% to 8%). So as to account for the fact that some of these estimates are not in terms of TNT equivalent mass (Section 3), we multiply W by a triangular distribution with mode of 1 and range of 0.5–1.8, based on the table of TNT equivalents in ASCE [20]. Furthermore, the bomb size is often estimated from the size of the bomb crater, and based on a study of craters from two hundred accidental explosions, Kinney & Graham [35] report a CoV of about one third for $W^{1/3}$ when estimated in this way, so we also multiply $W^{1/3}$ by a lognormal distribution with mean of 1 and standard deviation of 0.33 (both non-logged).

The implementation of the mitigation measures will presumably cause the attack probability to decrease. We account for this using a simple – but physically meaningful – approach. If θ (radians) is the angle between the line connecting the bomb and a point at some arbitrary height on the building's face (H_a) – an angle which necessarily decreases as R increases – the annual attack probability is specified as

$$p = (p_0 \cdot \theta) = p_0 \cdot \arctan\left(\frac{H_a}{R}\right) \tag{9}$$

where p_0 is what the annual attack probability would be in the case without the mitigation. From the boundary condition $p(R=R_0=3 \text{ m})=p_0$, it follows that $H_a=4.672$ (note that 3 m is the mean of R_0 – Section 5.8). However, the nature of the attack probably is highly uncertain, so we model it as a triangular distribution with mode from Equation 9 and range of zero (complete decay) to p_0 (no decay).

6. RISK AND COST-EFFECTIVENESS CALCULATIONS

The benefit of the mitigation measures is the net present value of the risk reduction, or

$$B_{NPV} = \sum_{n=1}^{life}\left[\frac{risk_0 - risk}{(1+dr)^n}\right] = \sum_{n=1}^{life}\left[\frac{(p_0 \cdot d_0)-(p \cdot d)}{(1+dr)^n}\right] \tag{10}$$

where *risk* is the product of the annual attack probability, *p*, and the (monetized) damages in the event of an attack, *d* ($; see below), *dr* is the annual discount rate, *life* is the building service life (we use 50 years), and a zero subscript denotes the base (no mitigation) case. As such, the model considers both the attack probability, and the damages in the event of an attack, and how both of these might change as mitigation is implemented. We use a 5% annual discount rate, but do not discount future deaths and injuries.

Injury severity is assessed using the Abbreviated Injury Scale (AIS) [36] (Table 4). For each AIS level, DOT [37] gives the quality-adjusted portion of remaining life lost, which is then multiplying by the value of a statistical life (VSL) (middle=$9.1 M; range=$5.4–$12.9 M) to obtain the monetary value of injury at each AIS level. These do not represent the value of any particular person's life or injury, but rather the willingness to pay to prevent one statistical death or injury of some level of severity. At each AIS level, we use a triangular distribution over the range of injury values in Table 4. We apply the geometric mean of AIS 1–2 to non-hospitalized injuries (on the basis that lower rather than higher AIS values are more likely to occur), the geometric mean of AIS 3–5 to hospitalized injuries, and AIS 6 to fatalities.

We evaluate the mitigation measures using various metrics: the breakeven attack probability, or the value of p_0 (the attack probability in the no mitigation case) at which $B_{NPV}=C_{NPV}$; the cost per statistical life saved, assuming an attack occurs at some point during the building's service lifetime; and the *deterrence value* and *mitigation measure effectiveness*, or the (absolute value of the) percentage change (decrease) in the attack probability (*p* – using Equation 9) and the (monetized) attack damages (*d*), respectively.

7. RESULTS

The cost of the mitigation is considerable. Even with only 5 m of setback distance, the average cost is $1.5 M (90% c.i. $0.3 M–$5.8 M); at *R*=25 m, the average cost is $8.9 M (90% c.i. $0.6 M–$38 M); and for *R*=50 m, the average cost is $23 M (90% c.i. $1.2 M–$100 M). At all setback distances, the majority of the total cost is land, and not the bollards themselves (59% of the total at *R*=5 m, steadily rising to 86% of total costs at *R*=50 m). The attack damages are summarized in Table 5. For the target, while all of the damages generally decrease as *R* increases, the drop in the number of deaths is the most marked. Table 5 also indicates that most of the damage reduction occurs within the first 25 m of setback distance.

Table 4: Monetary Values of Injury (2012$) for the Abbreviated Injury Scale (AIS) [36]

Severity Level		AIS 1	AIS 2	AIS 3	AIS 4	AIS 5	AIS 6 (fatal)
		Minor	Moderate	Serious	Severe	Critical	Maximum
Example Injuries[a]		Abrasion; laceration; contusion	Simple broken bone; serious strain/sprain	Compli-cated fracture; concussion	Heart laceration; loss of limb	Massive head injury	Usually (though not invariably) fatal
General Prognosis[a]		Treated and released	Follow-up and weeks to months to heal	Substantial follow-up; some minor disability	Hospital-ization; moderate long-term disability	Extended hospital-ization; significant long-term disability	
DOT[37] Injury Value	Low	$16,000	$240,000	$550,000	$1.4 M	$3.1 M	$5.2 M
	Middle	$27,000	$430,000	$960,000	$2.4 M	$5.4 M	$9.1 M
	High	$39,000	$610,000	$1.4 M	$3.4 M	$7.6 M	$12.9 M

[a] Source: Willis & LaTourrette [32].

Table 5: Predicted Mortality, Morbidity, and Property Damage in the Event of an Attack

	Target Building				Collateral
	Base case	R=5 m	R=25 m	R=50 m	(all R-values)
Killed	127 (1–493)	51 (0–202)	2 (0–8)	0 (0–1)	3 (1–6)
Hospitalized injuries	73 (5–269)	73 (7–261)	70 (7–250)	69 (6–248)	37 (26–53)
Non-hospitalized injuries	106 (9–371)	91 (9–310)	72 (7–257)	70 (70–252)	274 (116–529)
Property Damage	$150 M ($16 M–$540 M)	$150 M ($16 M–$540 M)	$94 M ($5 M–$390 M)	$52 M ($3 M–$210 M)	$270 M ($8 M–$1 B)

Notes: values in parenthesis are the 90% c.i.; property damage includes building contents and business interruption

A plot of the breakeven attack probability (that is, the *minimum* attack probability for the mitigation to be cost-justified) is given in Figure 1b. For all setback distances, the mean breakeven annual attack probability is on the order of 10^{-4}, which corresponds to a return period of around once every 10,000 years. The 5th percentile of the breakeven annual attack probability is on the order of 10^{-6} for R=5–35 m, and 10^{-5} for R=40–50 m, while the 95th percentile swings as high as 10^{-3} for R>20 m.

Figure 1c plots the cost-effectiveness of the mitigation ($/life-saved), assuming that the building is attacked at some point during its service lifetime (that is, an annual attack probability of 0.02 or greater). Also plotted in Figure 1c are dashed lines corresponding to the minimum, average, and maximum VSLs (see Section 6). The curve for the mean value of cost-effectiveness is below the mean VSL ($9.1 M) at all setback distances (it also decreases somewhat from R=5 m to R=10 m), and it only begins to exceed the minimum VSL ($5.2 M) between R=45m and R=50m. The 95th percentile curve begins to exceed the minimum VSL ($5.2 M) between R=25 and R=30, but does not exceed the mean VSL ($9.1 M) until R=40–45 m, and it never reaches the maximum VSL (the 5th percentile curve is essentially flush with the R-axis). These results suggest that this mitigation option is cost-effective for buildings that will be attacked during their lifetime.

Figure 1d gives the deterrence-effectiveness (D–E) plot (average values). The red dot at the origin represents the risk in the base case (that is, without the mitigation); points on the curve closer to the origin correspond to less setback distance; and the dashed lines along the top and right represent a state of zero risk. The D–E curve's position above the dashed diagonal line (that is, the line where D=E) suggests that most of the risk reduction of the mitigation derives from its deterrence value (that is, the reduced attack probability), rather than its effectiveness (that is, the reduction in damages in the event of an attack).

8. CONCLUSION

This paper presents a quantitative risk-cost model for using vehicle barriers to create setback distance around a building for protection from vehicle bomb attack. While highly empirical, and based on a relatively small number of bombing events, the model considers both the attack probability and the damages in the event of an attack, and how both of these quantities might change as mitigation is implemented to a greater degree. Results suggest that this mitigation option is worthwhile, but only in the case of buildings that would be at especially high probability of attack without the mitigation.

Acknowledgements

This work was supported by the John D. and Catherine T. MacArthur Foundation through grant #05-85373-0, the Department of Engineering and Public Policy at Carnegie Mellon University, the U.S. Department of Education Graduate Assistance in Areas of National Need (GAANN) program through

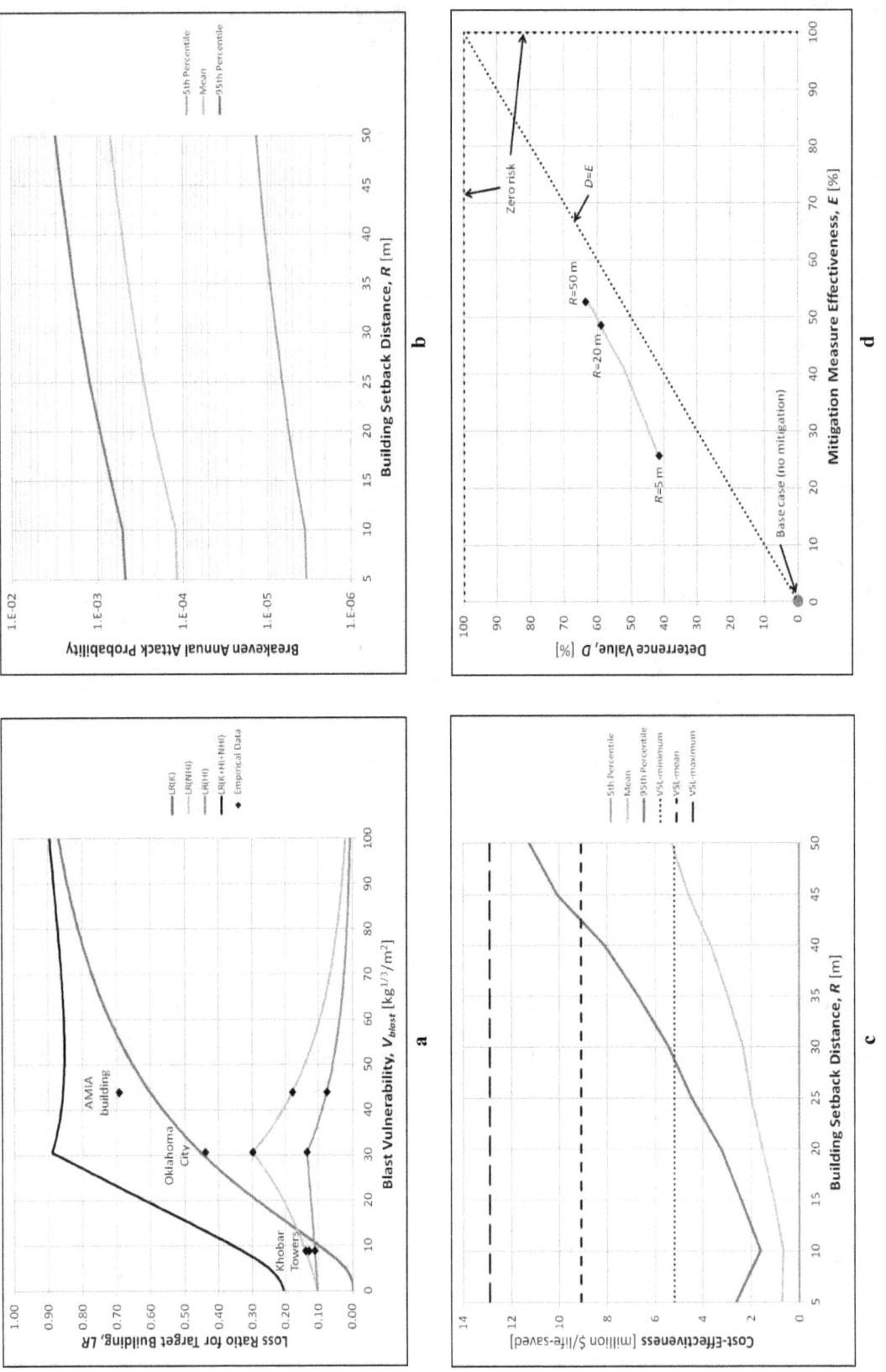

Figure 1: a) Estimated Loss Ratios for the Target Building Versus Blast Vulnerability (V_{blast}); b) Breakeven Attack Probability Versus Amount of Setback Distance; c) Cost Effectiveness Versus Amount of Setback Distance; d) Deterrence-Effectiveness Plot.

Probabilistic Safety Assessment and Management PSAM 12, June 2014, Honolulu, Hawaii grant #P200A090055, and the National Center for Risk and Economic Analysis of Terrorism Events at the University of Southern California. The author also thanks Daniel Ambrosini, Keith Florig, Eve Hinman, Matthew Kocoloski, Granger Morgan, Mitch Small, and Adam Rose for their assistance with various aspects of the study design, data, and analysis. However, all findings are those solely of the author, and do not necessarily represent those of the U.S. Department of Homeland Security or any of these individuals or organizations.

References

[1] M. Davis, *Buda's Wagon: A Brief History of the Car Bomb*, Verso, 2007, London, UK.

[2] U.S. Federal Emergency Management Agency and American Society of Civil Engineers, *The Oklahoma City Bombing: Improving Building Performance Through Multi-Hazard Mitigation*, 1996, Washington, DC.

[3] E.J. Conrath, T. Krauthammer, K.A. Marchand, P.F. Mlakar, eds, *Structural Design for Physical Security: State of the Practice,* American Society of Civil Engineers, 1999, Reston, VA.

[4] S.A. King, H.R. Adib, J. Drobny, and J. Buchanan, "Earthquake and Terrorism Risk Assessment: Similarities and Differences," in: J.E. Beavers, ed., *Advancing Mitigation Technologies and Disaster Response for Lifeline Systems*, American Society of Civil Engineers, 2003, Reston, VA, pp.789–798.

[5] U.S. Department of Defense, *DoD Minimum Antiterrorism Standards for Buildings*, UFC 4-010-01, 2012, Washington, DC.

[6] T. Krauthammer, *Modern Protective Structures*, CRC Press, 2008, Boca Raton, FL.

[7] U.S. Federal Emergency Management Agency, *Handbook for Rapid Visual Screening of Buildings to Evaluate Terrorism Risks,* FEMA 455, 2009, Washington, DC.

[8] U.S. Federal Emergency Management Agency, *Reference Manual to Mitigate Potential Terrorist Attacks Against Buildings*, FEMA 426, 2nd ed., 2011, Washington, DC.

[9] N.T. Heatwole, *Protecting buildings from vehicle bomb attacks: Towards more risk-based performance standards and a comprehensive strategy*, PhD dissertation, Carnegie Mellon University, 2011, Pittsburgh, PA.

[10] P.E. Gurvin, and A.C. Remson, "U.S. Embassy Designs Against Terrorism: A Historical Perspective," Proceedings of the ASCE Structures Congress, 2001, Washington, DC.

[11] M.G. Stewart, M.D. Netherton, and D.V. Rosowsky, "Terrorism Risks and Blast Damage to Built Infrastructure," *Natural Hazards Review*, 7(3), pp.114–122 (2006).

[12] J.E. Crawford, C. Hyung-Jin, and J.M. Magallanes, "Incorporating Uncertainties in the Blast-Resistant Design," Proceedings of the ASCE Structures Congress, 2007, Long Beach, CA.

[13] M.G. Stewart, "Cost Effectiveness of Risk Mitigation Strategies for Protection of Buildings Against Terrorist Attack," *J Performance of Constructed Facilities*, 22(2), pp.115–120 (2008).

[14] S. Foo, B. von Rosen, and E. Contestabile, "Risk assessment of buildings against blast effects," *Canadian J Civil Engineering*, 36(8), pp. 1285–1291 (2009).

[15] M.D. Netherton and M.G. Stewart, "Probabilistic modeling of safety and damage blast risks for window glazing," *Canadian J Civil Engineering*, 36(8), pp.1321–1331 (2009).

[16] R.G. Little, "Cost-effective strategies to address urban terrorism: a risk management approach," in: H.W. Richardson, P. Gordon, and J.E. Moore, eds, *The Economic Costs and Consequences of Terrorism*, Edward Elgar, 2007, Cheltenham, UK, pp.98–115.

[17] J.T. Baylot, and D.D. Rickman, "Uncertainties in Blast Loads on Structures," Proceedings of the ASCE Structures Congress, 2007, Long Beach, CA.

[18] U.S. Government Accountability Office, *Greater Attention to Key Practices Would Help Address Security Vulnerabilities at Federal Buildings*, GAO-10-236T, 2009, Washington, DC.

[19] W.E. Baker, P.A. Cox, P.S. Westine, J.J. Kulesz, and R.A. Strehlow, *Explosion Hazards and Evaluation*, Elsevier, 1983, Amsterdam, Netherlands.

[20] American Society of Civil Engineers, *Blast Protection of Buildings*, ASCE/SEI 59-11, 2011, Reston, VA.

[21] K.E. Case, "The value of land in the United States: 1975–2005," in: G.K. Ingram and Y-H Hong, eds, *Land Policies and their Outcomes*, Lincoln Institute of Land Policy, 2007, Cambridge, MA, pp. 127–146.

[22] U.S. Veterans Administration, *Cost Estimates for Physical Security Enhancements*, 2007, Washington, DC.

[23] A. Persily and J. Gorfain, *Analysis of Ventilation Data from the U.S. Environmental Protection Agency Building Assessment Survey and Evaluation (BASE) Study*, NISTIR 7145–revised, U.S. National Institute of Standards and Technology, 2008, Washington, DC.

[24] K. Jaiswal, D.J. Wald, and M. Hearne, *Estimating Casualties for Large Earthquakes Worldwide Using an Empirical Approach*, U.S. Geological Survey, Open-File Report 2009-1136, 2009, Reston, VA.

[25] K.S. Jaiswal, and D.J. Wald, *Rapid Estimation of the Economic Consequences of Global Earthquakes*, U.S. Geological Survey, Open-File Report 2011–1116, 2011, Reston, VA.

[26] *LexisNexis® Academic* database, http://www.lexisnexis.com/, 2013.

[27] S. Mallonee, S. Shariat, G. Stennies, R. Waxweiler, D. Hogan, and J. Jordan, "Physical Injuries and Fatalities Resulting From the Oklahoma City Bombing," *J of the American Medical Association*, 276(5), pp.382–387 (1996).

[28] C.A. Biancolini, C.G. Del Bosco, and M.A. Jorge, "Argentine Jewish Community Institution Bomb Explosion," *J Trauma*, 47(4), pp.728–732 (1999).

[29] D. Thompson, S. Brown, S. Mallonee, and D. Sunshine, "Fatal and Non-fatal Injuries among U.S. Air Force Personnel Resulting from the Terrorist Bombing of the Khobar Towers," *J Trauma*, 57, pp.208–215 (2004).

[30] C. Wilton, and B.L. Gabrielsen. *House Damage Assessment*. Report to Defense Nuclear Agency, URS Research, 1972, San Mateo, CA.

[31] R.S. Means Co., *Square Foot Cost Estimates*, http://www.reedconstructiondata.com/rsmeans/models/, 2014.

[32] H.H. Willis and T. LaTourrette, "Using Probabilistic Terrorism Risk Modeling for Regulatory Benefit-Cost Analysis: Application to the Western Hemisphere Travel Initiative in the Land Environment," *Risk Analysis*, 28(2), pp.325–339 (2008).

[33] U.S. Department of Defense, *Structures to Resist to Effects of Accidental Explosions*, UFC 3-340-02, 2008, Washington, DC.

[34] D. Terry, "Ceremonies Mark Passage of a Month Since Bomb," *New York Times*, May 21, 1995, p.16.

[35] G.F. Kinney and K.J. Graham, *Explosive Shocks in Air*, 2nd ed., Springer, 1985, New York, NY.

[36] Association for the Advancement of Automotive Medicine, *The Abbreviated Injury Scale 2005, Update 2008*, T.A. Gennarelli and E. Wodzin, eds., 2008, Barrington, IL.

[37] U.S. Department of Transportation, *Guidance on Treatment of the Economic Value of a Statistical Life in U.S. Department of Transportation Analyses*, 2012, Washington, DC.

From prescriptive arrival times to performance based fire service delivery – Parallels of Fire Service Planning and Fire Engineering

Adrian Ridder[*], Uli Barth

University of Wuppertal, Wuppertal, Germany

Abstract: The fire safety design process of buildings underwent a substantial shift in the last roughly two decades, switching from prescriptive building codes to performance-based, fire-engineered designs. A similar process can be observed with Strategic Fire Service Planning which defines "how much fire service" is necessary per municipality. The methods used there become more and more sophisticated as well. However, with increasing complexity it becomes harder to explain and interpret results to the decision-makers, which applies both to fire engineering and fire service planning. The need for further research is made clear as the major outcome of this paper.

Keywords: fire service, risk management, fire engineering, methodology

1. INTRODUCTION

For the decision-making process on how many resources, stations and personnel are necessary for the fire service of a local authority (town, city, county), several constrictions apply, the most important ones being limited available public funding and the statutory requirement to achieve a "deemed-to-satisfy" level of public safety (which usually is not quantitatively defined and therefore measurement of fulfillment is hard to achieve). A similar problem in quantifying the necessary level of safety exists in the design of buildings. This paper outlines some parallel developments in both fields, elaborates on similar problems in quantifying safety and delivering results that can both be validated and understood by the respective authority having jurisdiction.

2. FIRE ENGINEERING

In earlier times (and before the formal introduction of Fire Engineering in the second half of the 20th century), fire safety of buildings was based on prescriptive building-codes which contained the wisdom accumulated through lessons learned from fires over several decades or centuries in the respective country. Those concepts were hazard-based and prescribed in detail e.g. what building materials could be used, how far buildings must be separated and how thick a wall had to be. The cost of those very detailed, easy to apply rules in form of accepted solutions (called "deemed-to-satisfy" the objectives of the building codes) was limited flexibility and hampered progress in building design features.

With the emergence of Fire Engineering as respected science discipline and study course that changed. Deviations from prescriptive codes became allowed when the safety of a building could be assured otherwise. Building codes, e.g. the 2014 New Zealand Building Code [1], have been altered to include detailed quantitative "performance requirements". These can be achieved on numerous ways by architects and engineers and therefore allow for far more innovative designs than prescribed ready-made solutions where there remains not really any choice for architects and engineers except the one of implementing it or not.

To achieve the necessary safety level, the methods of Fire Engineering have been developed. In general they use acceptance criteria parameters to quantify general objectives like the protection of building occupants or easy fire service intervention. Through calculations and fire modelling the actual building in question is thoroughly analyzed as an individual case and the review shows if the building design can achieve parameters that are at least as safe as or safer than the performance requirements.

---

[*]Corresponding author: ridder@uni-wuppertal.de

This so-called performance-based concept is therefore risk-based, allows for more flexibility and at the same time – when carried out properly – ensures a high level of safety. Probably the findings of this process are more accurate than prescriptive solutions as the design is reviewed on an individual case basis and therefore "tailor-made", instead of generic prescriptive solutions which contain the risk of being not wholly applicable to the specific building. In accordance to that, it has been found that prescriptive solutions must not always be "safer" than performance-based ones. This discussion comes down to the fact that the level of safety of prescriptive requirements is not explicitly stated but it is rather assumed that by applying those measures, a design should be safe enough. That approach has lately been criticized as the level of safety of prescriptive requirements depends more on the building specifics than performance based designs do [2].

Methodologies and guidelines for the performance-based approach and relevant criteria parameters have been internationally developed, e.g. in the International Fire Engineering Guidelines IFEG [3], BS 7974 ([4] and the vfdb Fire Engineering Guideline [5]. Some of those documents also contain probabilistic risk assessment methodologies , e.g. [6].

3. STRATEGIC FIRE SERVICE PLANNING

Strategic Fire Service Planning has always existed where an organized approach existed to counteract the elementary hazard of fire. However, only very recently that area has been established as field of research. The resulting methodologies and algorithms strive to change fire service planning from experience and gut-feeling based towards a more scientific, fact-based and quantifiable approach (more often than not resulting from increasing pressures to justify fire service operations and the relating costs in the eye of the wider public and the politicians distributing funding).

3.1. Historical developments

In the past, many countries worldwide (e.g. Sweden, the Netherlands, UK, New Zealand, USA, Germany) chose prescriptive fire service arrival times as a substitute benchmark to answer the question if adequate safety has been achieved. That is, regulations or laws defined that the fire service had to arrive at the incident scene in a certain amount of time, usually somewhere between 8 and 15 minutes after the emergency call has been received by the dispatch center. Those times were universal or at best split in categories like "within city limits" and "outside city limits", therefore allowing for only small deviations [7], [8], [9], [10], [11].

In literature, that response or arrival time is seen as the dominating characteristic for quantification purposes of fire service delivery [12]. Though several attempts have been made, no validated correlation between a fire growth or fire toxicity curve to the specification of fire service arrival times can be found in international literature [7]. Existing time goals are therefore to be considered as expressions of the political will and the financial clout of the municipality at hand. However, it is obvious that shorter response times are favourable cp.[13].

The importance of the arrival time has to be put in context also with the question how much resources are put into action after that time, how well equipped and how competent they are in what they are required to do (cp. [13], [12], [14].

With the advent of increasing computation power, computer models have been used to calculate optimum locations for stations based on the existing workload (emergency calls). Also the other way around has been used, picking existing stations and calculating how far a fire truck can travel in a given time, with varying degree of detail (the type of road, quality of the road, traffic conditions, daytime, season etc.).

Today, the arrival time approach is still widely used in European countries and all over the world. The obvious advantage is the fact that it can easily be defined and controlled. But it comes with the disadvantage that it is not possible to derive what kind of equipment and trucks should be allocated to different stations. In that generic form it also fails to allow for more detailed responses, e.g. faster fire service response in areas with a higher risk (however, for example in the Netherlands approaches like that with varying response times in accordance to the type of object and time of day are already used).

In the EU, seven countries use such an approach, whereas seven others use a scenario-driven approach and nine countries use a risk-based approach (not all EU member countries have a written methodology for fire service planning)[15]. The so-called risk-based approach is a relevant recent development and started only in the early 2000s. It is hoped to include the probability of a fire incident in those methods and therefore allow for a more detailed allocation of resources (instead of spreading the fire service evenly over a city, some areas with less associated risk could get fewer stations or slower responses while high-risk areas would receive a faster response with probably more resources as well).

3.2. Current developments

The risk-based approach is to be considered the latest development in fire service planning. It is hoped that more sophisticated risk analysis methods and analysis of risk-related data will in the future allow for "tailor-made" allocation of fire service resources, without over-protecting (and therefore over-paying) but still guaranteeing an adequate safety level.

Methodologies used today are based on assumptions made based on the type of buildings in a given area (height, occupancy, construction type etc.), the number of population in the area (population density), the demand to the fire service measured in fire calls and rescues/injuries/deaths or a combination thereof [16], [17], [18], [19]).
However, more research needs to be done to create a more empiric foundation of those methods instead of pure assumptions and operational experiences. Newer research has shown a strong connection between socio-demographic factors and fire risk. A strong association between deprivation and other social factors with the occurrence of fire has been found [20].

Increasingly, different data sources from both inside and outside a fire service are used in conjunction with Geographical-Information-Systems (GIS), which allow for a convenient depiction of a large number of different datasets on a map. Data sources inside the fire service may be incident action reports containing the equipment used, the time spent at one incident, the number of injured persons etc.; data on availability of fire trucks due to maintenance and workshop-stays; availability of personnel (depending on percentage of sick leave, training, absolute number of personnel etc.); fire investigation reports and a lot more. Outside data sources may be the ambulance service (where separate from the fire service) contributing numbers of injuries of different severity and fatalities, Public Housing and Statistics departments contributing data on population distribution, type, age and occupation/zoning of buildings, Traffic departments delivering data on traffic density, blockages and road network changes, and many more. First works have been done on merging those different data sources in a uniform data warehouse per fire service [21] which facilitates the analysis enormously.
The move towards a more performance-based approach to fire service planning brings about and coincides with a similar movement to introduce performance measurement and performance management techniques and methodologies in the fire services [22], something private industry has done many years ago. Both need more and more sophisticated data collection pools and analysis tools to allow for the more detailed outcomes which are increasingly expected from authorities having jurisdiction. Therefore it has already been found that a risk based approach offered significant potential for the application of performance measurement in the fire services [20].

4. SIMILARITIES AND INTERDEPENDENCIES BETWEEN BOTH DISCIPLINES

Both Fire Engineering and Fire service planning share one common important question to answer: How safe is safe enough? And directly correlated to it is the more profane question, how much money must be spend to achieve an adequate safety level? Both questions have been widely discussed and elaborated on in the risk and safety literature [23], [24]. A more practical way to put that question is rather than considering absolute safety levels it should be considered what value is placed on a specified change in survival probability [24]. A holistic answer to that fundamental question of safety science has not been found yet, but it is important to communicate shortcomings in the risk assessment and risk management process to the relevant stakeholders, which mainly are the developer, owner and user of buildings in the case of Fire Engineering and the citizens and politicians in the case of Fire Service Planning. Uncertainties must be made public as to not create a false feeling of safety.

Both areas of research are directly linked when it comes to the question of how long a building's structural integrity must be maintained in order to allow for efficient fire service operations, an objective found in many building codes around the world. One approach for that is incorporated in the IFEG in form of a time-line approach for the fire service intervention based on the Australian Fire Brigade Intervention Model (FBIM, [25]). That shows how deeply interconnected both disciplines are: Building designers have to know how capable and fast the local fire service can deploy so that they can correctly calculate the buildings resistance against fire. Similar time-lines have been developed all over the world, e.g. [26]. In Germany a similar strong connection between fire brigade intervention and design of the building is only applied to industrial buildings, where the minimal fire resistance time determines the mandatory response time of the fire service [27].

Both disciplines also share common problems and shortcomings, which are discussed below.

4.1 Methodological Voids and Deficits

In both areas, the scientific work on the relevant topics is quite young, while Fire Engineering has at least some 50 years under its belt, Strategic Fire Service Planning in a methodological and scientific manner is even younger. Resulting from this fact, many assumptions are still used in state-of-the-art models, as simply the complex systems of compartment fires and fire service operations have not been fully understood yet. Additional fundamental research is necessary to understand the different parts and components of both systems, let alone the complex interactions between different subsystems (for Fire Engineering e.g. pyrolysis of wood and plastics, heat fluxes radiating back on surfaces, flashover-behavior in complex room geometries with various fuel-loads, toxicity of gases like HCN, CO, H_2S and their solitary and combination effects under varying ambient conditions; for Fire Service Planning e.g. the influence of competencies and training on outcomes of operations and the quantification of it, the efficiency and effectiveness of varying crew sizes and equipment as well as that of different firefighting agents (water, Class-A foam, compressed air foam etc.) and the influence of risk factors like building and population characteristics and many more).

Prescriptive arrival times did not consider local risk distributions, specific risk resulting from particular industrial complexes, residential buildings or transportation infrastructures. Therefore only rough estimations on the service delivery of a fire service could be calculated. The same can be said about prescriptive building-codes, which could only be used for off-the-shelf buildings and therefore hampered the development of more sophisticated designs. Hence both concepts became to rely on the risk concept.

But dealing with risks in both disciplines carries specific difficulties, too. Especially the probabilistic handling of fire service incidents in general and of structural fires as topic of fire engineering in particular highlight a methodological conflict associated with the risk concept. The important events of interest are of a low probability and high impact nature. Therefore it is difficult to not misrepresent those seldom but severe events as with simple mathematical multiplications those events are somewhat "adumbrated" by the far higher number of incidents with less or no damage at all (e.g. false automatic fire alarms). That makes those risks somewhat hard to statistically analyze and risk aversion factors and similar concepts must be used to counter that. The distinction between the probability of high risks and the pure frequency of all kinds of incidents also makes it hard to explain to laymen like citizens and politicians why some sort of "overhead" with the fire service resources has to be maintained in order to be able to respond to the larger, but more seldom events, although the daily call load could more often than not easily be handled with a fraction of fire service resources. Probabilistic models must therefore account for this specific fire service related aspects. Taking this approach one step further, fire services as the usually only "all-round" first responder agency in a town have to be able to deal with not only rare incidents but also totally unexpected ones. Therefore the quite new concepts of black swans [28], [29] and resilience-based contingency [30], [31] should be considered as well as.

Assuming a risk assessment could be executed, in both areas it is very tricky – and nonetheless important - to set the threshold for "acceptable" risks. Performance criteria contained in building codes can be a solution for this in terms of Fire Engineering. For Fire Service Planning suitable indicators and measuring parameters for acceptable risks have to be found. The author proposes the number of incidents where the loss could be stopped after arrival of the fire service as indicator for the protection of property and the percentage of rescued persons out of the total number of affected persons the fire service had a chance to rescue as performance indicator for life safety.

4.2 Validation & Verification and Credibility to Reviewers

Fire Engineering heavily relies on computational fluid dynamics, for which several software packages exist. The parameters and models used in those simulations must be validated and verified in order to prove that they actually represent realistic scenarios and can predict the probable course of events in case of a fire. It has been shown that actually the current state of development of the software packages in combination with the varying skills of the users (fire engineers, consultants, and technicians) does not allow for accurate a priori predictions of fire growth in realistic complex scenarios and contain a large margin of error [32].

Similarly to the CFD software used in Fire Engineering, increasingly GIS-based software systems and specific software packages receive attention by planning departments of fire services. However, looking at literature and current best practice one can easily get the impression that too much emphasis and trust is placed on the outcomes of those software packages, neglecting the application boundaries and working assumptions used in those models. One has to bear in mind that software packages can only be as good as the underlying models which should be based on empiric evidence rather on assumptions and "operational experience" only. However, operational experience comes in handy to cross-check the plausibility of simulation outcomes.

In terms of validation and verification, similar problems exist in both fields: All too often the authority having jurisdiction to decide on proposed fire service planning and fire engineering concepts lack the knowledge, manpower and capabilities to scrutinize the proposed solutions to the degree necessary. The increasing complexity of the used models drives that trend, as more and more specialists are also necessary to check the proposals. Especially with a fire service structure as in Germany, where roughly 11.000 independent municipalities exists which all have at least one independent fire department, one can imagine that is impossible to amass the necessary expertise at all of those municipalities and fire services. Besides more streamlined and more easily understandable models, probably also organizational changes towards larger fire services (as practiced currently in the UK, Australia and New Zealand for example) offer some advantages related to the economy of scale concept.

In addition, practical guidelines should be developed for both areas which authorities having jurisdiction could use for assessing the completeness, quality, and scientific validity of the models, parameters and assumptions used. For both areas those guidelines should be as uniform and transparent as possible in order to allow for comparisons and to reduce the additional workload for consultants and other involved stakeholders working for and with different authorities having jurisdiction. One best practice example for Fire Engineering is the Acceptable Solutions and Verification Methods of the NZ Building Code [1].

In both disciplines it is of paramount importance that the relevant outcomes of design and planning processes are not obstructed by a vast array of formulae, numbers, different models and codes etc.

Rather every consultant, engineer and planner should be able to explain his work in a way that is easily understandable by the relevant stakeholders and the authorities having jurisdiction.

5. CONCLUSION

Fire Engineering and Strategic Fire Service Planning face similar problems as identified in the literature review. Those problems are under scrutiny in a current work-in-progress project on developing a risk-based fire service planning methodology for Germany. That project called "TIBRO" (German acronym for tactical-strategic innovative fire service and risk-based optimizations) was started in February 2012 and is scheduled for its final report in the first quarter of 2015. The aim of the project is to outline the fundamentals of a comprehensive methodology to derive necessary fire service resources in accordance with the legal tasks to be fulfilled by the fire service.

It has been pointed out that more sophisticated models don't necessarily bring more clarity and accuracy in predicting the behavior of a building in case of fire or the operational performance of a fire service in the future. All current predictions and simulations are to be taken with a pinch of salt and should be compared to existing best practices and operational experiences where existent. More research in both areas is paramount to propel both young scientific disciplines to further levels of detail and validity.

Acknowledgements

The TIBRO research project on which this report is based has been funded from the German Federal Ministry of Education and Research (grant ID 13N12174). The authors are responsible for the content of this paper.

References

[1]   Ministry of Business, Innovation & Employment, New Zealand Building Code Handbook, Wellington/NZ, 14.2.14.

[2]   G. de Sanctis, Assessing the Level of Safety for Performance Based and Prescriptive Structural Fire Design of Steel Structures. 11th International Symposium of Fire Safety Science, 21.2.14, Christchurch, New Zealand, 21.2.14.

[3]   National Research Council of Canada (NRC), International Code Council (ICC), United States of America, Department of Building and Housing, New Zealand (DBH) and the Australian Building Codes Board (ABCB), International Fire Engineering Guidelines, 2005th ed., Australian Building Codes Board, Canberra, ACT, 2005.

[4]   British Standards, Application of fire safety engineering principles to the desgin of buildings. Code of practice, 2001.

[5]   Vereinigung zur Förderung des deutschen Brandschutzes e.V., Leitfaden Ingenieurmethoden des Brandschutzes. vfdb TB 04-01, Altenberge, 2013.

[6]   British Standards, Application of fire safety engineering principles to the design of buildings - Part 7: Probabilistic risk assessment, 2003 13.220.20; 91.120.

[7]   W.J. van Dijk, RemBrand Fase 1. Niet harder rijden, maar voorkomen en slimmer bestrijden - quick research scan naar een model voor de operationelen maatschappelijke prestaties gericht op brandveiligheid. TNO-rapport RA053.02994, TNO, Den Haag, 2013.

[8]   National Fire Protection Association, Standard for the Organization and Deployment of Fire Suppression Operations, Emergency Medical Operations, and Special Operations to the Public by Career Fire Departments, 2010 Edition, NFPA, Quincy/MA (USA), 2010.

[9] National Fire Protection Association, Standard for the Organization and Deployment of Fire Suppression Operations, Emergency Medical Operations, and Special Operations to the Public by Volunteer Fire Departments, 2010 Edition, NFPA, Quincy/MA (USA), 2010.

[10] S. Svensson, A shift in focus. Perspectives on Fire Cover. Second International Conference on Fire Servcie Deployment Analysis, 12 March 2002, Indianapolis, Indiana/USA, 2002.

[11] C. Reynolds, J. Pedroza, Fire cover modelling for brigades, Home Office - Fire Research and Development Group, London, 1998.

[12] S. Svensson, Quantifying efficiency in fire fighting operations, in: Interflam 10. Proceedings: 12th International Fire Science & Engineering Conference, London, 2010.

[13] H. Jaldell, Essays on the performance of fire and rescue services. Dissertation, Kompendiet, Göteborg, 2002.

[14] R. Ahlbrandt, Efficiency in the provision of fire services, Public Choice 16 (1973) 1–15. http://link.springer.com/content/pdf/10.1007%2FBF01718802.

[15] M. Weber, Brandschutzbedarfsplanung im europäischen Vergleich. Facharbeit in der Ausbildung zum höheren feuerwehrtechnischen Dienst, Lehmen, 2012.

[16] J. van der Schaaf, J. Jeulink, Handleiding Brandweerzorg. Systeem voor de beoordeling van de gemeentelijke brandweerzorg, Ministerie van Binnenlandse Zaken, Directie Brandweer, Den Haag, 1992.

[17] H.-G. Faasch, Strukturuntersuchung des Einsatzdienstes der Feuerwehr Hamburg, BrandSchutz - Deutsche Feuerwehrzeitung (1972) 355–358.

[18] R. Grabski, Erarbeitung einer Risikoanalyse für die Ausrüstung sowie die Anzahl der zu besetzenden Funktionen einer Gemeindefeuerwehr. Instituts-Bericht Nr. 437, Institut der Feuerwehr Sachsen-Anhalt, Heyrothsberge, 2007.

[19] DCLG, Fire Service Emergency Cover Toolkit. Executive Summary. Fire Research Report 01/2008. Fire Research Series 01/2008, Department for Communities and Local Government, London, 2008.

[20] DCLG, Risk Based Performance Measurement in the Fire and Rescue Services. Final Report. Fire Research Series 10/2008, Department for Communities and Local Government.

[21] von der Lieth, David, Neue Methode zur Erhebung und Analyse steuerungsrelevanter Kennzahlen. BRENNPUNKT, 25.10.12, Wuppertal, 25.10.12.

[22] D. Hilgers, Performance Management, Betriebswirtschaftlicher Verlag Dr. Th. Gabler / GWV Fachverlage GmbH Wiesbaden, Wiesbaden, 2008.

[23] C. Starr, Social Benefit versus Technological Risk, Science 165 (1969) 1232–1238.

[24] P. Slovic, The perception of risk, Earthscan Publications, London, 2002.

[25] Australasian Fire Authorities Council AFAC, Fire Brigade Intervention Model. V 2.2, Melbourne, 2004.

[26] AGBF Bund, Empfehlungen der Arbeitsgemeinschaft der Leiter der Berufsfeuerwehren für Qualitätskriterien für die Bedarfsplanung von Feuerwehren in Städten, Arbeitsgemeinschaft der Leiter der Berufsfeuerwehren in der Bundesrepublik Deutschland, 1998.

[27] Deutsches Institut für Normung, Baulicher Brandschutz im Industriebau - Teil 1: Rechnerisch erforderliche Feuerwiderstandsdauer, Beuth Verlag, Berlin, 2010 13.220.99.

[28] N. Taleb, The black swan. The impact of the highly improbable, 2nd ed., Penguin, London, 2010.

[29] T. Bjerga, R. Flage, On black swans in relation to some common uncertainty classification system, in: Safety, Reliability and Risk Management: Beyond the Horizon. ESREL 2013, Taylor & Francis Group Ltd, London, 2013, pp. 3197–3202.

[30] J. van Trijp, A. Breur, First overview of the relationship between quantitative dynamic operational resilience and the Dutch Fire Services occupational safety and quality management program Cicero, in: Safety, Reliability and Risk Management: Beyond the Horizon. ESREL 2013, Taylor & Francis Group Ltd, London, 2013, pp. 1671–1676.

[31] O. Renn, P.-J. Schweizer, M. Dreyer, A. Klinke, Risiko. Über den gesellschaftlichen Umgang mit Unsicherheit, Oekom-Verlag, München, 2007.

[32] G. Rein, J.L. Torero, W. Jahn, J. Stern-Gottfried, N.L. Ryder, S. Desanghere, M. Lázaro, F. Mowrer, A. Coles, D. Joyeux, D. Alvear, J.A. Capote, A. Jowsey, C. Abecassis-Empis, P. Reszka, Round-robin study of a priori modelling predictions of the Dalmarnock Fire Test One, Fire Safety Journal 44 (2009) 590–602.

Issues in Incorporating Probabilistic Safety Assessment (PSA) in the Design and Licensing Stages of Generation IV Reactors

Ibrahim A. Alrammah

ibrahim.alrammah@postgrad.manchester.ac.uk

School of Mechanical, Aerospace and Civil Engineering (MACE), University of Manchester, Manchester, United Kingdom

ABSTRACT

Probabilistic approaches has been used and are also highly recommended to be used from the very early stage of the reactor design process. So far, Probabilistic Safety Assessment (PSA) approach is increasingly being utilized in the demonstration of safety in combination with deterministic approaches (e.g. to justify the classification of situations, to determine the sequences of sophisticated failures) and used also to verify the systems and components reliability in order to satisfy safety targets. However, epistemic problems such as uncertainties due to lack of design information, unknown phenomena, plant-specific hazards, data, etc., are larger than that from existing reactors, and will impose a significant challenge to the decision makers. This paper will discuss some technical issues related to applying PSA in the design and licensing stages of Generation IV reactors. These aspects include: initiating events, passive systems modeling, reliability data, common cause failure (CCF), modeling of novel design features, modeling of preventive maintenance, technical specifications, human reliability analysis (HRA), systems interdependencies, modeling of instrumentation and control (I&C), external hazards, continuous design risk monitoring, supporting studies, interpretation of PSA results for new plants.

Keywords: PSA, Generation IV Reactors, CCF, HRA

1. INTRODUCTION

It has been widely accepted that nuclear power has a vital role to play in satisfying the increasing global energy needs. Most operating commercial NPPs around the world are of Generation-II category. The Gen-III reactors have just started operating, and Gen-III+ reactors are at the advanced phase of commercialization. The first reactors of the Generation IV concepts are foreseen to start operating in the period 2020-2030. However, nuclear safety has become an important issue of public concern, especially after the event of Fukushima in March 2011. In order to improve public perception, engineers and designers will have to show a satisfactory level of nuclear safety. Because of this, the Generation IV concepts will likely launch considerable innovative technological changes in comparison with current designs and these innovations will have to be at a higher level of safety.

Although the safety and reliability of these types of reactors are meeting high standards, Generation-IV reactor systems are targeting toward a joint target of providing safer, more reliable, proliferation-resistant and economically feasible nuclear power source. Six design systems have been nominated over others for particular research, development and deployment. They are: Gas-cooled Fast Reactor (GFR), Lead-cooled Fast Reactor (LFR), Molten Salt reactor (MSR), Sodium-cooled Fast Reactor (SFR), Very High Temperature Reactor (VHTR), Super Critical Water-cooled Reactor (SCWR). More safety enhancement for Generation IV concepts can be achieved through evolution in knowledge, technologies and the development of a solid safety methodologies in the early design stages. Such enhancements will particularly address the pathway to attain the safety level by the employment of safety concepts that would be "built-in" to the proposed design concept instead of "added on" to an existing system. The design process Generation IV concepts should be guided by a "risk-informed" methodology (i.e. utilizing both deterministic and probabilistic techniques). Safety of Generation IV concepts can be enhanced by properly implementing, as a supplement of the deterministic techniques, the probabilistic approaches such as PSA and other techniques as guiders of the design process. [1, 2]

One of the most effective and mature safety tools is the Probabilistic Safety Assessment (PSA). It is considered as an essential approach to achieve enhanced safety for Generation-IV reactor systems. In the past, the design of NPPs was mainly based on deterministic methods. PSA has been recently used to support deterministic criteria and analyses in the design process for new reactor concepts in several projects. In the framework of design, construction and licensing of innovative reactors, PSA plays a vital role as a supplement to traditional deterministic approaches. The importance of applying PSA in the development of new reactor designs is well recognized.

PSA has become a very complex method to identify scenarios of possible accident, quantitatively estimate their occurrence probabilities in a certain time period, and probabilistically estimate the consequences following postulated accidents based on a set of consequence parameters. Along with the conventional deterministic techniques, the approach has come to be broadly accepted as one of the foundations for confirming the safety of a NPP (as well as other installations) worldwide. Until recently, PSA technique was mainly utilized after the design was settled, or even after the plant was constructed. Applied in this stage, PSA was basically used as a tool of estimating the risk level associated with an operating plant. With the development of future evolutionary designs (such as Generation IV concepts), however, the significance of PSA as a vital driver for the design process is perceived. Concurrently, limitations and challenges have to be considered, mainly when the PSA tool is performed for innovative concepts characterized by great uncertainties; lack of precise knowledge and lack of empirical data about failure, provisions and degradation. [1,2]

The main advantages of applying a PSA during the design stage are related to the identifying of plant vulnerabilities, of inter-systems dependencies and potential Common Cause Failures (CCFs), and to the examination of risk levels from different design alternatives. The probabilistic insights will help with the design optimization of safety systems (particularly in terms of diversification and redundancy), and with the checking of the design homogeneity from safety standpoint and, in the near future, from cost to safety benefit concern. [3] Several published studies applies PSA to a reactor concept in the design stage, such as in: [4-12]

The application of PSA to novel systems faces some challenges. Applying a PSA for a plant in the design stage is quite different than applying it for an existing or operating plant in which most of PSA guidance and procedures are formulated. The key challenge in the application of PSA methodology to enhance the plant safety in the pre-conceptual design stage is the lack of information, which increases the uncertainties accompanying any quantitative risk measure. The absence of plant-specific operations procedures and operating experience data at the design stage lead to PSA results that do not reflect the future as-built, as-operated plant. A vague understanding of probable accident scenarios may become an obstacle to the building of risk-informed regulatory initiatives. The methodological challenges include the necessity to address a wide spectrum of systems and phenomena, the potential lack of key reliability and experimental data, the potential lack of knowledge on new main phenomena and the potential lack of accident analysis models. These challenges can influence a variety of factors (e.g. risk balanced concept, defense in depth assessment and plant safety level assessment, etc.) The technical challenges of the PSA for more advanced reactors, which are in research phase or in the early stages of conceptual design, as well as the aforementioned aspects, also comprise the potential requirement to consider very diverse systems and phenomenology. [13]

Currently operating NPPs had a deterministically-launched licensing basis before plant-specific or generic safety information and insights were made obtainable through PSAs. The PSAs generally proved that the original deterministic methodology to licensing was conservative (e.g., plants might respond to some failure scenarios in behaviors that were not attributed in the deterministic analyses) and additionally identified alterations that could enhance plant design and safety. Satisfying the deterministic requirements meant that application of their associated provisions embodied within the models of defense-in-depth, quality assurance, safety margins, conservative assumptions and analyses, and several other factors (numerous of which are not easily measurable within a PSA model) created a safety basis where the uncertainties were acceptable. However, PSA models have to depend on realistic data to ensure that the main risk insights are not hidden by falsely biased results resulting from the application of irregular conservatisms. Consequently, considerable care must be exercised in performing PSAs in the pre-conceptual design stage to ensure that the essential pillars of deterministic safety process are not unjustifiably compromised. Therefore, for future reactors, use of risk data can have a far more important impact on the safety foundation of the plant, including the ability to derive some main design decisions. [14] In the following sections, some technical issues in applying PSA to novel reactor designs will be discussed.

2. ISSUES IN INCORPORATING PSA IN THE DESIGN AND LICENSING STAGES OF GENERATION IV REACTORS

The main challenge in the utilization of PSA techniques to support the design of a NPP is the lack of information, which exponentially increases the uncertainties associated to any quantitative risk measure that can be associated to an early design. With the evolution of the design, as more information comes to be available, risk metrics become more realistic but they still need to be closely watched for the associated model uncertainties. The assumption/uncertainty database is going to track design alternatives in the form of different event trees or different fault trees explicitly modeled in the PSA, which will identify the uncertainty bounds for the damage and release frequency values. [15]

Another challenge to the use of PSA in design phase is the lack of a recognized standard against with to compare the technical adequacy of a PSA developed for a non-operating reactor, this in the view of the fact that the current PSA standard is dedicated to operating reactor and has requirements that are clearly not applicable until operation of the plant has commenced. [16,17] The following sections discuss issues related to applying PSA in the design and licensing stages of generation IV reactors.

2.1. Initiating Events

For Generation IV reactors, the initiating events list is typically established based on comparable existing reactors PSA and based on generic references like IAEA and NUREG handbooks. This list is then combined with some specific analysis to consider the unique characteristics of the new designs. The resulting initiating events list usually includes some new initiating events specific to the unique plant design features. [18] However, this method does not ensure completeness of the list, particularly if the definition of initial boundary conditions is not complete or not required for the goals of the design PSA (e.g., the loss of I&C initiators or loss of ventilation initiators might be excluded). [19]

2.2. Passive Systems Modeling

Several new NPP designs utilize passive safety systems. Owing to the specificities of passive systems that apply natural circulation (lack of data, small driving force, large uncertainties, etc.), there is a necessity for developing consistent approaches and methodologies for assessing their reliability. With the purpose of increasing confidence in the attained results, it is required to decrease the level of uncertainty associated with the passive system behavior, especially the phenomenological uncertainty. It is also required to determine the dependencies among the related parameters adopted to examine the system reliability. Another important issue is to study the dynamic aspects of the system performance. However, in many of the existing design PSAs, the passive systems models take into account only the failure of the systems components (e.g., pipe break, spurious valves actuation, etc.), and ignore the failure of the phenomena (such as natural circulation).

This issue may need to be addressed by modeling, for example, the scenario dependent situations which may result in a combination of conditions in which the passive system function cannot be executed. The modeling of passive systems in the PSA needs also to consider the impact on other PSA issues. For example, the functioning of the passive systems for extended term accident scenarios ought to be carefully studied. Another important matter is the treatment of the uncertainties of physical and thermal hydraulic data as well as of the uncertainties in the passive systems behavior.

For PSA supporting studies, the current thermal hydraulic codes might not be entirely applicable for the passive systems behavior analysis. Indeed, the main two issues with thermal hydraulic codes are:

1) When the input parameters are varied over their potential ranges, are the codes still within their domain of applicability or not?

2) Does the analysis consider the possibility of degraded conditions (e.g., subsequent to a seismic event) or not?

2.3. Reliability Data

The assumption that the evolutionary components have the same reliability as the existing ones might be a reasonable initial assumption. Therefore, reliability data and CCF parameters for the components included in the PSA for new reactors are extracted from the same sources as for the PSA for existing reactors. [18]

The method to select the most appropriate data sets depends on the assessment of similarity of the novel reactor components with the existing obtainable data. This approach is adequate in principle. Nevertheless, the similarity investigation between novel reactor components and the existing reactor which was used to calculate the existing reliability data is not a simple task. This evaluation has to take into account, besides the component category and safety level, also the operating conditions, the component population used to calculate the data, the surveillance requirements (test intervals), the recent operating experience trends, the operating environment and parameters (temperature, pressure, flow rates, ambient temperature and humidity), etc. The justification of selecting a given data has to be completely traceable and documented. [19]

Some analysts may perform PSA on a specific reactor without taking into account the applicability of generic reliability data, and suspicion about such assessments raised because of the absence of plant-specific reliability data. To remove this doubt, a study has investigated the applicability of generic reliability. It has been shown that reliability data from various sources does not distort the results of PSA if they are utilized in performing PSA for a specific reactor. [20]

In that study, a number of reliability data sets extracted from different sources were analyzed. The subsequent analysis evaluates a fault tree (FT) for a specific reactor, using a number of reliability data sets and demonstrates the variances in the results. Furthermore, a comparison is performed with a procedural analysis utilizing ranges of reliability data. The results revealed that the PSA for a specific reactor employing reliability data which are taken from different sources is acceptable.

The variances are slight for the majority of components, only a few crucial components should be given more attention and further study. In the time being, the lack of specific reliability data should not be a barrier for performing a PSA on a specific reactor. [20] However, it is better to attempt to get plant-specific reliability data to fully remove all doubts about their applicability. Nevertheless, in the meantime the absence of novel reactor specific reliability data should not be a barrier to conduct a PSA to improve plant safety, mainly when main initiating events are to be found out. In any situation, it is strongly recommended for the analyst who faces a lack of data to identify the main components in the system to pay them more attention. [21] For novel components or components with no operational experience, generic data for similar components are considered with supplementary reliability evaluations, manufacturers' information and expert judgment. [18]

2.4. Common Cause Failure (CCF)

CCFs are being considered as one of the most critical matters in the development of PSA, particularly within FT modeling. A growing number of studies to reliability and safety analyses of systems taking into account impact of CCF, considering the CCF uncertainties, valuation of CCF rates, are being introduced. In recent years, CCF have been an ongoing issue of investigation and arguments. Thus, CCFs have been given a great consideration within the PSA of NPPs. [22]

For a PSA, CCF data is generally extracted from existing operating experience issued by internationally available sources such as (IAEA) and (NUREG). Generic values may also be viable if it is considered that the available data is not relevant for the selected CCF group. Because there are generally no large inconsistencies between CCF parameters taken from different sources, this approach is acceptable in principle. However, it has to be traceable and documented.

Regarding the classification of CCF families, PSA applies assumptions in order to define the groups that contain redundant components for which CCF contributions should be considered. However, the assumption of complete diversification of certain redundant components (where it is assumed that CCF is not possible) has to be justified by a comprehensive analysis. This investigation has to cover all the CCFs and mechanisms (type, environment, manufacturer, maintenance, etc.) along with the long-term characteristics of these situations over the plant lifetime. This issue refers essentially to components of similar type, but made by different manufacturers, for cases where parts might be supplied by the same manufacturer or for components within a common maintenance program. Spare parts or maintenance materials may have an influence. Sensitivity studies will be useful in order to find out the potential CCF families for which thorough studies may be required. [10] It is a common practice not to model inter-system CCFs for existing plants because they are supposed to be insignificant contributors to large early release frequency, core damage frequency (CDF), etc. Nevertheless, for prospective reactors with inherent safety features needing to show compliance with reduced safety target values, special attention needs to be given to inter-system CCFs and CCFs associated with similarity in active sub-components (circuit breakers, motors, etc.). [23]

2.5. Modeling of Novel Design Features

Generally, in the PSA for novel reactors, the innovative design features result in decreasing the core damage frequency (CDF). Some new initiating events will be recognized, primarily associated with inadvertent actuation of the new automatic actions. The effects of these actions should be analyzed. Additional evaluations might be required in order to ensure that new design features are sufficiently addressed. These might include, as examples, studies showing the appropriate utilizing of conservatism in defining PSA success criteria, the employment of bounding parameters for PSA sensitivity studies and supporting calculations, and testing actions to validate calculations.

2.6. Modeling of Preventive Maintenance

Generally, because technical specifications and preventive maintenance detailed procedures are not available in the design stage, assumptions are made on preventive maintenance and on corrective maintenance intervals. These assumptions are built chiefly on the anticipated technical specifications and on the engineering experience. This method is generally accepted for a design phase PSA. However, if the preventive maintenance is predicted during power operation, comprehensive maintenance information, chiefly related to the configuration management, might be requested with the aim of ensuring that the maintenance configuration risk is appropriately addressed in the PSA.

2.7. Technical Specifications

The surveillance requirements and the technical specifications are generally not available during the design phase. The aspects should be modeled as accurate as possible since the PSA can be further used to define "risk-optimized" technical specifications and surveillance requirements. [15] In parallel with validation of the PSA, some safety authorities assess the design phase technical specifications to confirm that they will maintain the plant design validity by ensuring that the plant will be operated with the predefined design conditions, and with equipment that is crucial for preventing accidents and mitigating the accidents consequences. In some cases, complete design information, allowable values, equipment selection or further information are required to establish the basis for the technical specifications. These plant-specific values should be provided when a joint license application is submitted for a certain plant.

2.8. Human Reliability Analysis (HRA)

To make PSA more accurate, improving of human reliability analysis (HRA) is vital. Experience shows obviously that human interaction is one of the key contributors to operational disturbances and accidents. In a study, it was concluded that probability of human error has approximately 58% contribution to events leading to increasing in core damage frequency (CDF). [24] At the same time, humans can effect several components and systems and therefore present hidden coupling factors between systems. Consequently, a proper human error analysis is required even in the most reliable systems. [25]

Currently, the HRA approaches for the PSA for novel reactors are generally similar for the existing reactors PSA. These approaches are used to quantify the pre-accidental and post-accidental HRA. During the design stage, detailed accident procedures and the use of a simulator to enhance HRA quantification usually are not available. The HRA is often used to establish the operator plans for a variety of accident scenarios. Moreover, it is likely that the HRA qualitative and quantitative analyses will be used to help enhance the comprehensive simulator training scenarios and accident procedures. Current HRA approaches have to be improved due to the existed limitations including the lack of theoretical basis for human operators situation assessment, and lack of considerations on the interdependency between human operators and I&C systems. To solve these issues, new methods should be proposed for the quantitative safety assessment of human operators and I&C systems. [26]

In the future, the expanded application of HRA approaches is foreseen, as well as the wide spread use of simulators. The availability of detailed information of the accident procedures and the severe accident management guidelines is considered a vital issue by all innovative reactor project analysts.

2.9. Systems Interdependencies

The systems interdependencies represent a crucial point of the design of the new plants. The PSA is one of the most powerful tools to study the impact of different design solutions. Even if the complete design is not finalized, the interdependencies between the safety systems, i.e. functional dependencies or induced by the support systems (power supply, cooling, ventilation, I&C, etc.) should be modeled as detailed as possible, and conservative assumptions should be used if the information is not available. The omission of the dependency modeling, even the detailed design of support systems is not known, should be avoided. [15]

2.10. Modeling of Instrumentation and Control (I&C)

Digital I&C systems are the present design solution for innovative reactors. The many exclusive attributes of these systems, create challenges for PSA modeling. The main issues are the models ability to identify dependences created by digital I&C, specifically dependencies between an initiating event (such as a spurious signal) and failures of safety functions (theoretically, the FT modeling is a possible solution for this issue). The second issue concerns data, which is still challenging to find, particularly for software and CCFs. The digital I&C is not an exclusive issue to new plants, but due to higher safety expectations the role of I&C is growing and becoming a potential major issue. Even though there is no real practice consensus for the digital I&C modeling and quantification, some tentative methodologies are established and integrated in PSAs. [18]

2.11. External Hazards

The ability to identify the external hazards for the PSA is different for diverse new reactor projects. This is due to variations in the project development status, chiefly if the site is identified or not, and to the expected effect of the different external hazards on the prospective plant safety, which depends on the country and the site. Generally, nowadays only a few hazards for PSA are recognized for new reactors. Many external hazards are addressed using analysis or other simplified approaches that approximate the hazards contribution to overall prospective plant risk. The prospective possible hazards evolution (prompted by climate change for example) are generally not explicitly considered in the analysis (however, climate change is sometimes taken into account in the external hazard analyses normally to include bounding assessments that are meant to show the margin of design for these hazards). The combinations of hazards, in addition to the induced internal hazards, seem to have not been analytically considered in the performed assessments. [18]

In order to allow the assessment of the influence of the internal and external hazards on the plant safety, it is important that the design phase PSA incorporates useful information (like equipment location, fire compartments, etc.), even using simplified assumptions. A thoughtful verification should be done regarding the possible common mode failures of redundant trains, systems or functions. [15]

2.12. Continuous Design Risk Monitoring

The conventional use of PSA within the design phase is centered on a continuous monitoring of the design against established conventional quantitative risk metrics such as CDF and various release frequencies. To be able to enter in this phase of the PSA support to the design, a somehow complete preliminary design needs to be reached. Depending on the design stage, an extremely simplified fault tree (FT) modeling of support systems is used. The complete, even though simplified, PSA model allows at this point for a more comprehensive risk monitoring of the design by tracking intersystem dependencies that cannot be easily tracked in a single failure criterion approach.

The risk-informed design approach Generation IV concepts suggests continuous interaction between the design team and the PSA team, with a more structured feedback from the PSA to the design side. In this approach, PSA results have a more direct influence on the plant design, rather then simply following its development.

The main "drawback" of such an approach is that probabilistic studies need to be initiated at a very early stage of the design, when several required design information may only be partially or qualitatively available. This requires a more flexible approach to probabilistic analysis than used in the past and, especially, results in a relevant number of assumptions, which importance in the risk assessment is well beyond what is currently handled in a PSA for operating plants. A fundamental part of using PSA in the initial design stage was therefore the documentation and monitoring of all these assumptions for further analysis and confirmation of their actual applicability. [16]

2.13. PSA Supporting Studies

Specific support studies are usually conducted for the new reactors PSAs. This typically includes studies such as: thermal hydraulic analyses and system engineering analyses for defining mitigating systems' success criteria. The necessity for developing specific studies is identified according to the PSA standards and guidance. The design basis reports and safety report analysis represent other sources of information for the new reactors PSA development, along with the PSA reports of similar reactors.

2.14. Interpretation of PSA Results for New Plants

There are a number of ways that the results of the PSA are used to evaluate the design of a new plant, to identify the design weaknesses and to assess and rank potential alternatives for enhancing the design. Generally, these include:

- Safety metrics/indicators such as safety system reliability, core damage frequency, large early release frequency, etc. Safety metrics/indicators show whether the overall risk from the plant is low enough to start a license process.

- Lists of minimal cut-sets. The integrated list of top minimal cut-sets and lists of minimal cutsets generated for separate initiating event groups for different plant operating modes are reviewed. Both internal initiators and hazard-induced initiating events are considered. If a single order minimal cut-set representing an independent failure, e.g. a failure of a common support system component, appears in the list of minimal cut-sets provided within the internal event PSA, then, hence, the single failure criterion is not met, and redundancy of the system concerned has to be increased. If a similar finding is found in the internal hazard (e.g., fires and floods) PSA, then separation and segregation of safety related components is insufficient and needs to be improved.

- Importance functions for basic events, sets of basic events, sets of initiating event and safety systems. High importance of an independent failure event might be an indication of insufficient redundancy in some plant operating modes and the necessity for enhancement. In this situation, either system redundancy requires to be improved or limiting conditions for system operation should become tougher for this particular plant operating mode, if possible. High importance of a CCF could be an indicating of insufficient diversity to some safety functions. In this situation, a significant change in the basis of design might be required. High importance of a human error may indicate a poor man machine interface. Increasing automation of the plant can be considered as an additional design measure in this case.

These results are used to decide whether the proposed design is balanced or there is a need for additional measures to be integrated to reduce risk. The results of the PSA are being used as one of the inputs to a process of risk informed decision making respecting to the option to be incorporated into the design. The PSA is used to estimate the reduction in the risk for each of the options identified. [23]

3. FUTURE DEVELOPMENTS AND RESEARCH

While the PSA methodology is reasonably robust in most areas, additional research is needed and is in progress in several areas. In some cases this research is conducted to improve the efficiency of the PSA process. In other cases, it is performed to reduce the uncertainties associated with PSA results, thus making it easier to use the results and analyses in a regulatory environment or to change operational practices. Several activities are related to the development of new or advanced reactors.

Key areas of research in progress include the following: Development of PSA methods; PSA for internal and external hazards; Common cause failure (CCF) modeling; Human reliability analysis (HRA); Reliability data collection; PSA for passive systems; Reliability of digital systems; Level 2 and Level 3 PSAs; Uncertainties; Dynamic PSA; Modeling of ageing in PSA; Fuel route PSA; and Use of PSA in risk-informed decision making (RIDM).

It can be seen that the general areas of PSA research are not really new, but in each area substantial activities are ongoing. Of special note is research relating to severe accidents, to fire, and to human factors, which supports improved PSA modeling. Moreover, research relevant to problems relating to new plants (e.g., digital I&C and passive systems) is receiving high priority. [27] To improve the quality of the PSA, the following areas are suggested for future studies:

- Develop a systematic approach to estimate the reliability of a newly introduced system or component for the novel reactor.

- Establish a methodology for evaluating the reliability of a digital I&C in passive safety systems.

- Develop a methodology for estimating the CCF data of a newly introduced component.

- Establish a structural framework for HRA activities for novel reactor designs.

4. CONCLUSION

The use of the PSA from the early design for Generation IV reactors shows that the PSA is a very valuable tool to obtain an optimized and balanced design by taking into account the information provided by the risk assessment. On the other hand, the development and the use of PSA should take into account some specific methodological aspects of a design phase PSA. The decision making process should consider the fact that PSA for a novel plant concept may have substantial uncertainties. Extensive sensitivity studies should be performed and the uncertainties should be known and taken into account. [28]

In many engineering design activities, the use of PSA methodology is now accepted, but with controversy in some technical aspects. The main concern is with the misconception that PSA methodology is considered as a tool to conduct a 'risk study' only and not as a comprehensive 'probabilistic tool' for predicting the system design behavior and to optimize it respecting various goals (e.g. investment, safety, reliability, availability).

Modern designs typically consist of active and passive systems, controlled by computers and supervised by plant staffs. Therefore the PSA methodology need to be enhanced for considering in a proper way the passive components and computer software and hardware. It makes no sense if a system design is assessed by "traditional" PSA methodology and the effect of the computer system, which controls the system, is neglected. In this framework there exists a challenge for enhancing the models and the database in PSA methodology today.

For the new reactors, the PSA is being accepted as one of the key methodologies to justify safety-critical features in the conceptual and preliminary design phase and to address new operation conceptions. However, there still remain some aspects related to PSA in order to better consider the innovative reactors specific features. Most of the identified PSA matters are well recognized. Also, most of the subjects are relevant to all types of reactors in the design stage. However, there obviously are greater challenges in dealing with these matters when the plant is in the conceptual design stage (and complete design specifications have not yet been set).

REFERENCES

[1] *"Basis for the Safety Approach for the Design & the Assessment of Generation IV"*, Nuclear Systems, Generation IV International Forum (GIF) Risk and Safety Working Group (RSWG), Revision 1, GIF/RSWG/2007/002, The OECD Nuclear Energy Agency, November (2008).

[2] *"Proposal for a Technology-Neutral Safety Approach for New Reactor Designs"*, International Atomic Energy Agency (IAEA), IAEA-TECDOC-1570, September 2007, Vienna.

[3] F. Bertrand, *"Risk-informed analysis as a support to the preliminary design of the CEA GFR2400"*, Workshop on PSA for New and Advanced Reactors, OECD Conference Centre Paris, France, NEA/CSNI/R(2012)2, pp. 85-95, July (2012).

[4] C. Bassi et al., "Level 1 probabilistic safety assessment to support the design of the CEA 2400 MWth gas-cooled fast reactor", Nuclear Engineering and Design 240, pp. 3758–3780, (2010).

[5] K. Kurisaka, "Probabilistic Safety Assessment of Japanese Sodium Cooled Fast Reactor in Conceptual Design Stage", 15th Pacific Basin Nuclear Conference, Sydney, Australia, (2006).

[6] T. W. Kim et al., "Preliminary Level 1 PSA Results for SFR-600 Conceptual Design", Proceedings of the 18th International Conference on Nuclear Engineering ICONE18, ICONE18-30367, Xi'an, China, (2010).

[7] P. F. Nelson et al., "A design-phase PSA of a nuclear-powered hydrogen plant", Nuclear Engineering and Design 237, pp. 219–229, (2007).

[8] Yu. V. Shvyryaev et al., "Use of Probabilistic Analysis in Safety Validation of AES-2006 designed for the Novovoronezh Nuclear Power Plant Site", Atomic Energy 106, No. 3, (2009).

[9] J. Tong et al., "Development of Probabilistic Safety Assessment with respect to the first demonstration nuclear power plant of high temperature gas cooled reactor in China", Nuclear Engineering and Design, 251, pp. 385–390, (2012).

[10] M. H. PSAsad et al., "Level-1, -2 and -3 PSA for AHWR", Nuclear Engineering and Design, 241, pp. 3256– 3269, (2011).

[11] J. H. Lee et al., "Safety system consideration of a supercritical-water cooled fast reactor with simplified PSA", Reliability Engineering and System Safety, 64, pp. 327–338, (1999).

[12] H. Yamano et al., "Development of technical basis in the initiating and transition phases of unprotected events for Level-2 PSA methodology in sodium-cooled fast reactors", Nuclear Engineering and Design, 249, pp. 212– 227, (2012).

[13] "Workshop on PSA for New and Advanced Reactors", OECD Conference Centre Paris, Nuclear Energy Agency (NEA), NEA/CSNI/R(2012)2, July (2012).

[14] Kamiar Jamali, "Use of risk measures in design and licensing of future reactors", Reliability Engineering and System Safety 95, pp. 935–943, (2010).

[15] G. Georgescu et al., "Use of PSA at Institute for Radiological Protection and Nuclear Safety for EPR licensing purposes", Institute for Radiological Protection and Nuclear Safety (IRSN), Fontenay aux Roses, France.

[16] A. Maioli et al., "Use of PSA in the Development of SMRs, Workshop on PSA for New and Advanced Reactors", Workshop on PSA for New and Advanced Reactors, OECD Conference Centre Paris, NEA/CSNI/R(2012)2, pp. 241-261, July (2012).

[17] "Determining the quality of probabilistic safety assessment (PSA) for applications in nuclear power plants", International Atomic Energy Agency (IAEA), IAEA-TECDOC-1511, July 2006, Vienna.

[18] "A Joint Report on PSA for New and Advanced Reactors", Nuclear Energy Agency (NEA), NEA/CSNI/R(2012)17, (2012).

[19] G. Georgescu and F. Corenwinder, "Lessons learned from IRSN review of Flamanville 3 Level 1 PSA, Workshop on PSA for New and Advanced Reactors", OECD Conference Centre Paris, France, NEA/CSNI/R(2012)2, July (2012).

[20] Jia Ning, "Applicability Analysis of Generic Reliability Data for PSA on a Specific Reactor", 18th International Conference on Nuclear Engineering: Volume 3, Xi'an, China, 17–21 May, (2010).

[21] Ulrich Hauptmanns, "The Impact of Reliability Data on Probabilistic Safety Calculations", Journal of Loss Prevention in the Process Industries 21, pp. 38–49, (2008).

[22] Duško Kančeva and Marko Čepinb, "A New Method for Explicit Modeling of Single Failure Event within Different Common Cause Failure Groups", Reliability Engineering and System Safety 103, pp. 84–93, (2012).

[23] V. Morozov and G. Tokmachev, "Lessons Learnt from PSAs for New and Advanced Reactors in Russia", Workshop on PSA for New and Advanced Reactors, OECD Conference Centre Paris, France, NEA/CSNI/R(2012)2, July (2012).

[24]   In YH, *"Key risk concepts, in: Proceedings of the IAEA workshop on improvement of safety and economics of NPP"*, Daejeon, Korea, (2002).

[25]   Pekka Pyy and Bjorn Wahlstrom, *"Modeling the Human in PSA Studies"*, Reliability Engineering and System Safety 22, pp. 277-294, (1988).

[26]   Man Cheol Kim and Poong Hyun Seong, *"A Computational Method for Probabilistic Safety Assessment of I&C Systems and Human Operators in Nuclear Power Plants"*, Reliability Engineering and System Safety 91, pp. 580–593, (2006).

[27]   *"Use and Development of Probabilistic Safety Assessment: An Overview of the situation at the end of 2010"*, Nuclear Energy Agency (NEA), NEA/CSNI/R(2012)11, (2013).

[28]   Gabriel Georgescu et al., *"Use of PSA at Institute for Radiological Protection and Nuclear Safety for EPR licensing purposes"*, Institute for Radiological Protection and Nuclear Safety (IRSN), Fontenay aux Roses, France.

Need for PRA in the Oil and Gas Industry

Matt Johnson[a*], Nicholas Lovelace[a], and Michael Lloyd[b]
[a] Hughes Associates, Inc., Lincoln, NE, USA
[b] Risk Informed Solutions Consulting Services, Ball Ground, GA, USA

Abstract: Probabilistic Risk Assessment (PRA) is widely used in the nuclear industry to assess the risk from hazards to nuclear power plants. This paper discusses the application of PRA methods to the oil and gas industry, and, specifically, to assessing production platform safety and optimizing levels of hydrocarbon production. Oil and gas platform safety can be analyzed with a focus on potential loss of life to platform workers from internal hazards such as uncontained liquid or gas hydrocarbon releases with subsequent ignition. Additionally, platform production capabilities can be analyzed with a focus on reducing production downtimes. PRA methods can be effectively utilized to identify both safety and operating issues for typical platform alignments, maintenance and testing frequencies, and prioritization of enhancements to platform operation.

Keywords: PRA, PSA, Oil and Gas, Production Platform

1. INTRODUCTION

Probabilistic Risk Assessment (PRA) methodologies have been widely adopted and embraced by the nuclear power industry and the nuclear regulatory commission as an effective means of estimating the risk associated with operating nuclear power plants and with providing insights to the nuclear power plant response to hazards. Although the focus of nuclear power plant PRAs is primarily aimed at the core damage and release of fission products to the environment, the methodology can be applied to production risk and industrial (personnel) safety risk in both the nuclear industry and in other industries that operate large numbers of relatively complex facilities. This paper explores the potential for application of PRA methodology to the oil and gas industry, specifically to offshore hydrocarbon production facilities, with a focus on safety risk, environmental risk, and production risk. This paper also provides the framework for considering an integrated total platform risk assessment of each of these hazard types and consequences.

2. OIL AND GAS PLATFORM BACKGROUND

The oil and gas industry operates facilities on both land and sea to extract the hydrocarbons for storage and transport to market. The operations involve exploration, drilling, pumping, storing, and transporting the hydrocarbons. Each of these types of industrial operations entails risks associated with processing hydrocarbons.

The oil and gas industry operates hundreds of hydrocarbon production facilities worldwide. In general, major accidents are infrequent. The extent of local accidents that result in loss of life, environmental release of hydrocarbons, and loss of production is generally unknown to the greater public, since these events may not rise to the threshold of media attention. However, rarely major accidents do occur. A list of some of the most significant of these accidents [1] are listed below:

- A Blowout accident at Platform A offshore near Santa Barbara CA, resulted in a 100,000 barrel spill in 1969.
- The Pemex-operated Ixtoc I offshore well experienced a blowout, resulting in a 3 million barrel spill in 1979.
- The Alexander Kielland floating platform in the North Sea for off-duty workers capsized in 1980, resulting in 123 fatalities.

- The Ocean Ranger semi-submersible drilling rig sank off the coast of Newfoundland, Canada in 1982, resulting in 84 crew member fatalities.
- A blowout on the Enchova platform in the Campos Basin near Rio de Janeiro, Brazil caused an explosion and fire that resulted in dozens of fatalities in 1984.
- The Piper Alpha platform exploded and sank in the North Sea, resulting in 167 fatalities in 1988.
- An explosion on an offshore oil rig off the coast of Nigeria resulted in 13 fatalities and many more injuries in 1995.
- The P-36 offshore production platform sank off the coast of Rio de Janeiro five days after an explosion that killed 11 people. 10,000 barrels of fuel and crude spilled into the ocean in 2001.
- A fire destroyed the Mumbai High North Processing platform off India's west coast in 2005, resulting in loss of 15% of the country's production (123,000 barrels per day) and causing 12 fatalities.
- The Usumcinta rig collided with the Kab-101 platform off the coast of Mexico during stormy weather, resulting in 21 fatalities in 2007. The fatalities occurred when workers attempted to evacuate in life rafts.
- The West Atlas mobile drilling rig leaked oil and gas into the East Timor Sea near Australia, and later sank after a subsequent fire in 2009. The spill continued for months spilling millions of gallons of crude.
- An explosion and fire on the drilling rig Deepwater Horizon resulted in 11 fatalities and the spilling of roughly 5 million barrels into the Gulf of Mexico in 2010.
- A Venezuelan natural gas exploration rig sank in the Caribbean Sea in 2010.

The use of PRA methods is most effective for assessing low frequency, high consequence risk conditions.

3 PRA APPLICATION TO OIL AND GAS INDUSTRY

3.1. Hazards and Accidents

Oil and gas facilities are subject to hazards that impact the safety of the platform and personnel, environmental containment of the hydrocarbons, and production operations. Each of these hazards shares some underlying characteristic that can be assessed and managed beneficially for each hazard, although unique aspects of each type of hazard also exists. For example, the risk of loss of piping integrity for production piping is hazardous to both personnel and platform safety, environmental containment, and production, but the risk of loss of an operating pump may only impact production risk. Each individual hazard type is briefly described below. Note that the hazards can be classified as either "internal" (due to conditions or evolutions related to the processing operation) or "external" (due to weather, tsunami, earthquake, or other event that affects the processing operation from the environment).

3.1.1. Personnel Safety

Industrial accidents that result in injury or fatality occur during platform operations and are the result of equipment problems and/or human performance problems. Hazards include loss of hydrocarbon containment and subsequent ignition which can result in injuries or fatalities local to the source of hydrocarbon release. This type of risk has lower consequences assuming that mitigation strategies succeed in isolating or terminating the hydrocarbon release and/or ignition. Should mitigation fail to contain the release, the accident can result in higher loss of life or loss of platform, both are high consequence events. Typical safety hazards are as follows:
1. Hydrocarbon release and subsequent ignition (internal hazard)
2. Industrial hazards due to rotating equipment, systems under pressure, human error during maintenance/operation, etc. (internal hazards)
3. Severe weather events such as hurricanes/storms/rough seas (external hazard)

4. Seismic events (external hazard)
5. Transportation accidents resulting in fire/explosion (external hazard)

3.1.2. Environmental Impact

Accidents that cause hydrocarbon release to the environment result in fatalities, loss of production, and a decrease in public perception of the safety of the oil and gas industry. These events can be difficult to terminate (as seen during the Deepwater Horizon oil spill in the gulf of Mexico) and the economic impact can be significant if legal action is pursued against the oil and gas company and/or increases in regulation result from the event. Recent estimates put the cost of the Deepwater Horizon incident at over 40 billion dollars. Typical accident types are as follows:
1. Oil or gas well blowout
2. Pressure boundary ruptures in production facility
3. Transportation accident resulting in loss of crude

3.1.3. Production

Events that cause a loss of hydrocarbon production results in economic impact to the oil and gas company and significant losses can impact the oil and gas market of a country or the world as a whole. The following types of hazards impact production:
1. Equipment failures and human error leading to platform shutdown
2. Severe weather resulting in platform shutdown
3. Seismic events resulting in loss of well integrity
4. Transportation accidents

3.2. Risk Assessments

Current risk assessment methods employed by the oil and gas industry include qualitative assessment and quantitative risk assessments provided for safety accident risk (those that involve potential loss of life and loss of the platform facility). Some examples of existing tools that are applied to the industry are Layer-of-Protection Analysis (LOPA) [2] and Quantitative Risk Assessment (QRA) [4]. These tools are used and provide reasonable assessments of the risks associated with the safety aspects of platform operation:

- Layer of Protection Analysis: The LOPA combines the qualitative with some amount of quantitative approach to risk assessment. It identifies operations, practices, systems, and processes that do not have adequate safeguards and helps in deciding the layers of protection required for a process. It focuses on the most critical safety systems. This approach is essentially equivalent to the "Defense in Depth" approach in the nuclear industry.
- Quantitative Risk Assessment: A QRA is used to provide a fully quantitative risk assessment. This type of assessment is currently used to quantify safety risk of the oil platforms. It combines the frequency of the hazard with its consequences and provides a numerical representation of the risk generated via mathematical models of the hazard and consequences.

These types of assessments are generally effective at assessing the safety and accident risk for oil and gas platforms. However, given the common cause events that result in the safety, environmental, and production risk, as seen by the common impacts from the hazards described in Section 3.1, benefit can be gained from expanding the analysis to include a comprehensive, total integrated assessment of the risk to personnel and platform safety, risk of environmental release, and risk of production losses during platform operation by focusing on the assessment of the common hazards and accident types.

3.2.1. Personnel Safety

Significant focus is already given to personnel safety, but safety improvements are primarily focused on reducing the hazard and initiating events with relatively little focus on reducing the consequences of the accidents. Evaluation of the consequences beyond what is currently performed in the QRA would provide additional risk reduction benefit. An example of consequence reduction would be evaluating the distribution of personnel on the platforms and assessing strategy for reducing exposure to platform locations that carry more risk. This assessment can be provided by utilizing the maintenance schedule for internal equipment hazards to determine the risk of the platform configuration and then assessing the risk of hydrocarbon release and ignition that occurs during maintenance for each area of the platform. Personnel access and travel through the areas with elevated risk could be restricted. An integrated schedule tool/risk model can provide this information to the operating crew/maintenance staff prior to beginning maintenance activities on the platform.

Similarly, external hazards such as severe weather events and rough seas can have an impact on safety risk. The benefit of a PRA methodology and integrated tool would be to provide insights into combinations of internal and external events to assist in operations assessment of the personnel safety risk. This assessment could manifest itself as a deferral of maintenance activities as necessary to maintain pre-defined personnel safety risk levels that would be set in accordance with the company's safety goals.

3.2.2. Environmental Impact

Environmental impact events merit a risk assessment due to the high economic cost, potential for legal action, and negative public perception of these events. Often these events occur concurrent with personnel fatalities as well as loss of production [1] when originating from the platform. Other environmental impact events occur during transportation accidents.

One contributing cause to the Deepwater Horizon incident [5] cited weaknesses in the risk assessment of the annulus cement barrier which might have prevented the hydrocarbons from entering the wellbore annulus. Performing a risk assessment using an integrated model tool for this portion of the production operation may have identified mitigating features that could have been taken to prevent the loss of containment and also raised awareness of the risks of the operation, potentially allowing identification of the incident prior to events reaching a critical point in the accident timeline. An additional cause was identified as a misinterpretation of pressure test data. The interpretation may have been influenced by the lack of risk assessment which would have reinforced the awareness of the risks of the operation. A third contributing cause was identified that the venting of the leaking hydrocarbons through the mud gas separator vent line and subsequent communication of the hydrocarbons from electrically qualified areas through the HVAC system to unqualified areas created the potential for ignition. Each of these causes could have been prevented by performing a risk assessment of the hydrocarbon flow paths when well integrity is lost.

3.2.3. Production

The benefits of a production risk assessment are obvious; increased production has a beneficial impact on the oil and Gas Company's financial health as well as a benefit to the world markets by increasing supply. Application of a risk assessment to the production equipment, operational practices, and maintenance practices of the platform can provide insights into vulnerabilities that may result in a loss of production. The results of the assessment can be used to optimize platform operation and maintenance with goal of increasing production. The decrease in production loss risk would be balanced against the safety risk and the environmental risk.

3.3. Optimization

Given that industry currently treats these risks using established methods, the natural question that occurs is "how does the application of these methods and tools result in significant benefit beyond what is currently performed?" The answer lies in the ability to account for these hazards and goals with a combined assessment and with a risk assessment tool that can be used to understand the assessment. Because safety risk is typically the top priority for industrial facilities, improvements in safety can have a negative impact on production risk. Although this is the prudent way to approach industrial operations involving injury or fatality risk, the two do not necessarily have to be inversely related; the benefit of a combined assessment and tool would allow the oil and gas company to meet their safety goals and environmental impact goals while providing the ability to assess improvements that also decrease loss of production risks. At the very least, the integrated assessment and tool can be used to evaluate multiple options for decreasing the risk associated with safety and environmental impact while not increasing the risk of loss of production. This combined approach offers a best-of-all worlds approach to platform operation.

3.4. Scalability and Fleet Benefit

The oil and gas industry operates hundreds of offshore production facilities. Although there is some unique design and operating characteristics of each of these facilities, the generic approach using PRA methods and using an integrated model tool can be effectively scaled and applied to a fleet of facilities with much less effort than that needed to develop the first risk model. A comparison of results between platform models may yield additional platform-specific insights that general safety studies may not identify. This would benefit the operating crew by providing specific risk assessment information rather than providing general rules or strategies for reducing or mitigating risk associated with platform safety, environmental containment, and production loss.

4. CONCLUSION

This paper has provided an argument for application of a combined hazard and consequence risk assessment to offshore oil and gas platform operation. The combined assessment would focus on personnel safety, environmental containment of hydrocarbons, and risk of loss of platform hydrocarbon production. PRA methodology and tools can be effectively utilized to provide insights to all aspects of platform operation and provide significant global fleet benefit when applied across similar platforms.

References

[1]   TIMELINE-Major offshore accidents in the global oil industry, http://www.reuters.com/article/2010/05/13/venezuela-platform-idUSN1327238620100513, accessed 3/20/2014
[2]   The LOPA Method – Anton A. Frederickson, April 1, 2002.
[3]   A Guide to Quantitative Risk Assessment for Offshore Installations, John Spouge, 1999
[4]   BP Warns Gulf spill costs will exceed 42.4bn as compensation costs rise, http://www.telegraph.co.uk/finance/newsbysector/energy/oilandgas/10210318/BP-warns-Gulf-spill-costs-will-exceed-42.4bn-as-compensation-costs-rise.html, accessed 3/20/2014
[5]   Deepwater Horizon Accident Investigation Report, BP, September 8 2010.

Learning how to Learn from Failures:
The Case of Fukushima Nuclear Disaster

Ashraf Labib[a]

[a] University of Portsmouth, Portsmouth, United Kingdom

Abstract: In this work, it is argued that learning from failures and safety competence should be an important part of the curriculum of Engineering and Management students. The case of Fukushima will be used to illustrate how to learn about learning from failures using multi-models inspired by reliability and risk analysis in order to investigate disasters. This type of analysis can offer richness to our understanding of the root causes and provide insight into policy making and support decisions for resource allocations for prevention of such disasters. The analysis is based on a workshop related to learning from failures where students and practitioners were first given a brief about the related theory of reliability analysis and decision science, followed by introduction of the analytical techniques that can be used (such as FTA, RBD and AHP). They were then given a brief in the form of a narrative of the accident from investigation reports, and they were then divided into small groups with the task to perform an analysis of the disaster followed by presentation of recommendations in the form of a written report and an oral presentation. Finally, a set of generic lessons and recommendations are provided in order to prevent future system failure.

Keywords: Fukushima Nuclear Disaster, PSA, FTA, RBD, AHP.

1. INTRODUCTION

In the wake of Fukushima nuclear disaster, few investigation reports have been published in an attempt to explain the accident and outline lessons learnt. Most notably it was noted that probabilistic safety assessment is underutilized in nuclear industry. For example in a report of the Japanese Government to the IAEA [1], it was noted that *"Effective use of probabilistic safety assessment (PSA) in risk management PSA has not always been effectively utilized in the overall reviewing processes or in risk reduction efforts at nuclear power plants"*. Moreover, it was noted in a comment by the British Office of Nuclear Regulator's (ONR) final report on Fukushima [2] that *"This [under utilization of PSA in nuclear industry] is an important lesson, acknowledging that effective use of PSA could have helped help to prevent accidents like that at Fukushima escalating, and to help deal with them should they occur"*. In the same report the ONR's final recommendations include *"The circumstances of the Fukushima accident have heightened the importance of Probabilistic Safety Analysis for all nuclear facilities that could have accidents with significant off-site consequences"*. In this paper we analyze Fukushima disaster and develop a hybrid modelling approach using PSA related techniques.

2. BRIEF INTRODUCTION ABOUT THE ACCIDENT

In this section a narrative is provided in order to summarize abundant information in the literature reporting the incident. It is suggested that as the disaster happened a while ago, a primary data collection would be of lower quality as memories have faded and key persons may have disappeared. Therefore, it has been decided to use a secondary data analysis (which is a proven and widely used research method) for the problem structuring. A secondary data analysis of the disaster gives also the possibility to triangulate sources. Moreover, the case can be easily checked by others researchers. The same narrative was provided in the workshop conducted by the author. The delegates were then divided into two groups and each group was required to consult literature with respect to finding more evident about the disaster and utilize reliability engineering and decision science techniques in order to analyse the failure and make recommendations based on the analytic tools that have been used.

2.1 Logic sequence of the failure

On 11 March 2011 Japan suffered its worst recorded earthquake, known as the Tohuku event. It was classified as a seismic event magnitude 9.0, with maximum measured ground acceleration of 0.52g (5.07m/s^2). The epicentre was 110 miles E.N.E. from the Fukushima-1 site. Reactor Units 1, 2 and 3 on this site were operating at power before the event, and on detection of the earthquake they shut down safely. Initially, on-site power was used to provide essential post-trip cooling. About an hour after shutdown a massive tsunami, generated by the earthquake, swamped the site and took out the AC electrical power capability. Sometime later, alternative back-up cooling was also lost. With the loss of these cooling systems Reactor Units 1 to 3 overheated, as did a spent-fuel pond in the building containing Reactor Unit 4. This resulted in several disruptive explosions, because overheated zirconium fuel-cladding reacted with water and steam and generated a hydrogen cloud which, was then ignited. Major releases of radioactivity occurred, initially to air but later via leakage to the sea. The operators struggled to restore full control. This was a serious nuclear accident, provisionally estimated to be of Level 5 on the Nuclear Event Scale (INES), a figure which was later amended to a provisional Level 7 (the highest category). The Japanese authorities imposed a 20km radius evacuation zone, a 30km sheltering zone and other countermeasures. Governments across the world watched with concern and considered how best to protect those of their citizens who were residents in Japan from any major radioactive release that might occur [3].

Some have commented on reports of plant damage caused by the earthquake, concluding that the loss of effective cooling for the reactors stemmed directly from the earthquake rather than the subsequent tsunami. However, the information available on the emergency cooling systems and analysis of the circumstances does not support such a hypothesis [2].

This case study is a good example of a double-jeopardy, where the combination of earthquake and tsunami caused destruction on a scale that was not anticipated in the initial design specifications. For example, the plant was protected by a sea-wall - designed to withstand a tsunami of 5.7 meters (19 ft), but the wave that struck the plant on March 11 was estimated to have been more than twice that height, at 14 meters (46 ft). This, coupled with the now reported land movement of 2.4m experienced by much of Japan, ensured that the Tsunami caused enormous damage along the coast [4].

2.2 Consequences of the failure

The earthquake occurred under the sea near the north east coast of Japan. It lasted over 90 seconds, and caused widespread damage to property, although, due to the civil building design standards most properties did not collapse. As a result of the earthquake Japan has moved 2.4m laterally, and dropped 1m vertically. Also, the earth's axis has moved 0.17m and the length of the earth's day is now shorter by 1.8 microseconds [4]. This was by any measure a major global event.

The earthquake produced a tsunami 14m high that struck the coast of Japan, and travelled up to 10km inland, devastating infrastructure already weakened by the earthquake.

There were approximately 15,000 confirmed deaths and 10,000 people remain missing. It has been reported that the accident eventually cost Japan between 5-7% of its GDP, or US$300-600 billion [5].

The infrastructure affected included many different types of facility, such as houses, hospitals, electricity and water supplies, petrochemical and oil installations. However, it can be argued that the most significant damage in a global context, was to the Fukushima Nuclear Power Station at the town of Okuma. Fukushima is a city in the Tohoku Region of Japan. It lies 250km north of Tokyo, covering an area of 746.43km^2. As of May 2011, it had a population of 290,064.

The damaged caused by the earthquake and subsequent tsunami, which arrived at 15.41 JST [3], resulted in mandatory evacuation of the population within a 20Km radius around the site, loss of containment of radiological material to air, contamination in the sea (since detected in the Irish Sea) and of drinking water in Japan.

2.3 The Japanese Nuclear Industry

Japan is heavily dependent on its nuclear industry, with 54 nuclear reactors currently in operation consisting, of 30 Boiling Water (BWR) and 24 Pressurised Water (PWR) reactors. The industry is regulated by the Nuclear Safety Commission (NSC) through the Nuclear and Industrial Safety Agency (NISA), which are accountable to the government through the Ministry of Economy, Trade and Industry (METI) [3]. It was the stated goal of the Japanese government, prior to this event that, 50% of their electrical power should be nuclear power (although this, of course, may not continue to be the case). In the short to medium term the Japanese government has suspended operations at Tohoku until the sea defences are improved, which is estimated could take years to complete.

In an article in the Guardian Newspaper [6], Mr. Naomi Hirose, president of the Tokyo Electric Power Company (Tepco), which runs the stricken Fukushima plant, said "*nuclear managers should be prepared for the worst" in order to avoid repeating Japan's traumatic experience*", and then he continues to say "*... we have to keep thinking: what if..*" Hirose said that "*although the situation facing Fukushima Daiichi on 11 March was exceptional, measures could have been adopted in advance that might have mitigated the impact of the disaster. Tepco was at fault for failing to take these steps*". According to him, "*preventative measures included fitting waterproof seals on all the doors in the reactor building, or placing an electricity-generating turbine on the facility's roof, where the water might not have reached it. In addition, wrong assumptions were made*", he said. Finally he concluded with the following lesson: "*What happened at Fukushima was, yes, a warning to the world,"* he said. The resulting lesson was clear: "*Try to examine all the possibilities, no matter how small they are, and don't think any single counter-measure is foolproof. Think about all different kinds of small counter-measures, not just one big solution. There's not one single answer. We made a lot of excuses to ourselves ... Looking back, seals on the doors, one little thing, could have saved everything*".

2.4 Some basic information about risk assessment in nuclear industry

The International Nuclear and Radiological Event Scale (INES) was introduced in 1990 by the International Atomic Energy Agency (IAEA) in order to enable prompt communication of safety significant information in the event of nuclear accidents. The selection of a level, on the INES (Figure 1), for a given event is based on three parameters: whether people or the environment have been affected; whether any of the barriers to the release of radiation have been lost; whether any of the layers of safety systems are lost. Broadly speaking, events with consequences only within the affected facility itself are usually categorised as 'deviations' or 'incidents' and set below-scale or at levels 1, 2 or 3. Events with consequences outside the plant boundary are classified at levels 4, 5, 6 and 7 and are termed 'accidents'.

The scale is intended to be logarithmic, similar to the movement magnitude scale that is used to describe the comparative magnitude of earthquakes. Each increasing level represents an accident approximately ten times more severe than one on the previous level. Compared to earthquakes, where the event intensity can be quantitatively evaluated, the level of severity of a man-made disaster such as a nuclear accident, is more subject to interpretation. Because of this the INES level is assigned well after the incident of interest occurs. Therefore, the scale has a very limited ability to assist in disaster-aid deployment.

Nuclear reactor incidents/accidents are classified using the following scale (In descending order of criticality):
7 - Major Accident (Chernobyl, 1986 – USSR and Fukushima, 2011 - Japan)
6 - Serious Accident
5 - Accident With Wider Consequences (Three Mile Island, 1979 - USA)
4 - Accident with Local Consequences (Windscale, 1957 – UK)
3 - Serious Incident (2013; In a further incident of the Fukushima Daiichi nuclear disaster, 300 tonnes of heavily contaminated water had leaked from a storage tank.)
2 - Incident
1 - Anomaly
0 - Below Scale/No Safety Significance.

Figure 1: The INES scale of nuclear accidents

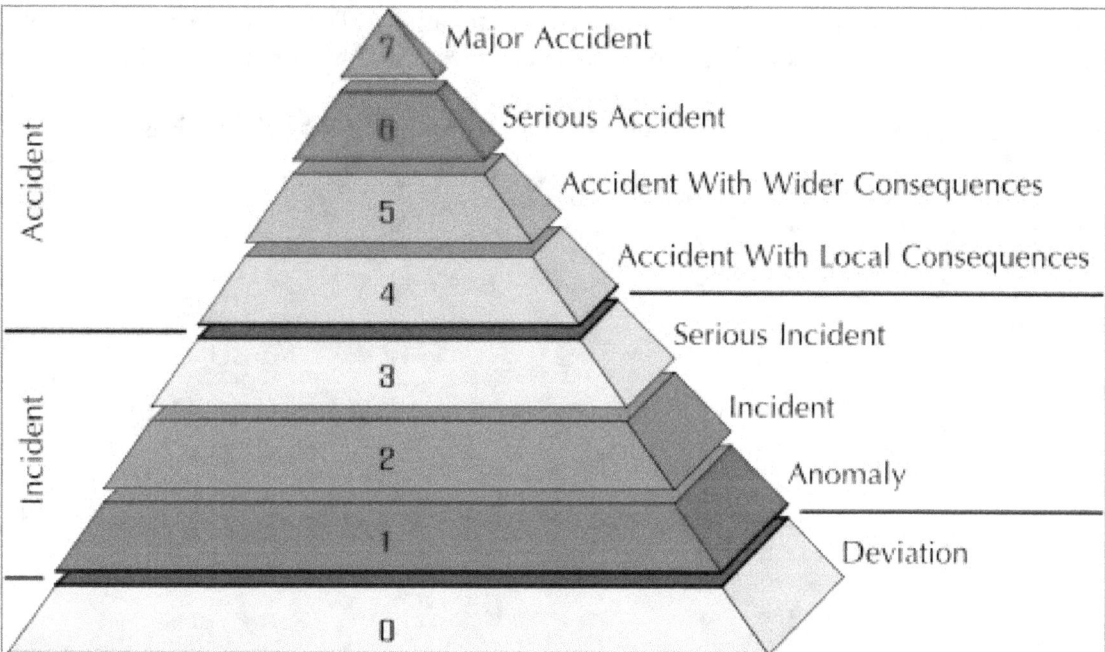

Note that up to level 3 on this scale the event is classified as an incident, whereas from level 4 onwards the event is classified as an accident.

3. ANALYSIS OF 1ST GROUP OF DELEGATES

3.1 Summary of the analysis of the causal factors of the disaster

The Fukushima Daiichi incident was fundamentally down to poor design, in that the normal mains and both backup power supplies were allowed to fail due to a single common mode failure, albeit from an extreme natural event. The mains supply integrity was such that the earthquake damaged it beyond repair and no diverse supply remained intact. The diesel generator system was located in a plant room likely to be swamped, and again no diverse connection point remained. And finally, neither could be repaired before the backup power supply was exhausted. The potential for an earthquake-generated tsunami in excess of the existing sea defences, and therefore capable of these effects, was not only realistic, it was actually foreseen in 2007 and calculated as likely in the lifetime of the plant [7]. But the plant continued to be operated and the sea defences were not improved, and the resilience of the cooling water system was not increased. Furthermore, the ability to provide cooling by other means was insufficient both with respect to the training and readiness to do so, and also as regards to the physical hardware to do so. Quite simply, without the installed pumps they could not provide enough water to prevent the pressure rising to unacceptable levels by any other installed emergency system [3]. The ability of the installed design to control, contain and direct excessive pressure was insufficient. Even if cooling could not be re-established there should be a means of safely directing the vented material to a suitable location. This was absent. In addition, the vent was not controlled from the point of view of fire suppression. In both cases, the venting to a suitably large installed system containing, for example, nitrogen blanketing would have limited the potential for the vented material exploding.

Fault Tree It is proposed that the process of evolution of the hydrogen explosion above Fukushima can be represented by a Fault Tree, as in Figure 2. The hydrogen explosion and the meltdown were due to three simultaneous factors; loss of coolant (ultimate heat sink), hydrogen built-up in cooling water and ignition (or detonation). The term 'ultimate heat sink' refers to the function of dissipation of residual heat after a shutdown or an accident.

Figure 2 (a,b,c): The Fault Tree Analysis of Fukushima Nuclear Disaster

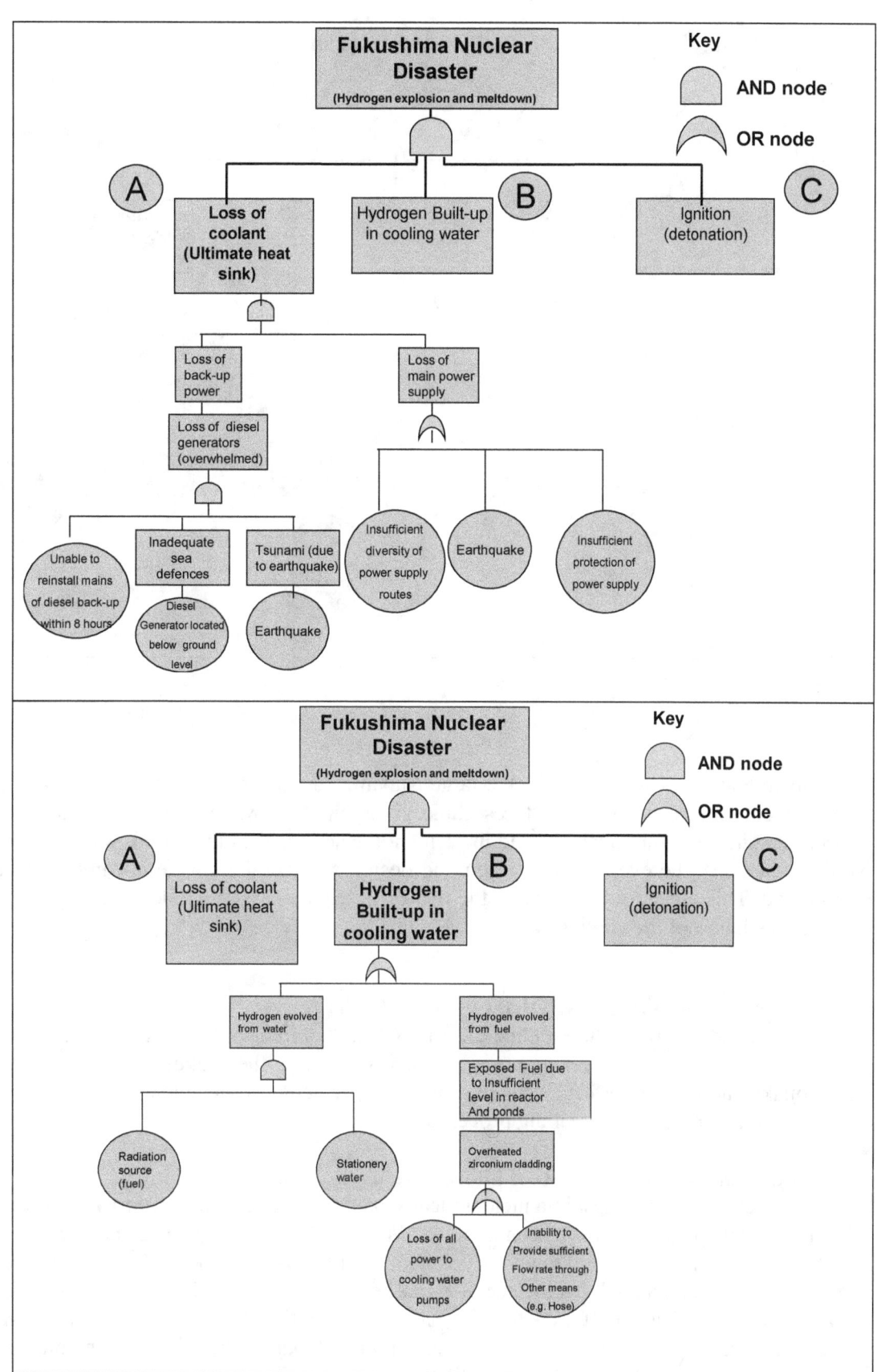

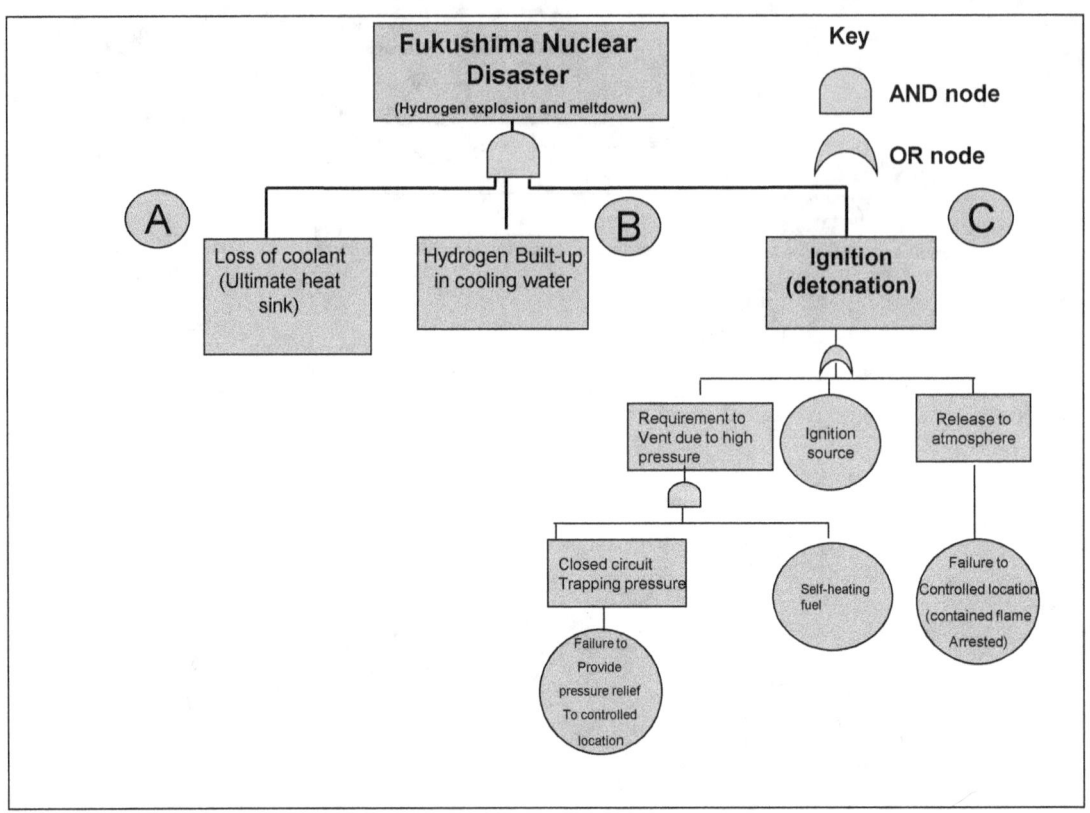

Given that the possibility of such a large tsunami was foreseen [7], it follows that through that the consequences were also foreseeable via a suitable FMEA. Therefore, the failure to carryout suitable hazard analysis, and implement the actions thus identified, was also a design failure. In short, the event was foreseen and design shortcomings were not investigated nor addressed. This aspect of the disaster, the hydrogen explosions, was fundamentally due to the lack of resilience of the cooling water circuit.

4. ANALYSIS OF 2nd GROUP OF DELEGATES

The second group part has applied the Analytical Hierarchical Process (AHP) to decide on the future of nuclear power usage in Japan following the Fukushima devastation by earthquake and tsunami effects. This part of the paper demonstrates the applicability of AHP in multiple criteria decision making processes.

4.1 Summary of the analysis of alternative nuclear power decisions for Japan

Exploration of the Fukushima incident leads back to the question; "*What went wrong?*" Could the station blackout have been avoided? Was it an engineering design and operations problem or a management and regulatory system failure? A Greenpeace International report [8] on the incident claimed that the accident marked the end of what it called the *'nuclear safety'* paradigm. The report drew the unusual conclusion that the notion of *nuclear safety* does not exist after what happened at Fukushima, but all that can be talked about concerning nuclear reactors are risks, unknown risks in the worst case. The report went on to say that, at any time, an unforeseen combination of technological failures, human errors or natural disasters at any one of the world's reactors could lead to a reactor quickly getting out of control. The report questioned the defence in depth of the engineering design barriers for nuclear power plants and disputed the PRA based postulation of only one core meltdown likely to occur in every 250 years. Being a humanitarian focussed organisation, Greenpeace did not consider the technicalities leading to the Fukushima accident, but rather focussed on the response both by the licensed operator, in this case TEPCO, and the Japanese regulatory authorities. It did not spare the IAEA in laying the blame and flaws on the agency's stance on the incidence. What becomes clear, one of the contributors to the report claimed, is that the weaknesses in the regulation and management of Japan´s nuclear power industry have not been 'hidden' faults in the system. On the contrary, people had been aware of, written and warned about them for decades [8]. So, from the humanitarian viewpoint of the report, the Fukushima accident was a regulatory system failure. Risks were known but no action was taken to address them. From a neutral perspective this does not justify the claim that safety in nuclear stations is non-existent. Rather, it points to the need to address some system deficiencies and suggests improvements that can make nuclear power even safer.

The IAEA report, on the other hand, conceded that Fukushima was an extremely unprecedented case and claimed that the response was the best that could be achieved considering the circumstances. However, it accepted that there were insufficient *defence-in-depth* provisions for tsunami hazards, in the sense that although these were considered both in the site evaluation and in the design of the Fukushima Daiichi NPP, and the expected tsunami height was increased to 5.7m after 2002, the tsunami hazard was actually underestimated. However, the view is that this was just a *black swan event* and does not invalidate the applicability of PRA postulates in nuclear power applications. In the Fukushima case the additional protective measures taken as result of the evaluation conducted after 2002 were not sufficient to cope with the high tsunami run up values and all associated hazardous phenomena. What comes out clearly from the IAEA report is that the design review underestimated the tsunami effect and this could therefore be classified as a design and reengineering failure. The nuclear authorities generally differ regarding the humanitarian view, in the sense that they see the incident as offering an opportunity for improvement in nuclear power probabilistic risk assessment, rather than a trumpet for propagating the message that nuclear power should be scrapped or be perceived as a public hazard. The general consensus at the *World Nuclear Fuel Cycle 2011 Conference* was in support of this view, the prevailing view at the conference seeming to be that nuclear energy will be providing utility power around the world for a long time, despite the accident at Fukushima Daiichi. This assertion was based on expert knowledge with minimal application of the decision making tools available at the time.

Faced with the foregoing two opposite views regarding the place for Japan's (and ultimately the world's) nuclear power usage, we shall now explore the available options for human safety driven improvement (or change) applicable to the Japan circumstances with respect to utility power after the Fukushima incident.

Option 1: Replace all nuclear power with alternative sources

This is a popular view among the environmental protection and humanitarian organisations. The Greenpeace report [8] suggested that a significant nuclear accident is bound to occur every decade, based on known incidences, and that puts a question mark over the applicability of nuclear power from the environmental safety perspective. The option to replace all NPPs in Japan is, however, based on the assumed existence of renewable energy sources, or other safer alternatives that could make up for the nuclear phase-out.

Option 2: Continue using NPP with improved barriers to external influences and better legislation

This is a popular view among the nuclear industry professionals. It is based on the belief that nuclear is one of the safest forms of energy and that PRA postulates on the probability of occurrence of catastrophic nuclear accidents are generally correct, i.e. the probability is remote. Failure in this case is an opportunity for learning, albeit that it comes at a great cost. Others have gone as far as proposing a review of how sites for nuclear power plants are selected by considering the historically based probability of natural occurrences.

Option 3: Continue with status quo

This option is based on the view that nuclear accidents of large magnitude are *black swan* incidents, one in every 250 years according to present probabilistic risk assessment theory. This *black swan* claim however, would appear to the environmental pressure group to be undermined by the much shorter time lapse between the Chernobyl and Fukushima disasters.

4.2 Application of MCDM

The three available options here are subjected to an MCDM process, viz. AHP based on the attributes of Safety, Environment, Economy, Image and Feasibility. The image criterion is considered from the legislature's point of view, i.e. that of the Japanese government and its nuclear regulatory agency, the Nuclear Industrial Safety Agency (NISA) and, on the extreme end, the IAEA and its affiliates. The AHP hierarchy thus developed is shown in Figure 3.

Figure 3: AHP Hierarchy

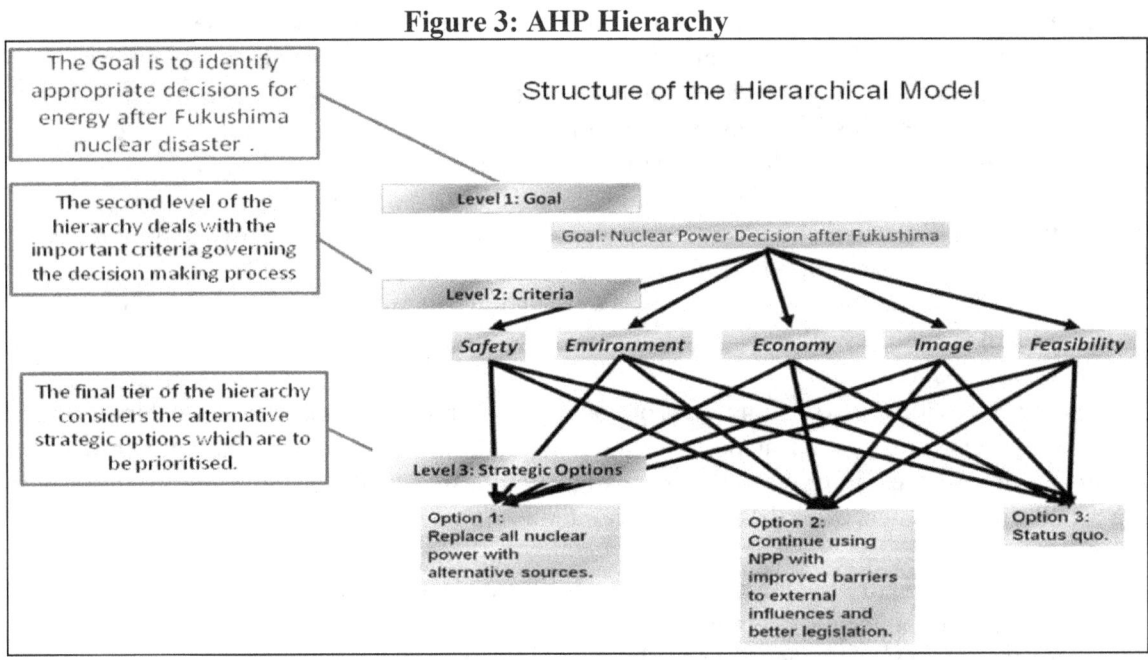

The AHP Results

Following traditional AHP guidelines the five attributes in the hierarchy were weighted, between 0 and 1. The attribute *"feasibility"* has the highest score, whatever alternative is to be chosen; first and foremost the alternative has to be feasible, then the rest can be considered, otherwise the analysis would be of no practical use and a waste of resources. Subsequent pairwise comparisons were done, more importantly the one with alternatives for the opposing sides i.e. environmentalists and IAEA.

Figure 4: AHP Synthesis and Sensitivity Analysis

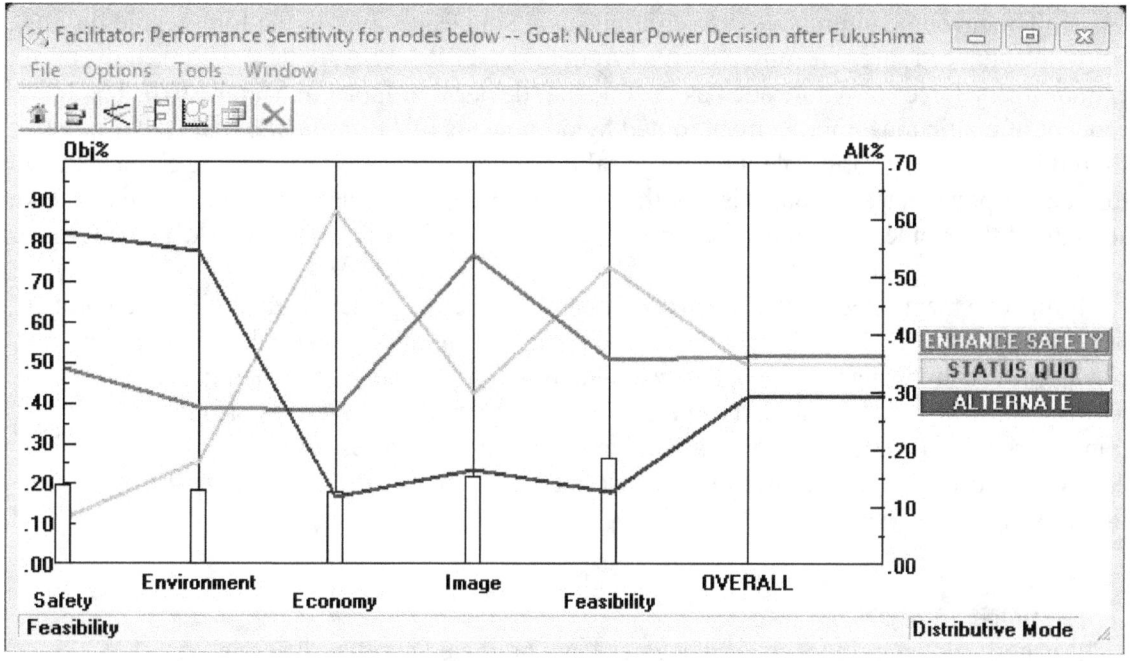

The performance sensitivity nodes representation in Figure 4 shows how the different alternatives rate with respect to the objective. It depicts the option *enhance nuclear safety* as the preferred option.

Figure 5: AHP Safety Sensitivity Graph

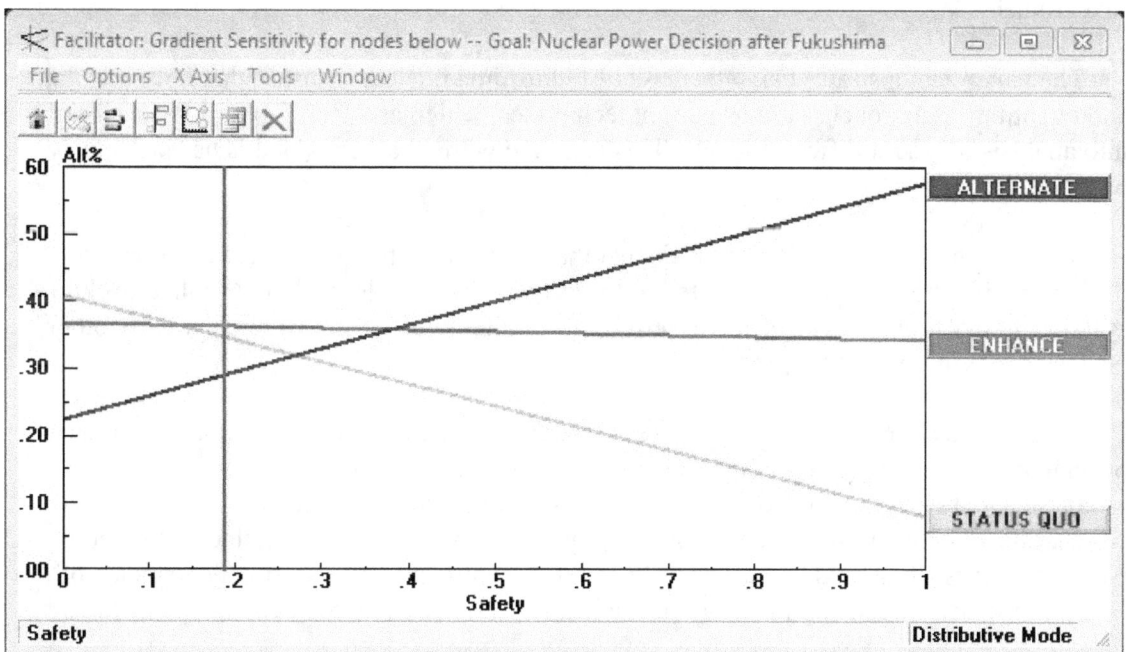

The gradient sensitivity with respect to the attribute *"safety"*, shown in Figure 5, indicates that, as far as safety is concerned, the use of alternative energy sources is the preferred option, while disregarding safety would result in the *status quo* option being preferred. This can be illustrated by moving the vertical line that indicates the importance of "safety" to the left, then the highest intersecting option (most preferred) becomes the one that belongs to the option *status quo*.

The favoured alternative is to continue using nuclear power in the foreseeable future, but with enhanced safety features, derived from revised PRA/PSA, to deal with the advent of extraordinary forces of nature such as the one that devastated Japan in March 2010. The concept of continuous improvement touted by proponents of *Probabilistic Risk Assessment* should be the guiding principle. One proposal is to possibly reconsider tectonic characteristics for nuclear power sites. Another is that the defence-in-depth structural design should also take account of the incidence of terrorist action such as the 7/11 attack on the World Trade Centre.

It must be mentioned, however, that normal AHP uses aggregate analyses from a number of people, presumably to reduce subjectivity, include all relevant stakeholders and promote consistency. But this has not been the case with this study as only one small group of participants was employed to carry out the analysis. Nevertheless, it does provide a framework which could be used by a number of participants to settle the nuclear power debate. The use of *expert knowledge* as prescribed by AHP could add more credibility to the findings of the analysis.

5. CONCLUSION

It is noticed that both teams of delegates have produced slightly different, yet complementary, mental models although they were exposed to the same narrative. The main differences were in the level of detail each group went into and techniques chosen where the fault tree analysis where used by the first group and it offered insight into the technical issues, whereas the second group used qualitative strategic analysis using the AHP approach. Nevertheless, on the whole, there were more agreements than otherwise in the findings of the two groups.

The paper demonstrates using the case of Fukushima nuclear disaster, that both qualitative and quantitative approaches are important techniques, which are useful to gain better insight into analysis of risk at different levels. This is in line with what Apostolakis has previously proposed [9].

It is clear from the fault tree analysis that the main causal factor was due to initial poor design specifications, especially related to the height of sea walls, and the installed backup systems, in that there was insufficient provision for alternate cooling water supply by other means or for controlled safe pressure relief.

Whereas on a more strategic level the nuclear power generation debate relates to the issue of regulation. According to [10 and 11], it is proposed that the time has come to introduce a Japanese and a global independent nuclear safety commission in order to separate national economic and political interests in promoting nuclear power from the regulatory function, which concerns all nations. Accordingly, it was recommended to elevate the mandate of the IAEA to include a licensing function for nuclear power plants, thereby changing its status from an advisory body to that of an international institution with authority to make legally binding decisions. This is in line with the findings of this paper, since it was noticed that the root causes for the disaster can be attributed to deficiencies in regulation and in setting design specifications based on risk assessment. In terms of setting design specifications based on risk assessment, it can be claimed that more research is needed in this field where the emphasis should shift from 'probability' assessment to 'possibility' identification. Mathematically, it is relatively easier to formalise the former than the latter.

To test for a cumulative probability of a one in ten million chance of a nuclear failure each year would require living for many years to prove its validity. But it could also mean to build 1,000 reactors and operate them for 10,000 years and expect a probability of one of them to fail during that period, which is a better proposition than the original one but still quite a long time.

Now, let us compare these ambitious estimates with the current state. Across the world there are about 435 nuclear power reactors operating, with over 140 in Europe, and 54 in Japan (Weightman, 2011), and around 100 in the USA. Fukushima is the third major nuclear accident (i.e. it was preceded by Three Mile Island and Chernobyl) and all three happened within less than half a century, which makes us question our models and original assumptions. So the current state suggests that the Mean Time Between Failures (MTBF) for the three major accidents (in 1979, 1986, and 2011) currently stands at just 10 years, which is very far from the ambitious 1 in a 10 million chance. This view is supported by [12] which also suggests a catastrophic accident to be expected every 12-15 years. Clearly, Three Mile Island, Chernobyl, and Fukushima each arose from very different circumstances, invalidating various modeling and risk assessment assumptions, and resisting assimilation into a single data set. It is difficult, with such a small sample size, to make generalizations about where current risk models fail, though we agree with the argument put forward by [10] which suggests that the original ambitious annual failure risk estimates were serious underestimations.

Acknowledgement

The author is grateful to his students at both University of Manchester (MSc Reliability Engineering) and University of Portsmouth (MBA). Specifically the author is grateful to the two groups involved in the Fukushima project and in particular, the leaders of the groups; Precious Katete and Michael Booth.

References

[1] Report of the Japanese Government to the IAEA Ministerial Conference on Nuclear Safety:The Accident at TEPCO's Fukushima Nuclear Power Stations Nuclear Emergency Response Headquarters Government of Japan June 2011.

[2] Weightman M., Japanese Earthquake and tsunami: Implications for the UK Nuclear Industry - Final Report, Office for Nuclear Regulation, September 2011

[3] Weightman M., Japanese Earthquake and tsunami: Implications for the UK Nuclear Industry - Interim Report, Office for Nuclear Regulation, May 2011

[4] IMechE (Institute of Mechanical Engineers) Report on Fukushima Daiichi Nuclear Power Plant & Tohoku Earthquake: What Happened When?, 11th March, 2011.

[5] Kashyap, A., et al, The Economic Consequences of the Earthquake in Japan, Web Publication. Freakonomics Web, 2011

[6] The Guardian, UK government must learn from Japan's catastrophe as it plans a new generation of plant, nuclear chief claims, Tisdall, S., 19 November 2013.

[7] Krolicki, K., DiSavino, S., Fuse, T., Engineers knew Tsunami Could Overwhelm Fukushima Plant, Insurance Journal, 2011.
http://www.insurancejournal.com/news/international/2011/03/30/192204.htm

[8] Morris – Suzuki T, et al, Lessons from Fukushima, Report, Greenpeace International, February 2012.

[9] Apostolakis GE. How useful is quantitative risk assessment? Risk Analysis 2004;24(3):515–20.

[10] Pfotenhauer, S.M., Jones, C.F., Saha, K., and Jasanoff, S, Learning from Fukushima, Issues on Science & Technology, University of Texas, 2012.

[11] Aoki, M., and Rothwell, G., A comparative institutional analysis of the Fukushima nuclear disaster: Lessons and policy implications, Energy Policy, 53, 240-247, 2013.

[12] Smythe, D., An objective nuclear accident magnitude scale for quantification of severe and catastrophic events, Physics Today: Points of view, December, 12, 2011.

Toward Demonstrating the Monetary Value of Probabilistic Risk Assessment for Nuclear Power Plants

Marzieh Abolhelm[a*], Justin Pence [a,1], Zahra Mohaghegh[a], and Ernie Kee[b]
[a] University of Illinois at Urbana-Champaign, IL, USA
[b] YK.risk, LLC, TX, USA

Abstract

Inefficiencies in the operation and maintenance of Nuclear Power Plants (NPPs) have caused unnecessary shutdowns, decreases in production, and increases in system risk. Probabilistic Risk Assessment (PRA), which guides risk-informed decision-making, helps expand the operational envelope by allowing more flexibility, adding to the efficiency of preventive and corrective actions and, therefore, generates more profit. However, the financial bottom line of PRA has not yet been formally estimated. This paper reports on the current status of first-of-its-kind research for estimating the monetary value of PRA. The proposed steps for this research include: (1) developing a Generic Financial Model (GFM) to estimate the Return On Investment (ROI) that results from profit generation or cost reduction associated with a typical PRA activity in an NPP (2) implementing GFM for one of the PRA programs and validating GFM, (3) conducting uncertainty quantification for the estimated ROI from Step 2, (4) identifying existing PRA programs at an NPP (i.e., South Texas Project Nuclear Operating Company; STPNOC), (5) obtaining ROI for all the PRA activities of STPNOC, running uncertainty analysis for the total ROI, providing a probabilistic monetary value of PRA, and (6) applying importance measure and sensitivity analyses to propose improvement approaches for PRA activities.

Keywords: Probabilistic Risk Assessment, Monetary Value of PRA, Socio-Technical Risk Analysis, Financial Modeling, and Business Case for PRA

1. INTRODUCTION

For most modern industries, safety is a goal that is given the same priority as efficient and economical production and, therefore, the connection between profitability and safety has long been an issue of interest to managers, business scholars, economists, and policy makers. However, the economic gains from using Probabilistic Risk Assessment (PRA) are yet to be discovered; thus, the goal of this work is to demonstrate the monetary value of PRA, and add another layer of justification to the advantages of PRA.

1.1 Probabilistic Risk Assessment

Methods to perform quantitative risk assessment in the U.S. aerospace and missile programs were developed and improved in the early 1960s. Later, the nuclear industry used PRA as a structured and formal method for identifying and assessing risk in nuclear technological systems. In most applications, PRAs have been utilized as tools to estimate risk as a function of equipment and operator performance. The process is used to identify potential accident scenarios, estimate the likelihood and consequences of accidents, and improve system safety designs and operations. This

---

1 Currently employed by Argonne National Laboratory

Probabilistic Safety Assessment and Management PSAM 12, June 2014, Honolulu, Hawaii
* abolhel2@illinois.edu

analytical technique has gradually improved and been applied over the last three decades and is now an important part of risk-informed regulation. At this time, PRA has been implemented in all of the 100 Nuclear Power Plants (NPPs) in the United States.

The risk-informed regulation and licensing practices of the Nuclear Regulatory Commission (NRC) have been improving, and new opportunities are becoming available for nuclear operating companies to use PRA to prove that, in some cases, alternative decision-making does not increase risk to unacceptable levels [1, 2]. The advancement in PRA methodologies, along with NRC support of risk-informed applications has created an opportunity for NPPs to advance their use of PRA applications. The purpose of this study is to uncover a new dimension of the benefits of PRA, i.e., the monetary value that PRA brings to the nuclear industry. By demonstrating the inherent financial return of PRA, this research aims to encourage the implementation of PRA programs at NPPs.

1.2 Socio-Technical Risk Analysis

Several investigations have recognized organizational factors as root contributors to technical and operational system risk. Numerous post-accident analyses reveal an urgent need to adopt new safety culture policies in order to reduce these contributors [3,4]. Socio-Technical Risk models have evolved from Reason's "Swiss Cheese" conceptual model [5], to the more recent Socio-Technical Risk Analysis (SoTeRiA) framework, introduced by Mohaghegh [6, 7] (Figure 1). The SoTeRiA framework is a theoretical foundation [8] for integrating organizational factors (e.g., safety culture and climate, organizational structure) with PRA. Mohaghegh, Kazemi, and Mosleh also proposed a hybrid modeling technique [9] to quantify this framework. The hybrid approach [10] quantifies the interactions of organizational safety risk factors using System Dynamics (SD) and the Bayesian Belief Network (BBN), and links them to Fault Trees (FTs) and Event Trees (ETs) in technical system PRAs. Advancing the measurement of factors of the SoTeRiA framework and further operationalization of this theory is the topic of other on-going research [11].

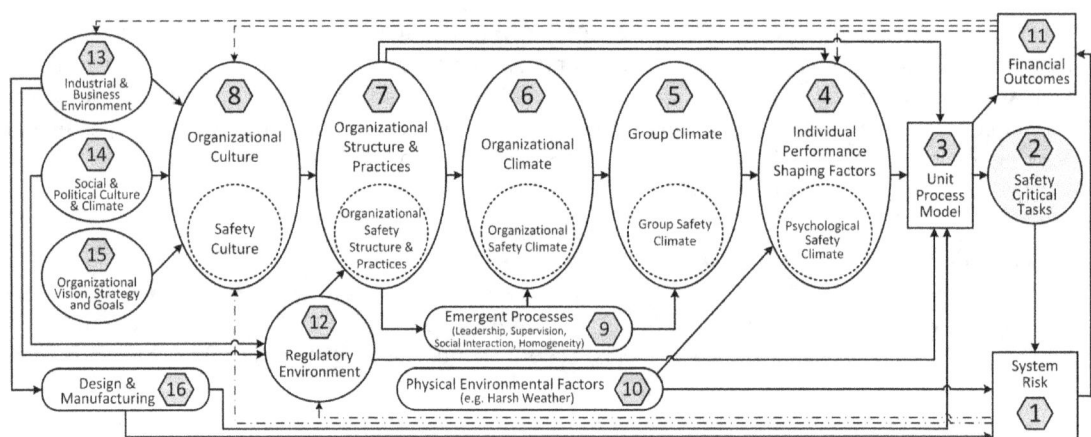

Figure 1 Socio-Technical Risk Analysis (SoTeRiA) Framework

While SoTeRiA has made a significant improvement on integrating organizational factors in technical system risk models, the relationship between "Financial Outcome" (Node 11 in Figure 1) and "System Risk" (Node 1 in Figure 1), as the two outputs of this framework, has yet to be quantified. This paper reports on the current status of on-going research that addresses the quantification of this relationship by evaluating the probabilistic monetary value of PRA. PRA-based programs at one nuclear operating company (South Texas Project Nuclear Operating Company; STPNOC) will be analyzed in order to estimate; (1) the Return On Investment (ROI) associated with PRA activities, and (2) the related uncertainties to provide a probabilistic representation of the monetary value of PRA.

1.3 System Risk and the Financial Outcome

"Risk assessment provides the ability for plant personnel to balance cost, power generation, and risk" Garrick mentions [12]. This paper reports on the current status of first-of-its-kind research for estimating the monetary value of PRA. This section summarizes the current status of the literature review related to the relationships between System Risk (Node 1 in Figure 1) and Financial Outcome (Node 11 in Figure 1). The associated literature is categorized as follows:

a. Correlations between safety and financial outcome:

   i. *Correlation between system safety and profit*: Through statistical analysis, several works have attempted to find a positive correlation between safety and profitability metrics, showing that the two outputs move in the same direction, regardless of the underlying causation behind their associations. The majority of studies correlating safety and financial outcome have taken place in the transportation industries. Examining the connection between profitability and safety in the U.S. railroad industry, Golbe [13] identified a positive association between contemporaneous profit and fewer accidents over the period of one year. Alternatively, in the aviation industry, Rose [14] reported a marginally significant positive relationship between the safety-profitability of most major U.S. airlines, with a stronger positive relationship for smaller airlines. In another work, Golbe [15] states that "the sign of the relationship between profit and safety is indeterminate and depends on risk preferences," suggesting that profitability affects a firm's propensity to take risks, and that safety affects this same propensity. In the U.S. nuclear power industry, there are inconsistent conclusions about the relationship between profitability and safety. For example, Feinstein [16] examines the relationship between the financial condition and contemporaneous regulatory compliance of firms, finding no relationship between the financial strength of a utility and its regulatory violations. In contrast, Marcus et al. [17] finds that safety is weaker in less profitable utilities, where the likelihood of a significant event occurring is higher. The number of correlation-based studies on the connection between system safety and profit exceeds the scope of this paper. Madsen [18] provides more examples on this topic.

   ii. *Correlation between occupational safety and profit*: Occupational safety has been extensively evaluated with respect to financial and operating profitability, in order to study the relationship between safety in the work place and the profitability of the business. Databased evaluations [19] of the relationship between occupational safety and operating performance support anecdotal evidence that good occupational safety is good for business. In other words, employees who do not feel safe in their jobs are not likely to do their jobs well; hence the economic productivity of the organization is associated with its safety conditions.

   It can be concluded that empirical perspectives from different domains of research provide partial support to the safety-profitability relationship; however, no existing research addresses the connection between system risk, calculated from PRA, and the financial output of the organization.

b. <u>Underlying mechanisms associated with the relationship between safety and financial outcome:</u>

   i. *The direct mechanisms through which system safety affects the financial outcome*: This topic refers to the impact of accidents and incidents on the financial status of organizations in high-risk industries such as nuclear, offshore oil platforms, aviation, aerospace, and healthcare. The hypothetical linkage is illustrated in Figure 1, connecting Node 1 to Node 11. The review of the literature under this topic is beyond the scope of this work, however, section 2 of the paper presents the methodology of this research, which is based on both the direct and indirect mechanisms through which system safety and financial output are connected. This method is used to develop a Generic Financial Model (GFM) in section 3.

   ii. *The indirect mechanisms influencing system safety and financial outcome*: Considering the SoTeRiA framework in Figure 1, safety and financial outcome can influence each other through a multitude of indirect paths of influence. For example, the external industrial and business environment (Node 13 in figure 1) can affect the financial output of an organization when there are disruptions in the global financial market (e.g., increases in the interest rate on investment loans, sudden changes in price). The financial status of an organization can influence the propensity of managerial decision-making towards risk or safety. In other words, financial distress in the organization, caused by the external business environment, could result in managers' failure to recognize and control the trade-offs between safety and short-term economic productivity [20].

   Focusing on the influence of the decision-making process on system safety, Baron and Pate-Cornell [21,22] illustrate the relationship between financial outcome and safety by analyzing the effects of alternative strategic decision-making for maintenance and operation based on long-term performance of the system. A number of other works [23,24,25] have focused on developing similar models that link managerial decision-making to system safety performance. Furthermore, as Starr and Whipple mention [26], the contingency of large financial losses due to a nuclear accident provide strong incentives for managers to focus on balancing the trade-off between system safety and economic productivity.

   Post-event analyses have demonstrated the influence of financial status on managerial decision-making. For example, before the Bhopal disaster, the operations and management staff had been reduced to half in order to save money, and the refrigeration unit, which could have mitigated the gas leak, had been shut down to reduce operating expenses [27]. Kurzman [28] reported that at Bhopal, "cuts...meant less stringent quality control and thus looser safety rules. A pipe leaked? Don't replace it; employees said they were told...MIC workers needed more training? They could do with less. Promotions were halted, seriously affecting employee morale and driving some of the most skilled...elsewhere". In the analysis of the Challenger disaster, a commonly cited contributing factor was the financial pressure on NASA, and its strong connection to the social and political climate, which led to management accelerating the launch despite concerns voiced by the engineering team [29]. Further analyses [30,31] have indicated that delaying upgrades and/or maintenance operations in order to meet production goals or deadlines played an important role in the magnitude of both the Challenger accident and the loss of the Piper Alpha oil platform.

c. Impacts of organizational factors and external environment (e.g., regulatory) on the financial output of high-risk organizations: Though the following topics do not explicitly focus on the connection between safety and financial performance, the internal organizational and external environmental factors have been defined by safety method research and, therefore, are considered as categories relevant to this research:

   i. *The connection between organizational factors (e.g., Node 7, 8 in Figure 1) and financial outcome (Node 11 in Figure 1):* In economics and finance, empirical studies have analyzed the relationship between a business' financial and organizational performance [32]. The findings of these studies affirm that businesses should focus on improving organizational factors (e.g., management and leadership, commitment to ethics, group culture, training, and organizational accountability) in order to achieve better results for their economic bottom line.

   ii. *The connection between regulatory environment (Node 12 in Figure 1) and financial outcome (Node 11 in Figure 1):* The NRC is moving towards a proposed risk management regulatory framework, which includes PRA and risk-informed decision-making in regulatory and oversight functions [33]. One of the proposed approaches in NUREG-2150 [33] is to add cost-benefit analysis to the already existing acceptance criteria of "as low as (is) reasonably achievable" (ALARA) for risk-informed decision-making. In the consideration of costs for regulatory decision-making and to determine ALARA thresholds, regulatory analysis, Severe Accident Mitigation Alternative analysis (SAMA) [34], rulemakings, risk-informed licensing actions, and backfits are used. According to Arrow et al. [35], with limited resources to spend on regulation, cost-benefit analysis is essential, as it can help clarify the trade-offs involved in making social investments. Furthermore, most economists would argue that economic efficiency, evaluated as the difference between benefits (e.g., the value of having a safer environment) and costs (i.e., the direct expenses of policy compliance as well as the indirect costs of time and training for PRA analysts), is a fundamental measure for testing proposed environmental, health, and safety regulations; and today, cost-benefit analysis is required for all major regulatory decision making. Currently, the NRC uses three guidance documents for cost-benefit analysis, NUREG/BR-0058, NUREG/BR-0184, and NUREG-1409. Cost estimates of public exposure (dollar per person-rem) and offsite property damage are considered in the analysis to justify the cost of safety enhancements and licensing actions. The analysis guidelines in NUREG/BR-0184 consider Three Mile Island as a low estimate of exposure level and Chernobyl as a high estimate. These estimates of exposure rates and cost are based partially on the Price-Anderson Act, which determines the liability insurance estimated from public claims or property damage claims. NRC's cost-benefit analysis guidelines are being updated to reflect new determinations of consequences, probabilities and uncertainties in the existing policies.

   There is great interest in using risk-informed regulation to allow more flexibility in the ways industry can reach compliance [33]. In our view, through risk-informed regulation and incentivizing PRA applications, innovations will emerge in all areas of NPP operation. For example, the Risk-Informed Asset Management (RIAM) process is where analysts review historical performance and develop predictive logic models and data analyses techniques to predict critical decision support figures-of-merit (or metrics) for managers of generating stations and electric utility company executives [36]. The RIAM metrics include (but are not limited to) profitability, projected revenue, projected costs, asset value, safety (i.e., catastrophic facility damage frequency and consequences, etc.), power production availability (i.e., capacity factor, etc.), efficiency (i.e., heat rate), etc. RIAM applies PRA techniques and generates predictions probabilistically so that metrics can be provided to managers in the form of probability distributions as well as point estimates. This application enables managers to apply the concept of "confidence levels" in their critical decision-making processes. There is an emerging list of other programs and activities that use PRA to promote the efficient functionality of a plant; risk-informed business modeling [37], on-line maintenance [38], safety assured maintenance scheduling and evaluation [39], Risk-Managed Technical Specifications (RMTS) [40], and risk-based project prioritization [41] to name a few. Significant work has been done to develop these applications, however, further research is needed to encourage the widespread use of these programs in high-risk industries.

2. RESEARCH STRATEGY

This section describes the proposed research strategy and demonstrates the necessary steps to reach the ultimate goal of determining the monetary value of PRA.

2.1 Selecting an Approach

As stated in the previous section, one method for calculating the monetary value of PRA is a purely probabilistic (correlation-based) approach that, through advanced regression analysis, finds the correlation between financial performance (as a measure of profitability and return on investment) and PRA. A statistically significant correlation would support the notion that the implementation of PRA would have a positive impact on the financial statements of the NPP operating organization. However, there are a number of deficiencies in this approach. First, due to the nature of regression analysis, finding redundant data is problematic. Secondly, due to the timelines of financial reporting, there would be a delay in the realization of the financial return on PRA programs. Finally, and most importantly, the change in the value of risk can be due to the change in the modeling approach, meaning that a change in risk does not necessarily represent a change in the real value of system risk or safety.

Considering these deficiencies, the monetary value of PRA could be redefined and looked at from another perspective. An inefficiently functioning NPP will cost more to operate, and will produce less energy. With a high frequency of component failure, an NPP will incur losses resulting from shutdowns. On the other hand, replacing and repairing the components too often will also result in a substantial increase in the cost of the operation [40]. There are various departments in an NPP carrying out different activities and programs that require an optimal financial management policy. Such a policy is cost-efficient in the way that it identifies and minimizes the associated costs and maximizes the benefits of each activity. Additionally, in complex industrial facilities such as NPPs, safety is a critical issue for decision-making; therefore, the optimal decision-making policy should consider both *financial* and *safety* factors.

The goal of this project is to show that using PRA not only advances the risk management of the NPP, but also improves the efficiency of operations and maintenance, resulting in considerable savings and added value to the economic bottom line (net earnings) of the organization. For example, in the maintenance department, events that are unimportant to safety can cause unnecessary plant shutdowns or significant resource allocations in order to determine a more reasonable Allowed Outage Time (AOT). A PRA-based program would bring more flexibility to maintenance (and operations) by expanding the operational envelope, and increasing the efficiency in estimating the times associated with surveillance frequencies, AOTs, etc., that, would help eliminate extra costs related to the deterministic methods of time estimation as well as to outage impacts. As an example, RMTS, a PRA-based program that has been implemented at STPNOC since 2007, allows the maintenance staff to exceed the front stop (or AOTs), and makes more time available to perform corrective maintenance without a significant concomitant increase in risk. This results in the prevention of unnecessary plant shutdowns that occur due to low-risk-in-service failures [42]. From this perspective, the monetary value of PRA would be the expected costs, as well as the anticipated profits generated, due to the implementation of PRA. The purpose of this research is to demonstrate that the use of insights from probabilistic risk assessment could be a significant aid in achieving risk-informed cost-efficient decision-making.

The proposed approach in this paper is an integration of *probabilistic* and *deterministic* approaches. The mechanisms and programs in an NPP that use PRA to reduce cost will be identified. In addition, the potential sources of uncertainty associated with cost saving in PRA-based programs will be identified in order to find a probabilistic monetary value of PRA.

2.2 Proposed Roadmap of the Research

The key steps in doing this research include:

1. *Developing a GFM to estimate the monetary value of a typical PRA-based activity in a nuclear power plant*: For the first step, a generic model that addresses and calculates the business return on a PRA project is developed. Considering PRA as an investment, one can calculate the ROI of PRA projects based on the modeled costs and benefits of a typical PRA activity. This model can then be modified for each PRA-based program at STPNOC and be used in later steps to evaluate the total return on PRA.

2. *Implementing the GFM for one of the PRA programs and validating the model*: Among the various PRA-based activities at STPNOC, the RMTS [40] is proposed as the basis of application for the first step of this research. The details related to RMTS are considered in order to modify GFM and to estimate the monetary return on PRA. This step aims to validate GFM.

3. *Conducting uncertainty quantification for the estimated ROI from Step 2*: All sources of uncertainty, according to the financial model, are identified and a probabilistic monetary value (ROI) of the PRA-based program of step 2 is presented.

4. *Identifying all the existing PRA programs at an NPP (e.g., STPNOC)*: The goal of PRA programs is to expand the operational envelope as widely as possible by increasing the robustness or resiliency of an operation. Therefore, such programs can bring a monetary *benefit* to the organization, in the form of profit generation or cost reduction. The maintenance and operations departments are the most likely places to implement such programs. Program examples include; RMTS, On-line Maintenance, Risk-Informed In-Service Inspections, Reactor Vessel Head Replacement [39], and RIAM.

5. *Estimating total monetary value of PRA*: This step includes obtaining the total return on PRA associated with all the PRA-based programs {i=1,...,k} identified in Step 2, running uncertainty analysis for the estimated value, and providing the probabilistic monetary value of PRA. The GFM developed in Step 1, details of each program and the associated sources of uncertainties would be considered in this step.

6. *Applying importance measure and sensitivity analyses to propose improvement approaches for PRA activities*: The total monetary value estimated in this project is based on the

current status of PRA activities at STPNOC. In implementing PRA for risk-informed decision-making, STPNOC has a very strong record of implementing successful PRA applications. Therefore, reporting the total monetary value that PRA has brought to STPNOC would be considered valuable to the industry, in order to realize the uncovered benefits of PRA that may otherwise be assumed to be expensive, luxury tools for NPPs. This derived value, however, is not the ultimate possible monetary value of PRA. Some of the PRA-based activities considered in this research might not yet have reached their maximum possible level of efficiency, and therefore, sensitivity analysis is used to present solutions for making these activities even more effective.

3. GENERIC FINANCIAL MODEL

This section explains Step 1 of the proposed approach (explained in Section 2.2.), which refers to the development of GFM. A definition for the total monetary value/return on PRA is shown as:

$$Return\ on\ PRA = ROI\_PRA = \sum \frac{Benefit_i - Cost_i}{Cost_i} \tag{1_a}$$

$\{i=1,...,k\}$ for all k PRA programs at STPNOC

The numerator in Equation 1_a, shows the Net Value of the PRA-program$_i$ and is based on the following definition:

$$NV\_PRA_i = Benefit_i - Cost_i \tag{1_b}$$

$\{i=1,...,k\}$ for all k PRA programs at STPNOC

Where **Benefit$_i$** is the monetary gain from the PRA Program$_i$, compared with not performing the program, or not implementing PRA.

Benefit$_i$ is defined as:

$$Benefit_i = Benefit_i^o + Benefit_i^r \tag{2}$$

In Equation 2, **Benefit$_i^o$** refers to the Ordinary monetary gains from program$_i$. This will be specified in each program based on its functionality. There are various resources for Benefit$_i^o$. When the number of unnecessary outages is reduced, or the plant uptime is expanded, there will be both more generated profit due to a growth in the production, and fewer incurred costs due to a drop in the operation. For instance, in the GSI-191 project, more cost-savings are realized from avoiding the need for changes in the design, insulations, and outage impact. Moreover, a PRA-based program$_i$ would help prevent a number of expected risks/accidents; therefore, it reduces the costs associated with those risks. As an example, in the GSI-191 project, the risk of worker radiation exposure is reduced because of the decreased outage impact; thus, there would be monetary gains realized due to reduction or elimination of the expected costs of accidents.

Considering this explanation, Benefit$_i^o$ is regarded as the combination of growth in the production, and a decline in the expected costs. The main principle for both is the increase in the plant uptime. The expected value of the increased uptime is a function of the expected average costs associated with *support organizations* (programs) had the plant been shutdown, plus the monetary value for the expected increase in the risk, plus the growth in the production (due to increased plant uptime). Eventually, the present value of this function must be calculated, in order to get the current value of the expected Benefit$_i^o$.

$$Benefit_i^o = F\,(c_j,\ g,\ x_i^o) \qquad (3)$$
$$\{i=1,\ldots,k\};\{j=1,\ldots,n\}$$

$\{c_1,\ldots,c_n,\ g\}$ = the expected average Costs, associated with support organizations, and Growth in the profit based on an increased production.

$$X_i^o = d*R_i^o \qquad (4)$$

The above calculates the dollar equivalent for the expected increase in the risk. This amount is associated with the condition of not conducting the PRA-Program$_i$. For instance, the worker radiation exposure dose would be monetized here. In order to calculate X_i^o, we use the product of **d**, the dollar conversion rate, and $\mathbf{R_i^o}$, the increase in risk, both from NRC standard guidelines and STPNOC records.

$$PV\_Benefit_i^o = [(1-\exp(r*t))/r]*F \qquad (5)$$

This is the present value of the expected ordinary benefits associated with the PRA-program$_i$, where r is the risk-free rate of return (by definition the interest rate on a three-month U.S. Treasury Bill), and t is the time-period analysis.

Obviously, there are certain resources of uncertainty and many what-if scenarios. Uncertainty quantification will eventually present a probabilistic range for **PV_ Benefit$_i^o$**, based on the expected changes in average future costs, future risk increases, and the sensitivity of the model to the dollar conversion rate. As mentioned in the previous section, this would be done in Step 5.

Returning to Equation 2, **Benefit^r_i** refers to the monetary gains associated with *rare* events. These gains are typically the benefits from mitigating economic impacts from severe accidents. Rare event severe accidents have expected economic/societal/human/health risks. Calculating the present value of the expected monetized impact of severe accidents provides us with the rare benefits associated with each program$_i$. To get the expected risk of severe accidents, PRA level 3 is needed, and this raises some concern due to the complexity of level 3 modeling. However, we can use generic values defined by the NRC in their standard guidelines [43].

$$Benefit_i^r = F\,(R_i^r) = d*R_i^r \qquad (6)$$

Benefit$_i^r$ is a function of $\mathbf{R_i^r}$, the expected risk of severe accident/rare event, and **d** is the dollar conversion rate. Again, calculation of the present value for the monetized rare benefits is as follows:

$$PV\_Benefit_i^r = [(1-\exp(r*t))/r]*F \qquad (7)$$

Where r can be the risk-free rate of return based on the return on US Treasury bills at the time of analysis, or it is assumed to be 7% (0.07/year) as recommended in NUREG/BR-0184. A value of 7% is conservative because cost estimates are usually performed by utilities using values between 11 and 15%. t is the time period. Time period analysis, t, can be defined based on either of the two options: the standard fiscal years at STPNOC, or the time remaining until the license term ends. Step 5 would then provide the corresponding probabilistic value.

Returning to Equation 1$_b$, **Cost$_i$** is defined as the total expenses associated with the program$_i$, which brings up equation 8:

$$Cost_i = Cost_i^p + Cost_i^b \tag{8}$$

The total cost associated with program$_i$ is the linear summation of two components: the Program Cost, and the Basic Cost. **Cost$_i^p$**, or the program cost, refers to the costs of initiating and conducting the program. An example of this is the research cost of the project. **Cost$_i^b$**, or the basic cost, refers to the costs of maintaining the PRA model of record at the plant or the forward cost of PRA. This basic cost includes costs such as the software license and maintenance, digital assets, and staffing labor. Financial statements of STPNOC (mainly the income statement) provide us with these amounts.

Now, the combination of Equations 1$_b$, 2, and 8, determines the Total Net Present Value of PRA:

$$\sum NV\_PRA_i = \sum Benefit_i^o + \sum Benefit_i^r - \sum Cost_i^p - \sum Cost_i^b \tag{9}$$

The focus of our research is on the monetary value of PRA; hence, the Severe Accident Benefit is assumed as $\sum$ **Benefit$_i^r$** = **Benefit** $_{SA\_Average}$. This is the average expected economic value of mitigating severe accidents (rare events) by conducting and implementing the PRA Programs. For the simpler calculations, this can be obtained from the NRC standard guidelines, and a generic risk value can be assumed [43]. In a similar approach, the total basic cost of PRA in the plant is defined as $\sum$ **Cost$_i^b$** = **Cost** $_{PRA\_Average}$. This represents the average PRA costs based on the STPNOC annual basic PRA expenses, which are related to performing basic PRA.

Rewriting Equation 9 results in:

$$\sum NV\_PRA_i = \sum Benefit_i^o - \sum Cost_i^p + Benefit_{SA\_Average} - Cost_{PRA\_Average} \tag{10}$$

The focus of this study is to find the return on PRA for the nuclear power plant as a complex socio-technical organization. In doing so, this research aims to uncover the positive effects of PRA on the financial statements of STPNOC. The positive effect is defined in comparison with not using PRA or not having the specific PRA-based program. This comparison narrows the focus of our model down to the first two components of Equation 10: the total Ordinary Benefits and the sum of PRA Program Costs. The Return On PRA would then be as follow:

$$ROI\_PRA = \sum \frac{Benefit_i^o - Cost_i^P}{Cost_i^P} \tag{11}$$

$\{i=1,...,k\}$ for all k PRA programs at STPNOC

Equation 11 gives us the monetary value of PRA at STPNOC. This value is compared with the cost of not implementing PRA or not conducting the program$_i$. The uncertainty quantification, which is addressed in Step 5, will provide us with a probabilistic representation of this value.

4. CONCLUSION

This paper reports on the current status of first-of-its-kind research for estimating the monetary value of PRA. Despite existing research that shows the importance of PRA in quantifying reliable values for risk, a review of the literature highlights a gap in the assessment of financial impacts of PRA implementation in complex socio-technical systems. Therefore, it is the goal of this research to provide, through an integration of probabilistic and deterministic approaches, an estimation of financial return on PRA in order to encourage the industry to innovate and implement new PRA activities and programs.

Demonstrating monetary value of PRA will help address (1) whether we can consider system safety/risk (calculated from PRA) as a competitive advantage that adds value to a business, (2) if there is a linkage between risk/safety and the financial performance, as two important outcomes of an organization, and (3) how the risk/safety status of a complex socio-technical system (such as an NPP) affects its financial outcome.

The proposed steps for this research include: (1) developing a Generic Financial Model (GFM) to estimate the Return On Investment (ROI) that results from profit generation or cost reduction associated with a typical PRA activity in an NPP (2) implementing GFM for one of the PRA programs and validating the model, (3) conducting uncertainty quantification for the estimated ROI from Step 2, (4) identifying the existing PRA programs at an NPP (e.g., South Texas Project Nuclear Operating Company; STPNOC), (5) Obtaining ROI for all the PRA activities of STPNOC, running uncertainty analysis for the total ROI, and providing a probabilistic monetary value of PRA, (6) applying importance measure and sensitivity analyses to propose improvement approaches for PRA activities.

This paper (a) summarizes the existing literature related to a linkage between financial performance and system risk, (b) justifies the proposed integrative probabilistic- deterministic approach for this research, and (c) explains the details regarding the development of GFM, which refers to Step 1 of the proposed approach.

5. REFERENCES

1. Mohaghegh, Z., Kee, E., Reihani, S., Kazemi, R., Johnson, D., Grantom, R., Fleming, K., Sande, T., Letellier, B., Zigler, G., Morton, D., Tejada, J., Howe, K., Leavitt, J., Yassin, H., Vaghetto, R., and Lee, S. *"Risk-Informed Resolution of Generic Safety Issue 191"*, International Topical Meeting on Probabilistic Safety Assessment and Analysis (PSA 2011), September 2013
2. Kee, E., Mohaghegh, Z., Kazemi, R., Reihani, S., Letellier, B., and Grantom, R. *"Risk-Informed Decision Making: Application in Nuclear Power Plant Design & Operation"*, Proceedings of 2013 American Nuclear Society, Risk Management Topical Meeting, November, 2013
3. Vaughan, D., *Autonomy, interdependence, and social control: NASA and the space shuttle Challenger.* Administrative Science Quarterly, 1990: p. 225-257.
4. Meshkati, N., *Lessons of the Chernobyl Nuclear Accident for Sustainable Energy Generation: Creation of the Safety Culture in Nuclear Power Plants Around the World.* Energy Sources, Part A: Recovery, Utilization, and Environmental Effects, 2007. **29**: p. 807-815.
5. Reason, J., *Human Error: Models and Management,* BJM Volume 320, 2000: P.769.
6. Mohaghegh, Z., *On the theoretical foundations and principles of organizational safety risk analysis.* 2007: ProQuest.
7. Mohaghegh, Z., *Socio-Technical Risk Analysis.* 2009, VDM Verlag. ISBN.
8. Mohaghegh, Z. and Mosleh, A., *Incorporating organizational factors into probabilistic risk assessment of complex socio-technical systems: Principles and theoretical foundations.* Safety Science, 2009. **47**: p. 1139-1158.

9. Mohaghegh, Z., Kazemi, R., and Mosleh, A., *Incorporating organizational factors into Probabilistic Risk Assessment (PRA) of complex socio-technical systems: A hybrid technique formalization.* Reliability Engineering & System Safety, 2009. **94**: p. 1000-1018.

10. Mohaghegh, Z. *Combining System Dynamics and Bayesian Belief Networks for Socio-Technical Risk Analysis.* in *Intelligence and Security Informatics (ISI), 2010 IEEE International Conference on.* 2010. IEEE.

11. Pence J., Mohaghegh Z., Ostroff, C., Kee E., Yilmaz, F., Johnson D. and Grantom R. , *"Toward an Internet of Organizational Safety Indicators by Integrating Probabilistic Risk Assessment, Socio-Technical Systems Theory, and Big Data Analytics"*, Proceedings of 12th International Topical Meeting on Probabilistic Safety Assessment and Analysis (PSAM12), 2014

12. Garrick B. J., *Quantifying and Controlling Catastrophic Risks,* Elsevier Inc. 2008

13. Golbe, D. L., *Product-Safety in a Regulated Industry: Evidence From the Railroads.* Economic Inquirty, 21, 1983: 39-52

14. Rose, N. L., *Profitability and Product Quality: Economic Determinants of Airline Safety Performance,* Journal of Political Economy, 1990: P. 944-964

15. Golbe, D. L., *Safety and Profits in the Airline Industry,* Journal of Industrial Economics, 1986: P. 305-318

16. Feinstein, J. S., *The Safety Regulation of US Nuclear Power Plants: Violations, Inspections, and Abnormal,* Journal of Political Economy, 1989: P.115-154

17. Marcus, A. Nichols, M., Bromiley, P., Olson, J., Osborn, R., Scott, W., Pelto, P., and Thurber, J., *Organization and Safety in Nuclear Power Plants,* NUREG/CR5437: NRC, 1990.

18. Madsen, P.M., *Perils and Profits: A Reexamination of the Link Between Profitability and Safety in U.S. Aviation,* Journal of Management, 2013, Vol. 39, No. 3, P. 763-791.

19. Veltry, A., Pagell, M., Behm, M., and Das, A., *A Data-Based Evaluation of the Relationship Between Occupational Safety and Operating Performance,* The American Society of Engineers, Vol.4, No.1, 2007.

20. Reason, J., *Human Error,* Cambridge University Press, 1990.

21. Baron, M.M., and Pate-Cornell, M.E., *Safety and Productivity Trade-offs: Managing Nuclear Reactor Outages,* Int. J. Technology Management, 2000, Vol. 19, Nos 3/4/5, P. 420-438.

22. Baron, M.M., Pate-Cornell, M.E., *Designing Risk Management Strategies for Critical Engineering Systems,* IEEE Trans Engng. Mgmt. 1999, 46(1): 87-100.

23. Murphy, D.M., *Incorporating Human and and Management Factors in Probabilistic Risk Analysis,* PhD. Dessertation, Dept. Industrial Eng. And Management, Stanford University, 1994.

24. Pate-Cornell, M.E., and Fischbeck, P., *PRA as a Management Tool: Organizational Factors and Risk-Based Priorities for Maintenance of the Tiles of the Space Shuttle Orbiter,* Reliability Eng. And Systems Safety, 1993, Vol. 40, P. 239-257.

25. Cowing, M.M., Pate-Cornell, M.E., and Glynn, P.W., *Dynamic Modeling of the Tradeoff Between Productivity and Safety in Critical Engineering Systems,* Reliability Engineering and System Safety, 2004, Vol. 86, P. 269-284.

26. Starr, C., Whipple, C., *Coping with Nuclear Power Risks: The Electric Utility Incentives,* Buchanan JR, editor. Nuclear Safety, Vol. 1, 1982: P. 1-7.

27. Scheberle, D., *The Night of the Gas: Why Bhopal Matters.* 2012.

28. Kurzman, D., *A Killing Wind: Inside the Bhopal Catastrophe.* New York: McGraw-Hill, 1987.

29. Presidential Commission, *Report on the Space Shuttle Challenger Accident,* Vol. 1, Washington DC, 1986.

30. Dunar, A. J., and Waring, S. P., *Power To Explore, History of Marshall Space Flight Center 1960-1990,* 1999. NASA.

31. Pate-Cornell, M.E., *Risk Assessment and Risk Management for Offshore Platforms: Lessons From the Piper Alpha Accident,* J Offshore Mech. Arctic Eng., 1993: 115: 179-90.

32. Smet, A. D., Palmer, R., and Schaninger, W. *The Missing Link, connecting Organizational and financial Performance,* Mckinsey & Company, 2007

33. Apostolakis, C. G., Lui, C., Cunningham, M., Pangburn, G., Reckley, W., *A Proposed Risk Management Regulatory Framework.* US NRC, NUREG-2150, 2012.

34. Severe Accident Mitigation Alternative (SAMA) Analysis, Guidance Document, NEI 05-01 [Rev A], 2005

35. Arrow, K. J., Cropper, M. L., Eads, G. C., Hahn, R. W., Lave, L. B., Noll, R. G., Portney, P. R., Russell, M., Schmalensee, R., Smith, V. K., and Stavins, R. N., *Is There a Role for Benefit-Cost Analysis in Environmental, Health, and Safety Regulation?* Science, Vol. 272, 1996: P. 221-222

36. Liming, J. K. and Kee, E. J., *Integrated Risk-Informed Asset Management for Commercial Nuclear Power Stations,* Proceedings of ICONE10 International Conference on Nuclear Engineering, 2002

37. Liming, J. K. and Grantom, C. R., *Risk-Informed Business Modeling for Nuclear Power Generation,* The 5th Conference on Probabilistic Safety Assessment and Management, 2000.

38. *Kee E., Richards, D., Disnard, R., Grantom, C.R., and Mikschl, T., Extensions to On-Line Maintenance Using BOP PRA Results: Initial Deployment in STPNOC Units 1 and 2.*

39. Ergunia V., *Safety Assured Financial Evaluation of Maintenance*, PhD Dissertation, Texas A&M University, May 2004.

40. Yilmaz F., Kee E., and Grantom R., *Risk-Managed Technical Specifications Application at STP: More Than Three Years of Experience,* ANS PSA 2011 International Topical Meeting Safety Assessment and Analysis, Wilmington, NC, 2011.

41. Koc, A., Morton, D. P., Popova, E., Hess, S. M., Kee E., and Richards D., *Prioritizing Project Selection*, The Engineering Economist: A Journal Devoted to the Problems of Capital Investment, 54:4, 267-297, 2009.

42. Belyi D. Damien P., Kee, E., Morton, D., Popova, E., and Richards, D., *Bayesian Nonparametric Analysis of Single Item Preventive Maintenance Strategies*, Proceedings of 17th International Conference on Nuclear Engineering, ICONE17-76050.

43. Generic Environmental Impact Statements for License Renewal of Nuclear Plants: Appendices (NUREG-1437, Volume 2), May 1996.

A methodology for determining of Plant Operating States of Low Power Shutdown Probabilistic Safety Assessment for the next-generation Nuclear Power Plants

Jae Gab Kim[a], Kwang Nam Lee[b], Hak Kyu Lim[a]

[a] KEPCO-ENC, Integrated Engineering Department, Korea, kjg@kepco-enc.com
[b] KEPCO-ENC, Power Engineering Research Institute, Korea, knlee@kepco-enc.com

Abstract: This paper outlines the Low Power Shutdown (LPSD) Probabilistic Safety Assessment (PSA) portion of a methodology for the determination of the Plant Operating States (POSs). This is to determine how best to characterize them for inclusion into the LPSD PSA. The characterization of POS will begin a review of available shutdown PSA studies for current generation plants. The next-generation Nuclear Power Plants (NPPs) provide useful references for POS development. Several sets of current and next-generation NPPs including NUREG/CR-6144 of Surry Unit 1 shutdown PSAs have been reviewed to identify potential POS. The POS defined for the next-generation NPP PSA must represent all conditions that can occur over the course of a fuel cycle. This paper considers all plant conditions except full power operation which is addressed with the internal events PSA. The development of POSs can lead to group plant states that require similar equipment, timing, and operator action to respond to an upset condition. POS Grouping is based on Technical Specifications (TS) requirement as well as key factors associated with the main shutdown risk contributors like RCS temperature, RCS pressure, RCS inventory, State of RCS pressure boundary, and Decay heat levels.

Keywords: PSA, LPSD Level 1, POSs

1. INTRODUCTION

The purpose of this paper is to identify unique plant operating states (POSs) during Low Power Shutdown (LPSD) operation. POSs will cover the LPSD evolution from full power operation to refueling conditions. During shutdown states, initial conditions, such as decay power and primary pressure, differ significantly from conditions during power operation.

The first step in evaluating each core damage sequence is the determination of POS. In the POS analysis, a thorough and systematic search was performed to define the spectrum of potential POS for the next-generation NPP. The available studies were reviewed to identify potential NPP shutdown states. In addition, the design control documents for the next-generation NPP were also reviewed to determine whether the current generation POS list was expected to remain applicable or not.

The POSs are expected operating conditions. These states are based on the existing outage practices common to all PWRs. It is expected that outage practices for the next-generation NPP will be similar to existing outage practices but POS for the next-generation NPP reflecting plant specific design feature can be developed. The developed POS can be used for the next-generation NPP PSA. The development includes planned shutdown refueling as well as unexpected shutdowns for unplanned maintenance and other events. As a result, the scope of potential states has been established. The task is to determine how to best characterize them for inclusion into the LPSD PSA.

The full power PSA is based on assumption that the plant power is 100%. But the POSs are various because the operational mode is changed as the process of planned outage. The LPSD operation encompasses low power operation, hot & cold shutdown process to cool RCS after reactor trip, disassembly work of reactor internals for refueling preparation, refueling, maintenance and test for equipment and components, assembly work of reactor internals. Plant equipment arrangements should be changed in order to do each process during LPSD. The success criteria of mitigating systems for abnormal accident like loss of Shutdown Cooling System (SCS) are dependent on the changing of plant arrangement. For example, for loss of SCS during LPSD operation, time to boiling and core damage is dependent on the level of decay heat and the area of RCS open part. Thus, in order to evaluate LPSD PSA, it needs to classify various plant configuration and operational conditions into several POSs and each POS should be applied by the same success criteria conservatively.

The next-generation technical specification (TS) can lead to group operational states into six operational modes based on reactor criticality, the temperature and pressure of RCS. But operational modes of TS are limited to reflect various and complicate plant operational states such as level changing of RCS water and refueling during LPSD. POSs are detailed plant arrangements based on six operational modes of TS and delivers basic information for LPSD PSA.

The POSs of the next-generation NPP are based on operational procedures and the POS classification of reference plants.

2. POS Characterizations for LPSD Level 1

The first step in LPSD PSA analysis is to identify POSs. Due to the continuously changing plant configuration in any outage, POSs are defined and characterized within each outage type. Each POS represents a unique set of operating conditions (e.g., temperature, pressure, and configuration).
In general, a POS is characterized in terms of TS mode(s), RCS conditions (RCS liquid inventory and SG availability), and RCS vent status. The TS mode determines the bounding RCS temperatures and pressures of the reactor core. The key safety functions during LPSD are considered to define and characterize each POS.

The POSs can be defined for the various purposes. For the evaluation of core damage frequency during the LPSD states, the parameters related to the RCS and core are focused on items as the POS classification factors.

2.1 Review of Potential Plant Operating States

The characterization of POS will begin with a review of available shutdown PSA studies for current generation plants. Several sets of current and next-generation NPPs shutdown PSAs have been reviewed to identify potential POS. These are summarized below.

○ Current-Generation Reactor Analyses
NUREG/CR-6144 (Reference 1) documented a shutdown PSA for Surry Unit 1 in 1994. It included a comprehensive set of POS that correlate well with those selected for the next-generation NPPs.

○ Certified Designs for Next-generation NPPs
The NRC has certified the Westinghouse System 80+, AP600 and AP1000 designs (Reference 2). The AP1000 report includes a shutdown PSA.

○ Next-generation NPPs Designs under Review
The NRC is reviewing Mitsubishi's US-APWR (Reference 3) and AREVA's EPR (Reference 4) for design certification. Both designs include LPSD PSA analyses. An AP1000 amendment is under review. These sources have the advantage of all being relatively recent (2011) documents.

Both the US-APWR and EPR analyses include a set of POSs that are very similar to those in NUREG/CR-6144. The published AP1000 discussion omits many details of the shutdown PSA, such that the POSs are not listed. The AP1000 discussion does note that drain-down evolutions and reduced inventory conditions are the dominant shutdown states. The AP1000 amendment does not propose any changes to the shutdown analysis as documented in the original, approved design control document.

○ Summary of POS Survey

The POS listed in the documents discussed above are summarized in Table 1. Comparing the POS horizontally across the rows, it is clear that the POS for each of the studies is similar, but POS reflecting plant specific design feature is developed.

○ Plant-Specific Experience

The plant-specific experience requires the consideration of plant-specific operating information, based on review and incorporation of plant-specific operating experience into the PSA, interviews with plant operators and other personnel, plant walkdowns, etc. Substantial industry experience with similar pressurized water reactors is available to confirm that the selected set of POSs is appropriate for next-generation NPP. However, as these reactors are currently in the design stage, no plant-specific experience has been acquired to date.

○ General Assumptions and Notes

The following are general assumptions and notes applicable to the POS development:

- Plant conditions that exist in forced outages result in no unique plant conditions not seen in refueling outages.
- The conduct of next-generation NPP outages will be similar to those performed for operating NPPs.

Table 1. POS Definitions from Various Sources

NUREG/CR-6144 Section 3.5 (June 1994) (Reference 1)	US-APWR Table 19.1-81 (Mar. 2011) (Reference 3)	EPR Table 19.1-87 (Aug. 2011) (Reference 4)
(1) Low Power Operation & Reactor Shutdown	(1) Low power operation (*)	(A) Power Operation (**)
(2) Cooldown with Steam Generators to 345°F	(2) Hot standby condition (*)	(B) Hot Standby to T > 248°F; (**)
(3) Cooldown with Residual Heat Removal to 200°F	(3) RHR cooling (RCS full)	(CAd1) RHR: RCS Normal Level with 2 RHR and SG (shutting down) (CAd2) RHR: RCS Solid with 4 RHR and SG (shutting down) (CAd3) RHR: RCS Solid 4 RHR (shutting down)
(4) Cooldown to Ambient Temperatures (using RHR)		
(5) Draining the RCS to Mid-loop	(4) RHR cooling (mid-loop operation)	(CBd) Mid-loop w/ RPV head on (shutting down) (Dd) Mid-loop w/ RPV head off (shutting down)
(6) Mid-loop Operation		
(7) Fill for Refueling		
(8) Refueling	(5) Refueling cavity is filled with water (6) No fuel in the core, or the core is partially offloaded (7) Refueling cavity filled (refueling)	(E) Cavity Flooded (fuel off load) (F) Core Off-load (E) Cavity Flooded (fuel onload)
(9) Draining the RCS to Mid-loop After Refueling	(8) RHR cooling (mid-loop operation)	(Du) RHR: Mid-loop w/ RPV head off

(10) Mid-loop Operations After Refueling		(starting up after refueling)
(11) Refill RCS Completely (After Mid-loop Operation)		(CBu) Mid-loop w/ RPV head on (starting up after refueling)
(12) RCS Heatup Solid and Draw Bubble	(9) RHR cooling (RCS full) (10) RCS leakage test (RHR isolated) (11) RHR cooling (RCS full)	(CAu) RHR: RCS Normal Level (starting up after refueling)
(13) RCS Heatup to 350°F		
(14) Startup with Steam Generators	(12) Hot standby condition	(B) Hot Standby (T > 248°F) (**)
(15) Reactor Startup and Low Power Operation	(13) Low power operation	(A) Power Operation (**)
Notes: (*) APWR POS 1 and 2 are analyzed with the at-power model. (**) EPR POS A and B are analyzed with the at-power model.		

2.2 POS Grouping

The POS defined for the next-generation NPP PSA must represent all conditions that can occur over the course of a fuel cycle. This document considers all plant conditions except full power operation which is addressed with the internal events PSA. The POS, therefore, represent the process or cycle from the reduction for power operation to refueling and back to power operation.

The POS for forced outages are subsets of those for a refueling outage. Furthermore, no unique plant conditions are expected for a forced outage that would not occur during a refueling outage. For example, a reactor trip will take the plant directly from power operation to hot standby bypassing low power operations on entry to the forced outage. From hot standby, the plant may be restarted or it may be cooled and depressurized further. If the plant is restarted, then forced outage will only involve two POSs. Although it may be argued that the decay heat load will be higher in hot standby following a reactor trip than after a planned shutdown, this difference is of negligible consequence. Some forced outages may take the plant down to cold shutdown for work which does not impact the reactor coolant system (RCS) pressure boundary. Other forced outages may require that the RCS pressure boundary be opened. Regardless of whether the RCS is maintained intact or not, all plant conditions that occur during a forced outage also occur during a refueling outage.

The duration of forced outages typically is shorter than refueling outages. Although past practices resulted in forced outages being significant contributors to plant unavailability, recent operating experience with existing reactors shows a very small contribution of forced outages to plant unavailability. In addition, improvements in design are expected to further reduce forced unavailability. With short duration of refueling outages expected for advance design reactors, the duration of any POS is the important consideration. Because the next-generation NPPs are under the design or construction stage, the length of any POS is postulated based on generic practices. Since the overall time spent in any outage is expected to be short for the next-generation NPPs design and since all plant conditions that are expected during forced outages are also expected during refueling outages, risk during forced outage is considered bounded by the risk of refueling outages and separate POS and analyses are not required for forced outages.

Development of POS groups plant states that require similar equipment, timing, and operator actions to respond to an upset condition. Groupings reflect Technical Specification (TS) requirements as well as key factors associated with the main shutdown risk contributors such as RCS temperature, RCS pressure, RCS inventory, State of RCS pressure boundary (vented or intact), Decay heat levels, and draining the RCS.

The POS scope Table 1 represents a reasonable division of potential states for use in LPSD PSAs. Some states could arguably be grouped together without losing resolution. On the other hand, the drain-down evolution can reasonably be divided into two states, one with and one without the pressurizer manway open, reflecting the presence of an open vent during drainage. However, the particular grouping is determined based on the expected conduct of plant outages and specific thermal-hydraulic features unique to a specific design.

The NUREG/CR-6144 analysis provides reasonable set of POS on which to base a LPSD PSA. These POS have been used as the basis for several industry analyses with little variation. The next-generation NPPs POS groupings will be based on those of NUREG/CR-6144. Details of the next-generation NPPs specific POS definition are provided.

2.3 Identification of Plant Operating States

The POSs defined for the next-generation NPP PSA are based on the 15 POSs defined in NUREG/CR-6144 with adjustments made to account for plant specific design feature. In addition, use of the POS defined in NUREG/CR-6144 is consistent with other recent analyses as summarized in Table 1.

The POSs defined for the next-generation NPP are segregated into two broad categories based on decay heat levels: high decay heat and low decay heat. High decay heat is when the reactor core contains only spent fuel, i.e., from shutdown before defueling. Low decay heat is when the reactor core contains some new fuel, i.e., from commencement of core reload to reactor restart.

Further distinction between POSs is based on the equipment available and needed to mitigate an accident sequence initiated while in each POS. For example, if the RCS is intact, use of feed and bleed cooling requires opening of valves to provide a vent path. However, if the RCS is not intact and the opening provides for adequate flow, then feed and bleed cooling can be accomplished without operator action to open valves.

The shutdown POSs for the next-generation NPP PSA are summarized in Table 2.

Table 2. The Plant Operating States for next-generation NPP

POS	Description	Primary System Water Level (1)	Primary System Pressure & Temperature	TS Mode
1	Reactor trip and Subcritical operation	In Pressurizer	2250 psia; 548-585°F	1, 2
2	Cooldown with Steam Generators to 350°F		2250-450 psia; 548–350°F	3
3A	Cooldown with Shutdown Cooling System to 212°F		450-15 psia; 350–212°F	4
3B	Cooldown with Shutdown Cooling System to 140°F		450-15 psia; 212–140°F	5
4A	Reactor Coolant System drain-down (pressurizer manway closed)	Below Reactor Flange	Slight positive pressure or depressurized; <140°F	5
4B	Reactor Coolant System drain-down (manway open)		Depressurized; <140°F	5
5	Reduced Inventory operation and nozzle dam installation			5
6	Fill for refueling			6
7	Offload	Cavity flooded		6
8	Defueled	N/A	N/A	Defueled
9	Onload	Cavity flooded	Depressurized or slight vacuum during refill;	6
10	Reactor Coolant System drain-down to Reduced Inventory	Below Reactor Flange		6

POS	Description	Primary System Water Level [1]	Primary System Pressure & Temperature	TS Mode
	after refueling		<140°F	
11	Reduced Inventory operation with steam generator manway closure			5
12A	Refill Reactor Coolant System (pressurizer manway open)			5
12B	Refill Reactor Coolant System (manway closed)		Depressurized, or at a slightly elevated pressure; <140°F	5
13	Reactor Coolant System heat-up with Shutdown Cooling System isolation at 350°F		15-450 psia 140–350°F	4
14	Reactor Coolant System heat-up with steam generators	In Pressurizer	450-2250 psia; 350–548°F	3
15	Reactor startup		2250 psia; 548-585°F	2, 1
(1) When level changes during a POS, the minimum level is listed.				

For the next-generation NPP, the plant operational parameters of 15 POSs are identified as follows;

- Technical Specifications (TS)
- The Level of Core Decay Power
- The level of RCS water and pressure
- The Primary temperature
- The States of RCS such as the RCS open part (i.e., pressurizer manway, SG manway, pressurizer vent valves, the heat of reactor vessel, and ICI tube)
- Plant Equipment Arrangements
- Success Criteria of Mitigating Systems for Abnormal Accident
- The Availability of Mitigating System
- The maintenance of front and auxiliary system
- System Design Feature
- The decay heat removal mechanisms
- Containment Status
- Before or after the Refueling
- The Outage Experience of Reference Plant.

It is assumed that the planned outage of the next-generation NPP is basically similar to that of reference plant as the same design concept of PWR. Thus, LPSD PSA of the next-generation has used the procedures and POSs reflected the experiences of the planned outages of reference plants.

The plant arrangements and RCS states are assumed as the same for each POS. The first 7 POSs from POS 1 to POS 7 show the progression of shutdown operation modes and before refueling process. The remained 7 POSs from POS 9 to POS 15 show the progression of the startup operation modes after refueling process. POS 8 shows defueled process.

2.4 POS Classification

Characterization according to the POS classification factors such as refueling, RCS status, and Containment status have been identified as follows:

Table 3. The POS Classification for the next-generation NPPs LPSD Level 1

POS	Description	Factors for POS Classification	
1	Reactor trip and Subcritical operation	The Level of Core Decay Power Mode 3 (Hot Standby)	Residual Heat Removal System Mode 4 (Hot Shutdown, SCS operation)
2	Cooldown with Steam Generators to 350°F		
3A	Cooldown with Shutdown Cooling System to 212°F	Mode 5 (Cold Shutdown, The Primary temperature of $\leq$ 99°C) Containment Isolation Status	
3B	Cooldown with Shutdown Cooling System to 140°F		The level of RCS pressure down to the Atmosphere
4A	Reactor Coolant System drain-down (pressurizer manway closed)	Pressurizer Manway Open	
4B	Reactor Coolant System drain-down (manway open)		The level of RCS water at reduced inventory operation Nozzle dam installation
5	Reduced Inventory operation and nozzle dam installation	Mode 6 (Refueling)	
6	Fill for refueling		
7	Offload	Core Location Change	Core Alterations
8	Defueled		
9	Onload		
10	Reactor Coolant System drain-down to Reduced Inventory after refueling	Mode 6 (Refueling)	The level of RCS water at reduced inventory operation Nozzle dam removal
11	Reduced Inventory operation with steam generator manway closure		
12A	Refill Reactor Coolant System (pressurizer manway open)	Pressurizer Manway Close	
12B	Refill Reactor Coolant System (manway closed)		The level of RCS pressure up from the Atmosphere)
13	Reactor Coolant System heat-up with Shutdown Cooling System to 212°F	Mode 4 (Hot Shutdown, The Primary temperature of > 99°C) Containment Isolation Status	Residual Heat Removal System Mode 3 (Hot Standby, SG operation)
13	Reactor Coolant System heat-up with Shutdown Cooling System to 350°F		
14	Reactor Coolant System heat-up with steam generators	The Level of Core Decay Power	
15	Reactor startup	Mode 2 (Startup)	

Conclusions

For the purpose of the development of LPSD Level 1 PSA of the next-generation NPP, POSs have been classified. This is based on the data of operating NPP and certified or under review for next-generation NPPs. The POSs for each of the studies are similar but each has been reclassified according to the TS requirements as well as key factors associated with LPSD risk contributors.

The POSs for the next-generation NPP have been classified into 15 POSs in total. By developing PSA Level 1 during outage according to the POSs for the next-generation NPP, it is possible to develop appropriate and reasonable risk assessments for LPSD.

References

[1]   Evaluation of Potential Severe Accidents During Low Power and Shutdown Operations at Surry, Unit 1, NUREG/CR-6144 (BNL-NUREG-52399), June 1994.

[2]   AP 1000 DESIGN CONTROL DOCUMENT, Chapter 19, Probabilistic Risk Assessment, Revision 11, certified January 2006.

[3]   DESIGN CONTROL DOCUMENT FOR THE US-APWR, Chapter 19, Probabilistic Risk Assessment and Severe Accident Evaluation, MUAP-DC019, Revision 3, March 2011.

[4]   U.S. EPR FINAL SAFETY ANALYSIS REPORT, Revision 3, Chapter 19, August 2011.

[5]   NUMARC 91-06, "Guidelines for Industry Actions to Assess Shutdown Management," December 1991.

Shutdown PSA for Ringhals NPP Unit 1. Insights, overview and results.

Stefan Eriksson[a]*, Marie Gryte[a], and Erik Cederhorn[b]

[a] Ringhals AB, Väröbacka, SWEDEN
[b] Risk Pilot, Stockholm, SWEDEN

Abstract: During 2011, 2012 and 2013 a Shutdown PSA (SPSA) has been developed for Ringhals NPP unit 1. Ringhals 1 is a Boling Water Reactor (BWR) made by ASEA-Atom situated at the West coast of Sweden. The SPSA supplement the existing PSA Level 1 and 2 for Ringhals 1 and the final outcome will give a complete risk profile for the unit, providing support for verification of plant safety and upgrades. This paper gives an overview of the level 1 SPSA. A description is made of the basic conditions for identification of Plant Operating States (POS), analysis of initiating events, sequence analysis and system analysis. The result for level 1 SPSA of R1 is briefly discussed.

Keywords: PSA, NPP, Shutdown conditions.

1. INTRODUCTION

Ringhals 1 is a Boiling Water Reactor (BWR) at a four-reactor site in the West coast of Sweden. During 2005 - 2009, Ringhals 1 has been undertaken a large modernization program including an additional I/C system, new diesel generators and a new cooling water supply chain. The program was initiated partly by findings in the previous Probabilistic Safety Assessment (PSA) analysis and partly by new regulations and demands from the regulatory body.

Several PSA studies have been made for Ringhals 1. The present study originates back to 2000 but has been complemented, revised and updated several times. The R1 PSA for at power is a full-scope PSA Level 1 & 2 covering both internal, external and area events. In the work with the upgrading of the reactor, the PSA model has been fully revised, e.g. the modeling of a Digital RPS complementing the old analogue RPS. For more information about findings and results see PSAM10-paper No. 14 - *Use of PSA in a Modernization Program. Findings and Results from the Ringhals 1 PSA*. Concerning details about the digital I&C refer to PSAM10-paper No. 110 - *Development of the Ringhals 1 PSA with regard to Implementation of a Digital Reactor Protection System*.

In 2011, Ringhals AB decided that an updated analysis of the remaining plant operating modes (POM) should be developed that would be integrated with the existing PSA. During 2011 to 2013 a PSA has been developed for shutdown operation. Today the shutdown study only includes PSA Level 1 and internal and external events. At the end of this year the SPSA will include a full-scope PSA Level 1 & 2 covering internal, external and area events.

---

* stefan.x.eriksson@vattenfall.com

2. OVERVIEW OF THE ANALYSIS

The analysis follow the main task in a SPSA Level 1 and 2, see figure 1. At Ringhals AB, the general procedure of performing a PSA is described in figure 1. The SPSA follow that procedure in all aspects.

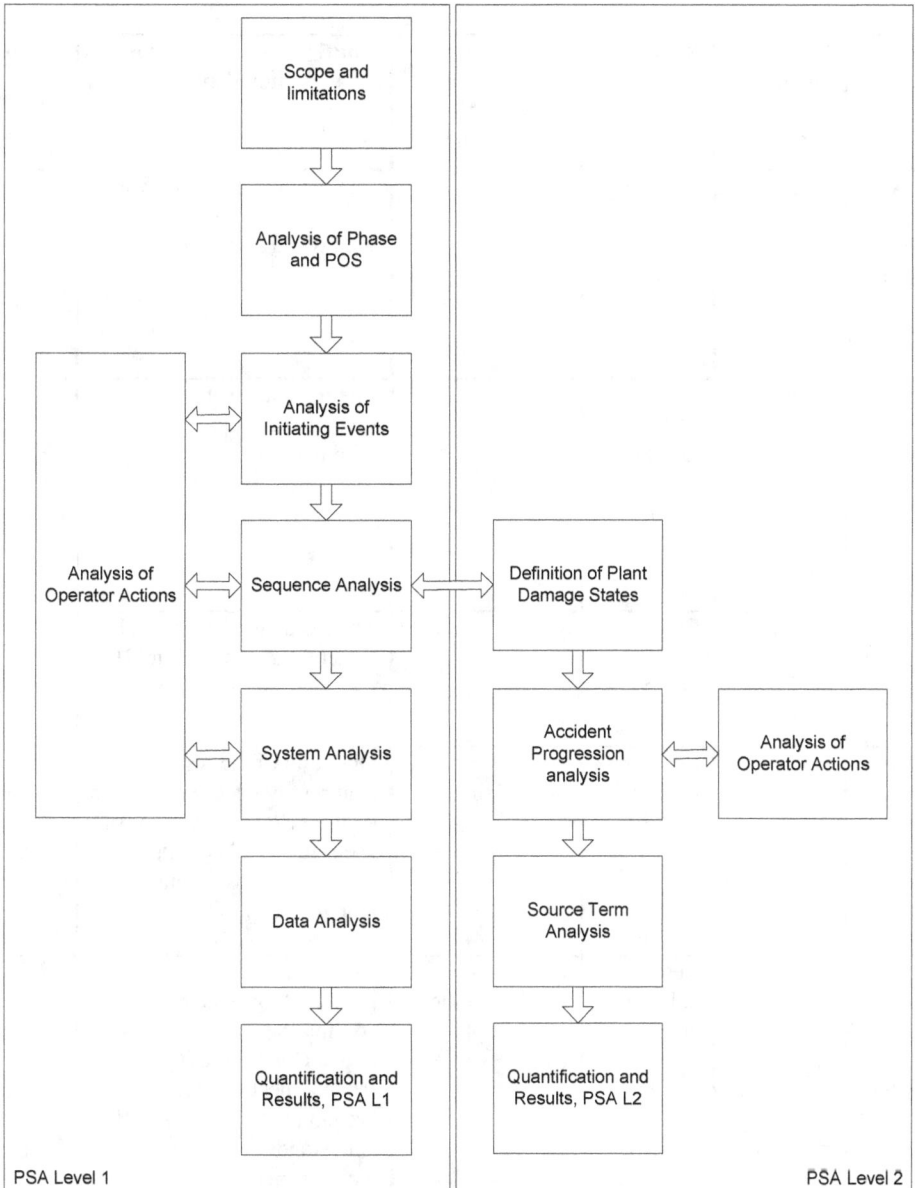

Figure 1   **The main tasks in a SPSA Level 1 and 2**

This paper describes main task for PSA Level 1. The procedure follows [1].

3. PLANT OPERATING STATES (POS)

Based on the shutdown procedure for Ringhals 1, Plant Operating States (POS) were defined as described in table 1.

Table 1: Ringhals 1, Plant Operating States (POS)

Phase	Description	Closed/ Open Primary System	Reactor Vessel Level/C-pool	Power supply unavailable because of maintenance	Configuration Residual Heat Removal system	Total time [h]
K1	Cold shutdown – Reactor Vessel Head mounted, water level under streamlines	Closed	Normal	-	The residual heat removal system (321) is cooling RPV with two trains.	20
K2	Cold shutdown – Reactor Vessel Head mounted, water level above streamlines	Closed	Top filled/ top filled above steam lines	-	The residual heat removal system (321) is cooling RPV with two trains.	36
K3	Cold shutdown – Open Reactor Vessel	Opened	Empty reactor hall pools	-	The residual heat removal system (321) is cooling RPV with two trains.	41
K4	Cold shutdown – Open Reactor Vessel. 40 h -7 days. B-side unavailable	Opened	Reactor hall pools are filled	Power supply B unavailable	One train of the residual heat removal system (321) is cooling RPV. Two trains of spent fuel pool cooling system (324) is cooling the reactor hall pools.	169
K5:1	Cold shutdown – Open Reactor Vessel. 7-14 days. B-side unavailable.	Opened	Reactor hall pools are filled	Power supply B. 50% of the time (in phase K5)	Two trains of spent fuel pool cooling system (324) is cooling the reactor hall pools. One train of the residual heat removal system (321) is cooling RPV is in standby, but maintenance on 321 possible.	253
K5:2	Cold shutdown – Open Reactor Vessel. 7-14 days. A-side unavailable.	Opened	Reactor hall pools are filled	Power supply A. 50% of the time (in phase K5)	Two trains of spent fuel pool cooling system (324) is cooling the reactor hall pools. One train of the residual heat removal system (321) is cooling RPV is in standby, but maintenance on 321 possible.	337

Phase	Description	Closed/ Open Primary System	Reactor Vessel Level/C-pool	Power supply unavailable because of maintenance	Configuration Residual Heat Removal system	Total time [h]
K6:1	Cold shutdown – Open Reactor Vessel. 14+ days. B-side unavailable.	Opened	Reactor hall pools are filled	Power supply B. 50% of the time (in phase K5)	Two trains of spent fuel pool cooling system (324) is cooling the reactor hall pools. One train of the residual heat removal system (321) is cooling RPV is in standby, but maintenance on one 321 and one 324 possible.	561
K6:2	Cold shutdown – Open Reactor Vessel. 14+ days. A-side unavailable.	Opened	Reactor hall pools are filled	Power supply A. 50% of the time (in phase K5)	Two trains of spent fuel pool cooling system (324) is cooling the reactor hall pools. One train of the residual heat removal system (321) is cooling RPV is in standby, but maintenance on one 321 and one 324 possible.	785
K7	Cold shutdown – Open Reactor Vessel. , 1 bar.	Opened	Empty reactor hall pools	-	The residual heat removal system (321) is cooling RPV with two trains.	920
K8	Cold shutdown – Reactor Tank idle on flange	Closed	Normal	-	The residual heat removal system (321) is cooling RPV with two trains.	1016

Cold shutdown is defined according to Technical Specification, as a subcritical reactor with water temperature below 100°C and the two operation mode switches turned to state "0".

4. INITIATING EVENTS

Identification of initiating events was made with the same condition as for the power operation PSA, i.e. the cladding temperature will reach above 1204°C due to loss of water inventory, loss of cooling, or reactivity transients (defined as BS1 for core damage in the RPV, BS2 for core damage in the spent fuel pit and BS 3 for core damage due to exposure of fuel rod during load/unloading because of outage LOCA. A time frame of 20 hours is defined. To handle or distinguish cored damage after 20 hours separate consequences are defined. Other consequences that are analyzed are:

- Exceeding of HTG (Highest accepted limit for the Pressure Vessel), primarily cold over pressurization.
- Exceeding of HTG for the temperature in the fuel pool (> 60°C)

Sequences where residual heat removal has been effective during this time frame, are considered to have a stable safe end state.

The sources of radioactivity considered in the analysis are:
- Reactor Pressure Vessel (RPV)
- The Spent Fuel Pit (SFP)
- Exposure of fuel rod during load/unloading because of outage LOCA

An initiating event in this analysis is an event with potential for leading to any of the unwanted end states and that may require functions for:

- cooling of the fuel in the reactor vessel/spent fuel pit
- maintaining applicable parameters as pressure, level and temperature in the reactor vessel and in the spent fuel pit within allowed limits
- reactivity control

An initiating event in the PSA model for cold shutdown is defined as an event that requires one or more manual alternatively automatically initiates actions to bring the plant to a safe end state. A screening value of 1·10-7 per year is used. This means that events with a frequency lower than the screening value are screened out from further consideration in the analysis.

The following initiating event categories are considered:

- Internal events (process related)
- Area events
- External events

Reference reports and background material forming the basis of identification and analyses of initiating events were:

- Ringhals Licensee Event Reports (LERs)
- R1 Safety Analysis Report (SAR)
- Nordic Owner Group report regarding safety during shutdown conditions [2]
- Previous PSA analyses at Ringhals
- Previous PSA analyses in Sweden (especially earlier shutdown studies at Forsmark NPP)
- Reference literature
- Specific work groups at the NPP (experts) identifying events to occur during shutdown

The Master Logical Diagram which describes the initiating event process is presented in figure 2. The categorization of initiating events follow [2]. Observe that CCI is added to list of initiating events.

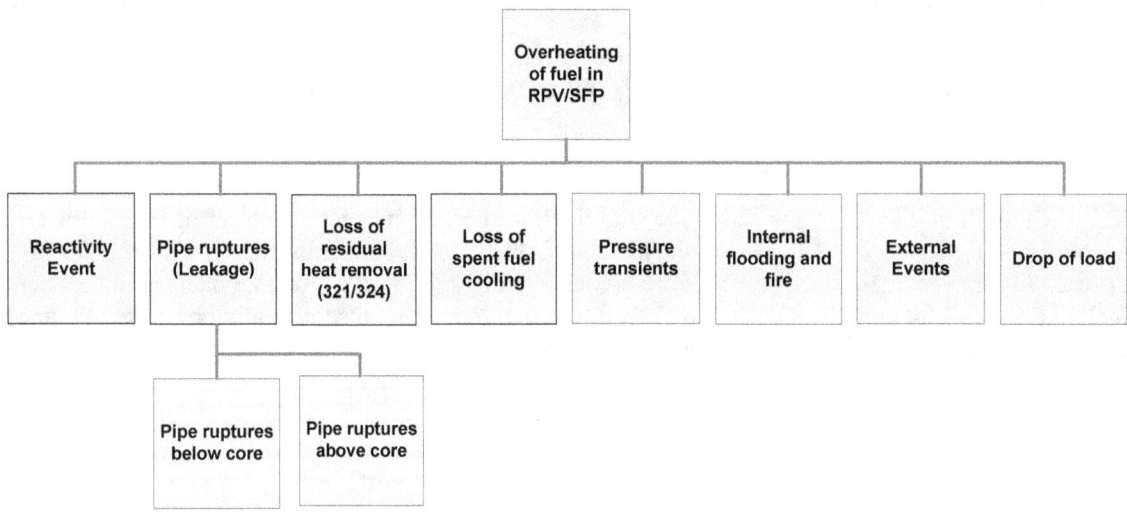

Figure 2: General Master Logic Diagram for overheating of fuel

5. SEQUENCE ANALYSIS

The sequence analysis follows the same model as for the power operation, thus it describes each sequence with a Success Block Diagram. All functions given in the Success Block Diagram (and subsequently in the event tree) are thoroughly described. The end state in the Success Block diagrams for the SPSA, Level 1 PSA will be some of the core damage consequences listed before (BSX) or safe state (OK). As far as possible, the structure of the full power PSA has been followed, but with focus on following functions:

- Pressure Relief and depressurization with system 314 and 326 (2 events for SPSA)
- Release of water to condensation pool through system 324 (1 event for SPSA)
- Water injection in Containment with system 733, 367 or 323 (3 events for SPSA)
- Closing of door between reactor pool and spent fuel pool (1 event for SPSA)
- Core cooling/Water injection in RC with system 416, 329, 733, 342, 322, 762, 323 (6 events for SPSA)
- Residual Heat Removal in RC/Containment with system 322, 321 or 324 (3 events for SPSA)
- Isolation functions leakage (3 events for SPSA)

For each of the identified initiating events, given in the previous chapter, a description is given as follows:

- Which POS are affected
- General success criteria
- Activation signals and time aspects

In all there are about 77 success block diagrams:

1. 14 for LOCA below the core
2. 18 for LOCA above the core
3. 4 for external LOCA below the core
4. 20 for external LOCA above the core
5. 4 for loss of residual heat removal due to loss of system 321 and/or 324
6. 4 for loss of residual heat removal due to CCI
7. 4 for loss of residual heat removal due to external events (loss of offsite power)
8. 4 for loss of residual heat removal due LOCA
9. 4 for loss of residual heat removal for spent fuel pool due to LOCA
10. 1 due to exposure of fuel rod during load/unloading because of outage LOCA

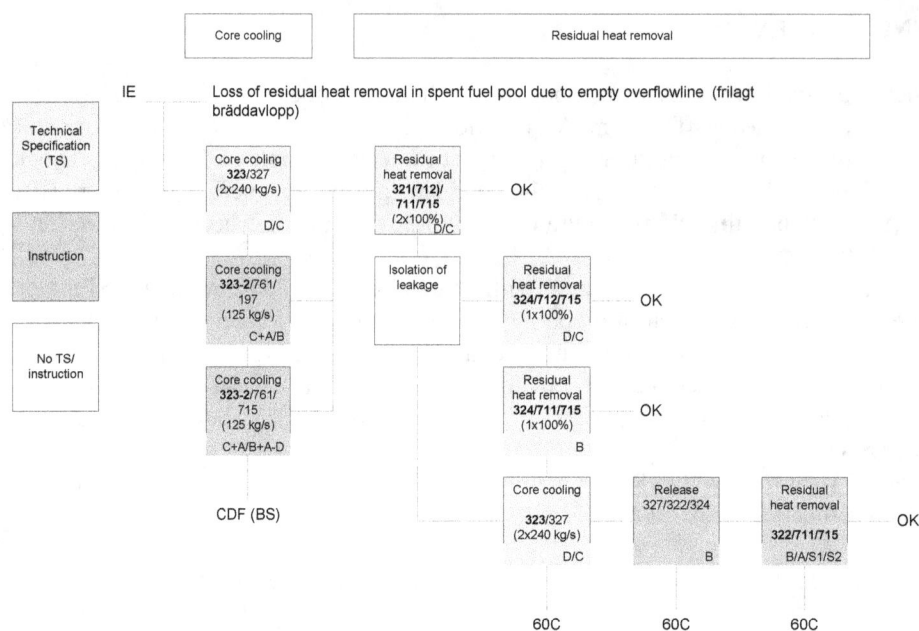

Figure 3: **Success diagram for LOCA_221: LOCA 221 control rod drive mechanism (leakage below core, 76 mm)**

6. SYSTEM ANALYSIS

The most important front line systems during shutdown conditions are
- System 314 – Pressure Relief System
- RH – System 321: Residual Heat Removal System
- SP – System 322: Containment Spray System
- SI – System 323: Safety Injection System
- System 326 – Reactor Vessel Head Spray System
- AF – System 329/416: Auxiliary Feedwater System
- SF – System 324: Spent Fuel Pool Cooling and Cleanup System
- System 367 – Mobile Pump for Containment Spray

In addition, the following support systems are covered by the systems analysis:
- CC – System 711: Cooling System for 321 and 322 (RHR and Containment Spray)
- SW – System 715: Salt Water System
- System 733 – Demineralised Water Storage and Distribution System
- FP – System 762: Fire Protection System
- Electrical system (overview)

For each system a description is given of:
- System Overview
- System tasks during shutdown
- System functions
- Assumptions and limitations
- Human actions related to system functions
- Fault tree modelling

7. HUMAN RELIABLITY ANALYSIS (HRA)

The included human interactions were divided according to IAEA-praxis [3], i.e. the categories A, B and C, where:
- Category A - Pre-incident tasks and errors
- Category B - Incident initiating errors
- Category C - Post-incident actions

Both screening analyses and detailed analyses were performed. The qualitative descriptions of the manual interactions, both for the screening- and detailed analyses, were emphasized.

7.1   Screening Analysis for Pre-Incident Tasks and Errors - Category A

The amount of category A actions are significantly lower in the outage analysis compared to the full power analysis. However, some category A actions are included in the outage analysis, and were identified departing from detailed descriptions of different scenarios. Screening values for category A actions were calculated based upon tables from e.g. THERP [4]. One example of an important category A action is the correct lining up of system 323 when performing maintenance on components which are critical from an outage-LOCA point of view.

7.2   Screening Analysis for Initiating Events – Category B

Manually initiated events for e.g. outage LOCA, loss of residual heat removal, and drops of heavy loads were analysed. For these three types of initiated events, different approaches were applied related to the explicit modelling. The most explicit modelling was done for the analysis of outage LOCA. For drops of heavy loads a semi-detailed analysis was done.

7.2.1   Screening Analysis for outage LOCA

In figure 4 the general model for analyzing outage LOCA is presented. For each of the sections in the figure a further subdivision is made. As an example, the section "possibilities for leakage when dismantling the component" consist of four different characterizations (D – Dismantling):
- D1 – A continuous and gradually increasing leakage will always occur in connection with the task, even if the task is not correctly performed. In order for a total (full scale) leakage to occur, the dismantling should be conducted for more than a minute despite of the increasing leakage.
- D2 – An abrupt leakage can occur even if the work procedures for dismantling the component are performed correctly.
- D3 – If the procedures for dismantling are followed on an overall level, but a couple of important steps in the procedure are not followed, there might be a leakage.
- D4 - If the dismantling is performed and a rather large deviation is done compared to what is stated in the procedures, there is a possibility that at leakage will occur.

For each characterization criteria (D1, D2, D3, D4, L1, L2 etc.) probabilities are assigned.

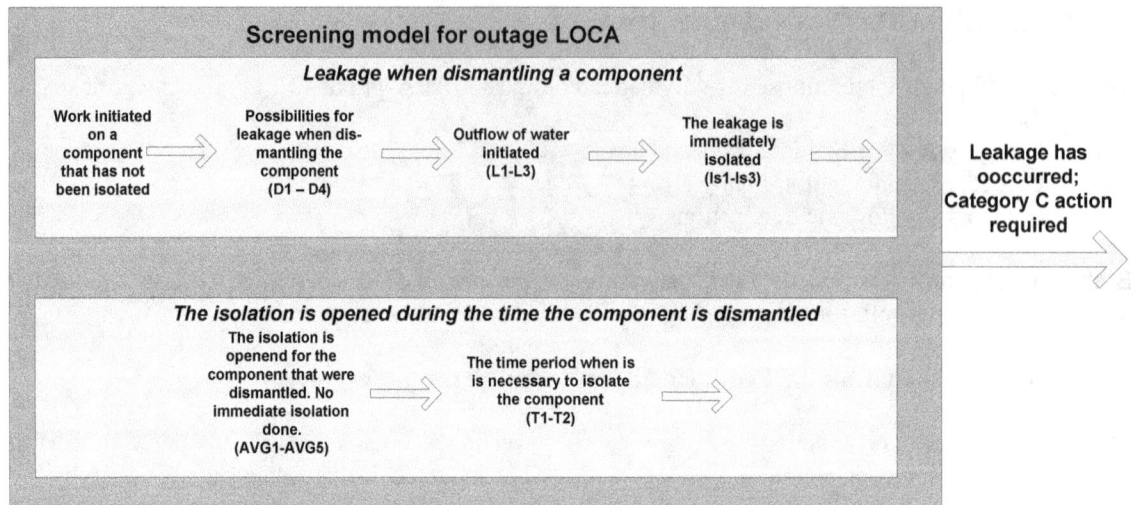

Figure 4: Screening model for outage LOCA

The general formula for estimating the probability for manually initiated outage LOCA for a specific component thus is:

$$P(\text{leakage}) = f(D/R, L, Is) + (AVG, T)$$

7.2.2 Screening Analysis for Loss of Residual Heat Removal

The probabilities for loss of residual heat removal due to manual interaction were estimated by using an expert judgment process (Delphi-influenced), in which three subject matter experts (SME) participated. They estimated the probabilities for losing systems that could either directly or indirectly lead to the initiating event, and both recoverable and unrecoverable loss of the systems were estimated. In table 2 two examples are shown.

Table 2: Example of probabilities for the loss of residual heat removal, on a systems level. For 90% of the cases the lost system can be repaired, for the other 10 % it is assumed that the lost system is unrecoverable.

System	Phase 1	Phase 2	Phase 3	Phase 4	Phase 5	Phase 6	Phase 7	Phase 8
7xx	9,8E-4	7,9E-4	2,5E-4	6,3E-3	8,3E-3	2,2E-2	6,6E-3	4,8E-3
6xx	5,6E-4	4,5E-4	1,4E-4	3,6E-3	4,7E-3	1,3E-2	3,8E-3	2,7E-3

7.2.3 Detailed analysis for initiating events

For a number of manually induced initiating events that contributed significantly to the core damage frequency, detailed analysis were done. Based upon several interviews, procedures, drawings and in some cases inspection of the actual work environment hierarchical task analysis (HTA) were made. These HTAs were then complemented with tabular task analysis in which for example possible errors, error mechanisms, consequences and barriers were identified. Performing shaping factors relevant for the respective works were identified and estimated, i.e. on a five grade scale ranging from "very bad support" to "very strong support" for the work. Finally, error probabilities for possible human errors were quantified based upon a Delphi-influenced expert judgement process. Three different estimates were made, i.e. the median value for the human error, as well as the values for the 5th and the 95th percentile. A triangular shaped distribution was assumed.

7.3   Screening Analysis for Post-Incident Actions - Category C

For manual actions that aim to prevent the initiating events from leading to core damage, or exceeding of HTG, an approach departing from THERP's time-diagnosis curve were used [4, 5, 6]. Five calibration factors (performing shaping factors) were applied [5, 6]. Based upon the result for these calibration factors and the available time for resolving the problem (i.e. primarily based upon the time from the initiating event to core damage of exceeding HTG, subtracting times for e.g. implementing actions) a probability for failing with the manual action were calculated. The following calibration factors were used and their values were estimated by SMEs:

1.   Quality and importance of procedures
2.   Quality and importance of training
3.   Feedback from process, quality of MMI
4.   Mental load
5.   Communication and coordination

In a few cases the time-reliability curve were not used. These cases consisted of mitigation actions (including observation, diagnosis and decision) that were only marginally cognitively demanding. In these cases THERP's ARM model [4] was used.

As a basis for all category C actions a rather detailed qualitative analysis were made, based upon interviews, procedures, HTAs etc.

7.4   Dependences

For assessing dependencies THERPS model [4] were applied.

7.5   Uncertainty Estimates and Reasonableness

Both uncertainty estimates and estimates of reasonableness were made for most of the included human interactions. The uncertainty estimates were done either by estimating the Beta-factors (primarily for the screening analysis), or by using Monte-Carlo simulations when performing Delphi-based expert judgments. When estimating the reasonableness of the results, the SMS reviewed the final results, primarily focusing on the internal ranking of e.g. the probabilities for manually initiated outage LOCAs for different components. Comparisons with actual data were done when such data existed. On a general level, the results were found to be reasonable.

In some cases the actual reasonableness for the actual human error probabilities were made. One example of the outcome from this was that the human error probability was not reasonable. In this case, the time reliability curve had been used when it was more justified to use the ARM-model.

8. RESULTS

The modelling of SPSA has been done in the same PSA model as for the R1 power operation model, i.e. the same Risk-Spectrum model has been used. The quantification has been done for internal events (including man-made initiating events) and the external event Loss of Offsite Power.

The preliminary results for level 1 SPSA of Ringhals 1 (to be finalized later this year) shows that the core damage frequency for the shutdown period is lower than for the full power operation mode but not significantly. The Plant Operating States 1 (cold shutdown. Reactor Vessel Head mounted, water level under steam lines) gives the largest contribution to the core damage frequency.

The preliminary results also show that there are no dominating sequences. The contribution from the sequence of highest order is just below 35 %. The relatively low core damage frequencies are probably due to:

- Events leading to cored damage after 20 hours not included in the results are sent to authorities
- Another reason for the low results is that events in the spent fuel pool are not considered in the results sent to the authorities
- Regulated and restricted instructions for which systems are to be operated during shutdown conditions

9. CONCLUSION

As for the preliminary results, the Level 1 SPSA indicates that the unit has large safety barriers during shutdown conditions. However, the final result must be studied before any final conclusion is made. From a PSA standpoint, the development of a Shutdown PSA does not differ much from an ordinary PSA. It is the author's belief that it is important to have an integrated model for all power operation modes. It is also important to have one common structure of the documentation. Having done that, it is fairly easy to progress with the PSA analysis. The difficulties lay, as always, in finding proper calculations (e.g. thermo hydraulics calculations of drain down events) in order to have the proper time frames. But that is probably the ever-remaining task for a PSA analyst, what is the available time for recovery actions?

An extensive amount of work is focused on a complete mapping of initiating events, even more compared to most other shutdown studies in Sweden. For example, every component (pumps and valves) in system connect to the reactor vessel and out to the isolation valve is mapped and analyzed regarding leakage rate, possibilities to isolate, level of studs and initiating frequency. Also the method for screening of possible initiating LOCA events inside containment due to faulty manual actions according is unique.

Because of the focus on a complete mapping of events showed that there were some events were core damage occurred after 20h (normal focus is on sequences with core damage occurred before 20h). To handle this type of event separate consequences was created. This feature makes it possible to analyze events with core damage after 20h.

Another unique feature compared to other shutdown studies is that the model also evaluate following events:

- Loss of residual heat removal for spent fuel pool due to outage LOCA
- Exposure of fuel rod during load/unloading because of outage LOCA

The biggest advantage of the new updated shutdown PSA for Ringhals NPP Unit 1 is that the model will support the possibility to analyze and plan future outages in a thoroughly and complete risk perspective.

Acknowledgements

The authors wish to acknowledge the support from Lennart Isaksson, Stefan Johansson and Stefan Peterson, operation support at Ringhals 1, for their support and wise words during the development and performance of SPSA.

References

[1] *A IAEA TECDOC-1144 - Probabilistic Safety Assessments of Nuclear Power Plants for Low Power and Shutdown Modes*, March 2000
[2] *NOG - Säkerhet och Miljö. Säkerhet under revisionsavställning. Delprojekt B - Komplettering av befintlig säkerhetsredovisning*, 2003-11-21

[3]   *Human Reliability Analysis in Probabilistic Safety Assessment for Nuclear Power Plants: A Safety Practice.* IAEA Safety Series No. 50-P-10

[4]   Swain, A.D., & H.E. Guttman, *"Handbook of Human Reliability Analysis with Emphasis on Nuclear Power Plant Applications",* NUREG/CR-1278/SAND80-0200, Sandia National Laboratories for the U.S. Nuclear Regulatory Commission, Washington, DC, August 1983

[5]   Holmberg, J.-E., Kent Bladh, K., Oxtrand, J., Pyy, P. *Enhanced Bayesian THERP — Lessons learnt from HRA benchmarking.* Proc. of PSAM 10 — International Probabilistic Safety Assessment & Management Conference, 7–11 June 2010, Seattle, Washington, USA, IAPSAM — International Association of Probabilistic Safety Assessment and Management, paper 52.

[6]   Holmberg, J.E. & Pyy, P. *An Expert Judgement Based Method for Human Reliability Analysis of Forsmark 1 and 2 Probabilistic Safety Assessment.* PSAM 5, International Conference on Probabilistic Safety Assessment and Management, November 27 – December 1, 2000, Osaka, Japan.

Developing a Low Power/Shutdown PRA for a Small Modular Reactor

Nathan Wahlgren
NuScale Power, LLC, Corvallis, OR, USA

Abstract: A growing area of interest in the field of nuclear risk analysis is the application of PRA techniques to low power and shutdown configurations when the availability of systems and components may differ significantly from normal operation. Many operating plants have performed (or are in the process of performing) a PRA for low power operations, and new reactor designs are required to complete one as part of the design certification process.

NuScale Power is developing a natural-circulation small modular reactor, and certain features of the design require refueling and maintenance procedures different from any in the industry. This uniqueness eliminates some sources of risk traditionally addressed in a shutdown PRA, but also introduces entirely new areas of risk. One major challenge is that all modules in the plant share a common refueling area, so each module must be lifted and moved from its operating location with fuel in the core. The module is completely disconnected and most systems credited in the full power PRA are unavailable when the module is in transit.

This paper will give an overview of NuScale's design and refueling process and discuss some of the challenges involved with developing a shutdown PRA for a reactor that is designed to be moved with fuel assemblies in place. Special attention is paid to determining a failure probability for a single-failure-proof crane with little directly applicable publicly available data.

Keywords: PRA, Low Power/Shutdown, Nuclear, Small Modular Reactor, NuScale

1. INTRODUCTION

NuScale Power, LLC is developing a small modular reactor that seeks to incorporate proven light water reactor technology with revolutionary design concepts to provide a modular approach to nuclear power that is both innovative and exceedingly safe. The design draws upon proven technology and materials while incorporating new design features to enhance operability and safety.

2. NUSCALE DESIGN OVERVIEW

A NuScale module is a self-contained assembly composed of a reactor core, a pressurizer, and two steam generators integrated within the reactor pressure vessel and housed in a high-pressure compact steel containment vessel. Each module uses traditional light water reactor fuel assemblies to produce 160 MWth, and a dedicated steam turbine to produce 45 MWe (net). Coolant flows through the RPV by natural circulation, with no reactor coolant pumps required for either normal operation or shutdown cooling.

A NuScale plant combines 12 reactor modules into a common reactor building to produce a total of 540 MWe (net). Each module operates independently, but all modules are managed from a single control room. The modules are submerged in a below-grade reactor pool that includes the spent fuel pool and a common refueling area. The pool functions as the ultimate heat sink for the backup cooling systems and also provides radiation shielding.

2.1. Safety Systems

Safety cooling systems are passively operated and can be passively actuated, with no power required for either function. The containment vessel on a NuScale module is a high-pressure steel vessel that functions as an integral part of the safety systems, conducting heat to the surrounding reactor pool using the simple physical processes of convection and conduction.

The decay heat removal system is analogous to the auxiliary feedwater system in a traditional plant, providing cooling through the steam generators when normal feedwater is not available. Valves on the main steam and feedwater lines redirect the flow of secondary coolant from the steam generators through a pair of closed loop two-phase heat exchangers mounted on the outside of the containment vessel, with the reactor pool acting as heat sink. Each heat exchanger is independent and capable of removing 100% of reactor decay heat.

NuScale's emergency core cooling system is unique to the industry, providing passive cooling in the event that normal feedwater and both trains of the decay heat removal system are unavailable. Steam exits the RPV through vent valves in the head of the RPV, condensing on the inside of containment and collecting in the bottom of the containment vessel. Recirculation valves mounted on the side of the RPV allow water to flow back into the RPV and are positioned at a height that maintains the water level in the core above the top of active fuel. The coolant in containment is cooled by the containment conducting heat directly to the reactor pool.

The volume of the reactor pool is sufficient to provide cooling for thirty days, by which time the decay heat has been reduced to a low enough level to allow the module to be air-cooled indefinitely. The emphasis on passive cooling, combined with a large volume of water in the ultimate heat sink, allows a NuScale plant to safely shut down and indefinitely maintain cooling with no operator action, no AC or DC power, and no additional water.

The emphasis on passive safety systems has enabled NuScale to achieve a Level 1 core damage frequency (CDF) for internal events less than 1E-7 per module critical year. Analyses of Level 2 and Level 3 internal and external events are currently ongoing.

3. REFUELING PROCEDURE

The unique design of a NuScale plant requires a refueling procedure different from any in the industry. The most obvious difference is that modules are not refueled in place, requiring that modules be transported while fueled. In addition, water is never removed from the RPV, eliminating drain-down events, and the reactor pool ensures that the module never occupies a condition that could be consider mid-loop. Cooling throughout the refueling procedure is maintained by the reactor pool, first by conduction through the containment vessel, and then by direct submersion when the RPV is opened. The planned refueling cycle for one module is 24 months, with outages staggered to allow other modules in the plant to continue operating.

After shutdown, the module is cooled using normal secondary cooling, then the containment is flooded and the reactor vent and recirculation valves opened to establish passive cooling by convection and conduction to the reactor pool. The module is lifted from its operating bay using a single-failure-proof (SFP) crane and transported to the refueling area, where it is disassembled. The lower portion of the containment vessel and RPV, including the core, remain in their stands in the refueling area while the upper vessels are transported to a dry dock area for maintenance and inspection.

3.1. Single-Failure-Proof Reactor Building Crane

The Nuclear Regulatory Commission (NRC) requires an SFP crane be used when lifting critical loads; a critical load is defined as a load that can be a direct or indirect cause of a release of radioactivity [1]. This is not limited to loads that contain radioactive material, but also loads that are lifted over or transported above safe shutdown equipment, where dropping a heavy load may damage systems or components relied upon to prevent core damage. A NuScale plant is laid out in such a way that modules do not pass over safe shutdown equipment at any time, eliminating that source of risk and leaving only the possibility of damage incurred by dropping the module.

General requirements for SFP cranes are given in NUREG-0554, Single-Failure-Proof Cranes for Nuclear Power Plants [1]. The criteria are that the system be designed so that a single failure will not result in the loss of the capability of the system to safely retain the load. Also required is that the crane must retain control of the load upon loss of electrical power and allow it to be lowered in a controlled manner. This is accomplished with a combination of redundant components, large safety margins, and rigorous procedures for both operation and maintenance.

4. DEVELOPING A LOW POWER/SHUTDOWN PRA

The low power/shutdown (LP/SD) PRA is a required part of the application for design certification, and an important tool in understanding risk present during refueling procedures, especially for a plant with no operating experience. The process involves identification of plant operating states, a screening process for existing initiating events, identification of new initiating events, modification of existing event trees and addition of new ones to construct a model that accurately depicts the module configuration during refueling operations.

4.1. Plant Operating States

It is standard practice for an LP/SD PRA to define a plant operating state (POS) for each configuration that occurs during an outage. Each POS has distinct initiating events, each with its own event tree. NuScale's LP/SD includes a POS for initial cooling, cooling with flooded containment, module disconnection and reconnection, transport to and from the refueling area, module disassembly and reassembly, and restart; the event trees are populated with systems that are available during that POS.

4.2. Initiating Events and Initiating Event Frequency

Initiating events for the LP/SD PRA are identified as those events that will cause a disruption to the critical safety functions of decay heat removal, coolant inventory, or reactivity control and require a response, either automated or by operators, to restore the stable condition of the plant.

When normal secondary cooling is taken offline, initiating events such as loss of feedwater, loss of condenser heat sink, and steam generator tube rupture can be screened out. Loss of coolant inside containment events can be screened out once containment is flooded, and loss of coolant outside containment events can be screened when active systems are removed from service and the containment is isolated. At this point the module is in cold shutdown, effectively immune to effectively all internal initiating events, including internal fires, internal floods, and loss of power. The module can occupy this state indefinitely without electrical power or further action from operators.

4.2.1. Initiating Event Frequency

For initiating events from the Full-Power PRA that are applicable to one or more POSs, a simple unit conversion is used to adjust the frequency. The adjusted frequency is used to account for the amount of time the frequency and duration of the POS, and also converts from units of per reactor critical year to per calendar year. The uncertainty distributions and parameters are not changed. The following equation is used to perform the adjustment:

$$f_{LP} = \frac{f_{FP}}{CF} \times f_{POS} \frac{d}{8760}$$

Where
- f_{LP} low power frequency, per calendar year
- f_{FP} full power frequency, per reactor critical year
- CF   module capacity factor, dimensionless
- f_{POS} frequency with which module enters POS, per calendar year
- d duration of POS, hours

For conservatism, the initial value of the module capacity factor is taken to be 0.844, the industry average for 2012[*] as calculated from the NRC's plant status data [3]. The f_{POS} term is estimated as the sum of the frequency of controlled shutdowns plus the refueling outage frequency, accounting for the fact that certain POSs will be applicable during each shutdown while others only apply to a refueling outage.

Representative frequency calculations are shown in Table 1 for three initiating events for POS1 (initial cooldown), POS2 (cooling with flooded containment), and POS7 (restart). For the purposes of this calculation, full-power frequencies are taken from generic values from the NRC Operating Experience Database [4] and expressed in units of per reactor critical year (rcry),. Two of these events are not applicable during POS2, as during that POS the module does not rely on secondary cooling or any system that requires electrical power.

Table 1: Sample Frequency Calculation for Initiating Events

Initiating Event	POS	Duration (hours)	f_{FP} (per rcry)	f_{POS} (per year)	f_{LP} (per year)
LOCA outside containment	1	10	3.67E-4	2.5	1.24E-6
Loss of secondary cooling	1	10	1.28E-1	2.5	4.33E-4
Loss of offsite power	1	10	6.14E-2	2.5	2.08E-4
LOCA outside containment	2	15	3.67E-4	1.5	1.12E-6
Loss of secondary cooling	2	15	1.28E-1	1.5	N/A
Loss of offsite power	2	15	6.14E-2	1.5	N/A
LOCA outside containment	7	20	3.67E-4	2.5	2.48E-6
Loss of secondary cooling	7	20	1.28E-1	2.5	8.65E-4
Loss of offsite power	7	20	6.14E-2	2.5	4.15E-4

4.3 Event Trees

Event trees in the LP/SD PRA are based on event trees in the Full Power PRA, especially for existing initiating events that are applicable to one or more POSs. The major change for all shutdown POSs is the removal of sequences that include a failure of the control rods to shut down the module. Several other changes are implemented to ensure that the top events reflect only those events that are applicable to the POS. For example, the definition of POS2 is that the containment is flooded with the vent and recirculation valves open; since opening these valves actuates the emergency core cooling system, any sequence that includes a failure of the emergency core cooling system to actuate are removed.

Quantifying the modified event trees for all POS not involving module transport gives a CDF that is approximately two orders of magnitude lower than that of the Level 1 PRA.

4.4. Reactor Building Crane in the LP/SD PRA

Due to the role that the crane plays in a NuScale plant, it is receiving special attention from both design and safety analysis engineers. Crane failure has been added to the LP/SD PRA as an initiating event, though the associated event trees are still in preliminary form as analyses of the potential effects of a crane failure are still in development.

---

[*] Although NuScale plants have no operating history, the design, including the power conversion system, is far simpler than existing design and therefore not subject to many of the upset events that can disrupt operations in the more complex plants that are currently operating. The industry average is therefore expected to be conservative.

4.4.1   Crane Failure Probability Estimation

The crane failure probability is estimated using operating experience data for cranes, which is compiled in NUREG-1774, A Survey Of Crane Operating Experience At U.S. Nuclear Power Plants From 1968 Through 2002 [2]. Cranes at nuclear power plants are used so frequently that it is difficult to find data of the total number of lifts performed, but the category of loads classified as "very heavy" (greater than 30 tons) was studied more closely by the authors of NUREG-1774; with a weight in excess of 500 tons, a NuScale module is certainly in this category. It was estimated that 54,000 very heavy load lifts were performed at nuclear power plants between 1980 and 2002, during which time nine failure events (six load slips and three load drops) were recorded. Note that most of these failures did not occur in SFP cranes.

Calculating a point estimate with these data gives a failure probability of 9/54,000 = 1.67E-4 per lift, however this is not a good indication of the failure rate of NuScale's crane. The narratives of the nine failure events suggest that none of the events are directly relevant to the NuScale design, due to the fact that the cranes involved in most of the failures were not SFP, or temporary rigging straps were not connected properly or failed, or the load was not dropped. A load drop caused by the mechanical failure of a single component in the temporary rigging system is not credible for NuScale crane due to the single-failure-proof crane and the dedicated coupling mechanism it uses to interface with the module, whereas a load dropped by a SFP crane caused by human error is more relevant.

A weighting system was developed to adjust each failure event for relevance. The narrative of each event was used to identify the consequence (slip or drop), the cause (human error, mechanical failure, or rigging), and the crane used (SFP or non-SFP). A weighting factor was assigned to each category, and the product of these weighting factors was used as the equivalent number of failures for that event. The sum of all nine equivalent failures is used to calculate the failure probability.

Weighting factors were determined by engineering judgment. A slip is assigned a consequence factor of 0.5, implying that two load slips have the same impact as one drop. A drop is assigned a consequence factor of 1.0. Human error is assigned a cause factor of 1.0, and mechanical and rigging failures are each assigned a cause factor of 0.1. The crane is designed to prevent mechanical failures from causing a drop, and the module is lifted with a purpose-built and permanent rigging device that attaches to the same points on the module each time, eliminating the need for temporary moveable rigging that is reattached at each lift. A failure involving a non-SFP crane is assigned a crane factor of 0.1 and those involving an SFP are assigned a crane factor of 1.0. By this system, the most relevant events will be counted as one failure, with each factor reducing the worth to less than that of a full failure.

The weighting factors are shown in Table 2 and application to the operating experience data is shown in Table 3.

Table 2: Weighting Factors for Crane Failure Events

Consequence	Factor	Cause	Factor	Crane	Factor
Slip	0.5	Human	1.0	SFP	1.0
Drop	1.0	Mechanical	0.1	Non-SFP	0.1
		Rigging	0.1		

Table 3: Applying Weighting Factors to Operating Experience Data

Date	Plant	Consequence	Cause	Crane	Equiv. Failures
11/1985	St. Lucie 1	Slip	Mechanical	Non-SFP	0.005
4/1990	Fort Calhoun	Slip	Rigging	SFP	0.050
9/1993	Arkansas Nuclear One 1	Slip	Human	SFP	0.500
12/1997	Byron	Slip	Human	Non-SFP	0.050
10/1999	Comanche Peak	Slip	Mechanical	Non-SFP	0.005
11/1999	Crystal River 3	Slip	Rigging	SFP	0.050
12/1997	Byron	Drop	Human	Non-SFP	0.100
5/2001	San Onofre	Drop	Rigging	Non-SFP	0.010
6/2001	Turkey Point 4	Drop	Rigging	Non-SFP	0.010
				Total	0.780

The 0.780 equivalent failures are used to estimate a failure probability of 0.780/54,000 = 1.44E-5 per lift, reducing the original estimate by an order of magnitude to approximately one failure is 70,000 lifts.

The uncertainty for this event is assigned a lognormal distribution with a error factor of 10 to account for the uncertainty in engineering judgment. OpenBUGS was used to perform uncertainty sampling calculations, resulting in a 90% confidence interval of 5.37E-7 to 5.38E-5, as shown in Table 4; the script used to generate these numbers is given in the Appendix.

Table 4: Summary of Uncertainty Sampling

Mean	Standard Deviation	5% Value	Median	95% Value
1.437E-5	3.411E-5	5.368E-7	5.383E-6	5.350E-5

4. CONCLUSION

NuScale's innovative design has proven to be exceedingly safe in the realm of normal operations, and the preliminary LP/SD analysis indicates the refueling process can be executed safely as well, with the CDF due to internal events approximately two orders of magnitude below the full power CDF. Future work will involve a more detailed examination of the crane that incorporates the results of analyses currently underway, as well as an expansion of the LP/SD PRA to include internal fires, internal floods, and external events.

APPENDIX

OpenBUGS script used to perform uncertainty sampling. The script was written by Sara Misic of NuScale Power, LLC.

```
Component : Crane
Failure Mode: Crane failure
Model: Lognormal distribution fit to data with error factor = 10
Analyst: Sara Misic
Date: 02/03/2014

Model {
lambda ~ dlnorm(mu, tau)
mu <- log(1.44E-5) - pow(log(EF)/1.645,2)/2
tau <- pow(log(EF)/1.645,-2)
}

Data
list(EF= 10)

end
```

References

[1]   U.S. Nuclear Regulatory Commission, "Single-Failure-Proof Cranes for Nuclear Power Plants," NUREG-0554, May 1979.

[2]   U.S. Nuclear Regulatory Commission, "A Survey Of Crane Operating Experience At U.S. Nuclear Power Plants From 1968 Through 2002," NUREG-1774, July 2003.

[3]   U.S. Nuclear Regulatory Commission, *Power Reactor Status Reports for 2012*, 1/30/2014, http://www.nrc.gov/reading-rm/doc-collections/event-status/reactor-status/2012.

[4]   U.S. Nuclear Regulatory Commission, "Industry-Average Performance for Components and Initiating Events at U.S. Commercial Nuclear Power Plants," NUREG/CR-6928, February 2007 (2011 data update).

[5]   NuScale Power, LLC, *NuScale Power Technology*, 2/3/2014, http://www.nuscalepower.com/ourtechnology.aspx.

Risk-Informed Design Changes of an Advanced Reactor in Low Power and Shutdown Operation

Ji-Yong Oh*[a], Ho-Rim Moon[a], Han-Gon Kim[a] and Myung-Ki Kim[a]

[a] Korea Hydro and Nuclear Power Co. Ltd, Central Research Institute, Deajeon, Korea

Abstract: APR+ has been developed in Korea since 2007. APR+ adopts various advanced safety features including passive auxiliary feedwater system, four emergency diesel generators. Through the implementation of the advanced designs, APR+ increased the safety to the world best level of evolutionary reactors. The full power core damage frequency or containment failure frequency decreased significantly comparing to APR1400 that is base model of APR+. However, low-power shutdown risk has not been improved substantially. This paper suggests several design changes that optimize low-power shutdown risk. Based on the design alternatives, this paper discusses risk effectiveness of the proposed design including various factors, e.g. equipment reliability, human error, training, procedure and so on.

Keywords: PRA, LPSD, Mid-loop, SDC

1. INTRODUCTION

Before 1980s, people in nuclear industry believed that the level of reactor decay heat during low power and shutdown (LPSD) state is very low comparing to the case of normal operation mode. People also believed that operator could have enough time to manage the accident in LPSD state so that the associated risk might not be significant and might be ignorable in comparison with full power state. In 1987, Diablo canyon nuclear power plant experience the accident that result from the loss of residual heat removal function during mid-loop operation. The industry and regulatory had focused on the safety issue for the shutdown sate, particularly the hazard on mid-loop operation. Nuclear Regulatory Commission (NRC) had investigated the accident in Diablo canyon and inspected the readiness of the same kind of accident for other plants in the U.S. The industry and NRC investigation resulted that the decay heat level might not be so small and the available resources of safety systems during LPSD might not be sufficient so that the associated risk for LPSD would not be ignorable comparing to that of full power. On the following, NRC issued a generic letter [1] urging the industries to implement the expeditious actions as well as programmed enhancements for LPSD. On the risk perspective, the industry provided the guidance [2] for industry actions to assess shutdown management.

The APR+ design adopts various advanced safety features including passive auxiliary feedwater system, four emergency diesel generators, rigorous containment design preparing for aircraft impact, and so on. Particularly, the design has four independent trains separating with the concept of mechanical, electrical and physical. Note that these four independent features enable the APR+ design to implement on-line maintenance with an effective manner. Through the implementation of the advanced features, the risk level of APR+ design for full power is the world best level of evolutionary reactors. However, in terms of LPSD, the efforts improving the safety have not been made a prominent progress during the stage of standard design approval [3]. In order to reduce the overall risk associated with all modes and all hazards, the appropriate process or method should be developed to deal with the LPSD risk.

This paper discusses the LPSD risk of the APR+ design. First, the characteristics of LPSD risk are presented including the concept of plant operational state, mid-loop operation, and special initiating events in section 2. In section 3, a new process is proposed to identify the design alternatives

* Corresponding Author, teslar@khnp.co.kr

improving LPSD risk. This process also performs the conceptual level designs for the alternatives. Moreover, the sensitivity analyses associated with the design alternatives are evaluated and the best design option is determined based on the process. In conclusion, this paper discusses the positive and negative effect when the proposed design is implemented.

2. CHARACTERISTICS OF LOW POWER SHUTDOWN RIKS

2.1. Plant Operational State (POS) Definition

According to the continuously changing plant configuration in any outage, plant operational states (POSs) are defined and characterized. Each POS represents a unique set of operating conditions (e.g., temperature, pressure, and configuration). For the typical refueling outage of PWR, up to 15 POSs are usually used, representing the evolution of the plant throughout a refueling from low power down to cold shutdown and refueling, and back-up to low power [4].

The LPSD risk model associated with APR+ design adopts 15 POSs including two more sub states in POS 4, 10. The 17 POSs are divided based on six operating modes in technical specification, reactor coolant system (RCS) water level, RCS opening (pressurizer manway, SG manway), active core in reactor, and maintenance schedule of main safety systems and supporting systems. Plant configurations of POS 1,2,14 and 15 are the same as these of full power and POS 5, 11 are correspond to mid-loop operation. POS 7, 8 are associated with fuel loading and reloading. The durations of POSs are referencing to the outage practice of Shin-Gori 3&4 plant [5].

2.2. Mid-Loop Operation

In mid-loop operation mode, operators decrease RCS level to mid-level of hot leg to install SG nozzle dams and replace seal or journal bearing of reactor coolant pumps. After that, operators increase RCS level up to the top of hot leg. Through the mid-loop operation, SG eddy current test (ECT) or maintenance work can be performed in parallel with core alteration so that the significant amount of the outage time can be save impacting on the plant economies.

On the contrary, RCS level is too low and its associated time to boil or core uncover is also very short that caused relatively high risk state during mid-loop operation. Particularly, the loss of shutdown cooling function during mid-loop is one of the most vulnerable events that some evolutionary plants have experienced. In addition, during outage including mid-loop, the available critical safety functions can be limited, for instance SGs and some safety injection pumps may not be available after RCS is drained and breakers are racked-out for the purpose of preventing inadvertent injections. In general, mid-loop operation is the most important stated during LPSD on the perspective of risk so that the cautious RCS level control as well as the continuous monitoring of shutdown cooling function are essential in this operating.

2.3. Initiating Event for Low Power Shutdown Operation

NUREG/CR-6144 [4] documented a shutdown PRA for Surry Unit 1 in 1994. It provided a sound analysis of a comprehensive set of initiating events and then-current data. LPSD risk model of APR+ adopts NUREG/CR-6144 [4] as the basis of initiating event. In addition to that, Standard Safety Analysis Report (SSAR) [6] of APR+, specifically LPSD parts and the PSA report of Sin-Kori 3&4 [5] are reviewed to identify LPSD initiating events.

There are two major categories for IE, i.e. Loss of Coolant Accidents (LOCAs) and transient events. LOCAs are divided into more detailed levels, i.e. unrecoverable LOCA (CVCS letdown line), LTOP safety valves fails to reclose, and so on. The categories of transients are almost the same as that of full power mode. The loss of shutdown cooling (LOSC) is one of LPSD specific initiators and the most important risk contributors in LPSD operation modes. LOSC is divided into four sub-level categories as following:

- S1 - Recoverable Loss of Shutdown Cooling System
- S2 - Unrecoverable Loss of Shutdown Cooling System
- SO – Over-drainage during Reduced inventory Operation
- SL – Failure to Maintain Water Level during Reduced Inventory Operation

The former two IEs are initiated by the mechanical failure of shutdown cooling (SCP) and the latter two IEs are caused by SCPs failure due to the cavitation by insufficient suctions. The most dominant IE during LPSD is over-drainage during reduced inventory operation, which results from inadequate level control of operators or the failure of level instruments.

3. DESIGN ALTERNATIVES

The design alternatives are identified by the overall process that is developed in this paper. The overall process uses the LPSD PSA model of APR+ design and identifies the design alternatives that provide additional safety functions and prevent plausible accidents during LPSD operation modes. Conceptual designs for the alternatives are performed and the design effectiveness is evaluated by the sensitivity analysis with the associated risk parameters. With the outcome for the alternatives, the priorities are determined with a comprehensive evaluation considering the negative influences for other areas.

3.1. Overall Process

In order to identify appropriate design alternatives that possibly minimize LPSD risk, developing stage PRA model of APR+ and its results, for instance CDF lists depending on POS and IE, the importance values, minimal cutsets (MCSs) are used [7]. In the first step, important operational mode, accidents and accident mitigation functions are identified by reviewing CDF lists for POSs, IEs and importance values. Important sequences are discussed with the MCS analysis of APR+ design. In the second step, the preventive measures and design alternatives are elicited based on the result of the first step. The conceptual designs for the alternatives are performed with the cooperation of risk analyst and system and component designers. Specifically, the interpretation of MCSs identifying the importance accident sequences on the stand point of risk and the findings of their remedial measures are very important in this step. The third step evaluates the effectiveness for the proposed design alternatives through the sensitivity analysis of associated risk parameters. The purpose of this step is not to obtain explicit risk value but to acquire the risk insights from the proposed designs. Therefore, more comprehensive interpretation should be done in the determination of risk parameters. In the fourth step, the priorities are determined and the negative influences are discussed with more comprehensive perspectives including industrial safeties, licensing issues and design varying controls. The final alternative selected by the former steps proceeds to the detailed design and risk evaluation step. Particularly, an iterative process is carried out throughout the first step and the third step in the process. The actual applications of the process are implemented with APR+ design in the followings.

Figure 1: Design Optimization Process for LPSD risk

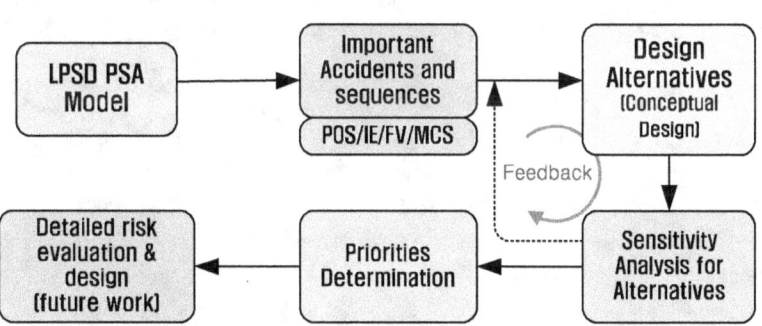

3.2. POS/IE/Importance/MCS analysis

The result of LPSD PRA for the APR+ shows that the most risk values are concentrated on the mid-loop and associated drain operation that correspond to POS 5&11 (mid-loop), and POS 4B&10 (drain operation). POS 5&11 and POS 4B&10 take 55.9% and 23.9% portions of the total LPSD CDF, respectively. Particularly, POS 5 (first mid-loop operation) takes 44.8% of the total LPSD CDF and identified as the most vulnerable operation state in LPSD.

In terms of the LPSD initiating events, SO (over drain), S1 (recoverable SCP fail) and S2 (unrecoverable SCP fail) are dominant and take 28.3%, 18.7% and 14.4% portions of the total LPSD CDF. These events categorizing the loss of shutdown cooling are major initiating events that lead to core damage with high frequencies. Except these kinds of IEs, the noticeable events are the loss of component cooling water, station block out, and stuck open of pilot operated safety relief valves. Particularly, some of AC sources e.g. EDG, UAT, and SAT might not be available even in the mid-loop operations since the component maintenance activities.

With the view of the importance analysis, the SO initiating event has the highest value of Fussel-Vessely (FV) [8] where the events contribute to the system reliability. Following the events, the common caused failure (CCF) of essential chillers, sump plugging event under feed & bleed operation, and operator action to RCS inventory recover ranks the top four of the FV values. The events associated with high importance event, e.g. dependent events are ranked high in the FV lists.

In the MCS analysis, the most probable sequence is occurred in POS5 and the sequence takes 13% portion of the total LPSD CDF. The sequence can be demonstrated by the following detailed scenario description. During the start of mid-loop operation, operator errors or level indication failures lead to the loss of shutdown cooling. Operator actions for recovering RCS inventory using available SCPs are attempted but failed. And then, Operators tried to feed & bleed operation using the available safety injection pumps. When the trial also failed, operators attempted the feed Y steaming operation using the charging pumps. All these mitigation systems with associated operator actions are failed sequentially and finally the core reached the success criterion temperature of CDF. The scenario of the second high sequence is similar to that of the first one except the initiating event and first mitigation action. The sequence is also occurred in POS5 with the recoverable SCP fail and operators attempt to recover SCP as the first mitigation action but failed. The other progress is pretty similar to that of the first one. The combination events of the loss of shutdown cooling and CCF of essential chillers (WOCHKQ4) are shown in the high rank MCS. In case of WOCHKQ4 leads to core damage with a simple combination with several initiating events, representing that the common caused failure of safety injection system might become a very critical event with some IEs in the several POSs.

Table 1: Minimal Cutset for APR+ LPSD

Rank	Mean (/year)	(%)	Minimal Cutset			
1	3.58E-07	13	%SOP05	HR-FB-SOP05-02-DE	HR-FS-SOP05-02-DE2	HR-MK-SOP05
2	2.03E-07	7.4	%S1P05	HR-FB-S1P05	HR-FS-S1P05-DE	HR-RS-S1P05
3	1.48E-07	5.4	%SOP11	HR-FB-SOP11-02-DE	HR-FS-SOP11-02-DE2	HR-MK-SOP11
4	1.25E-07	4.5	%S2P05	HR-FB-S2P05	HR-FS-S2P05-DE	HR-RS-S2P05
5	8.73E-08	3.2	%CCP05	HR-FB-CCP05	HR-FS-CCP05-DE	HR-RS-CCP05
6	7.97E-08	2.9	%PLP02	WOCHKQ4-CH01A/B/C/D		
7	7.86E-08	2.8	%S1P05	HR-FS-S1P05	HR-RS-S1P05	SISPP-P456
8	6.13E-08	2.2	%LPP10	WOCHWQ4-CH01A/B/C/D		
9	4.85E-08	1.8	%S2P05	HR-FS-S2P05	HR-RS-S2P05	SISPP-P456
10	4.79E-08	1.7	%SOP05	WOCHKQ4-CH01A/B/C/D		
11	4.79E-08	1.7	%SOP11	WOCHKQ4-CH01A/B/C/D		
12	4.53E-08	1.6	%S1P10	WOCHKQ4-CH01A/B/C/D		

※ SO : loss of SCP due to over drain , HR-: human action, WOCHWQ4 : CCF of essential chillers
S1 : recoverable SCP mechanical random fail , F&B : feed & bleed, F&S : feed & steaming
S2 : unrecoverable SCP mechanical random fail

Through the risk analysis depending on POSs, IEs, importance, and MCSs, three design alternatives are identified. The first alternative is associated with the prevention of the initiating event caused by the loss of shutdown cooling including over drain (SO) and failure of maintain water level (SL). The second alternative is corresponds to the reliability enhancement for the operator manual actions in feed and bleed operation (HR-FB-**). An alternative design is proposed with automated injections by hot leg level signal. In the third alternative, an alternative associating with the diverse function for room cooling of safety pumps is elicited. By the implementation of the alternative, the probability of essential chiller CCF can be mitigated. The first and second design alternatives are applied to only LPSD operations. However, the third item is applicable to both full power and LPSD. Since the various safety functions my not be available due to the maintenance, the CCF of essential chiller would be critical during LPSD.

3.3. Design Alternatives

3.3.1. Prevention of loss of shutdown cooling (alternative 1)

RCS levels during mid-loop operation are controlled and balanced by shutdown cooling system (SCS) and chemical & volume control system (CVCS). The heat removal loop (loop A in figure 2) is established by shutdown cooling system with a closed loop shape. The purification loop (loop B in figure 2) starts from the discharge line of shutdown cooling (SDC) heat exchanger (HX) and connected to volume control tank (VCT) of CVCS. In order to prevent over drain or inadvertent level loss during mid-loop, the first design alternative is proposed with the installment of automatic isolation of MOVs (red dot line in figure 2) in the purification line when RCS level decrease to Lo level (figure 2). The design corresponding level sensors, signal transmitters and logic devices are included in the alternative. This design prevents the RCS inventory loss during the accident with an automated manner. Through the implementation of the design alternative, the initiating event associated with over drain or failure of maintain water level can be significantly reduced.

Figure 2: Design alternative 1&2 for reducing LPSD Risk (Automatic purification line isolation and SIP injection)

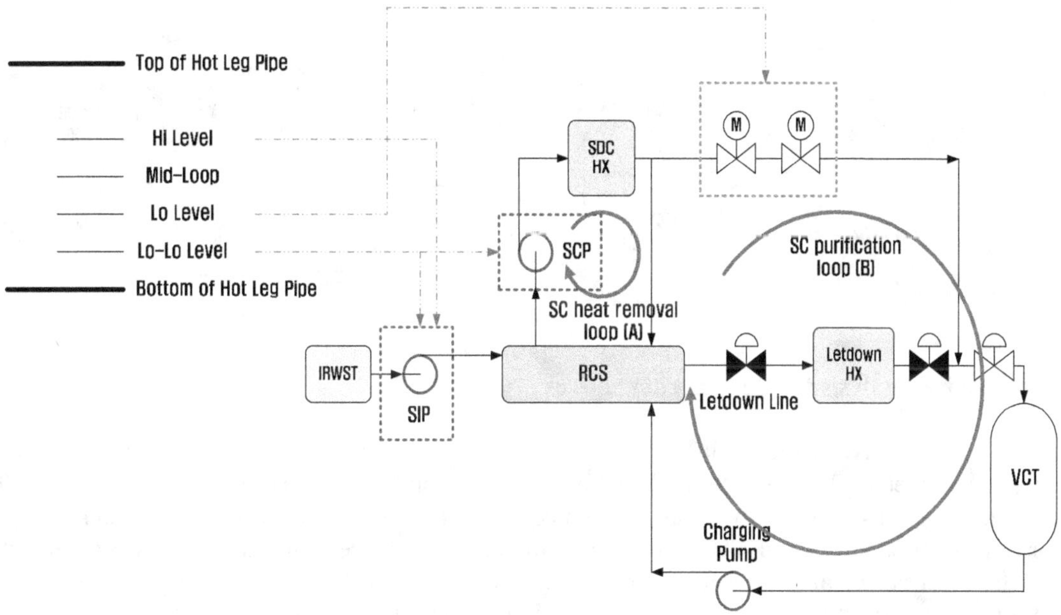

3.3.2. Automated safety injection under RCS level reduced (alternative 2)

When an event associating with over drain (SO) and failure of maintain water level (SL) is initiated, immediate action to stop the running SCP and to start safety injection pumps (blue dot line in figure 2) should be taken to mitigate the accident. The second design alternative automates the injections by the level signal from hot leg. When RCS level decreases continuously below Lo level and reaches Lo-Lo level, the automated systems are activated. When RCS level is recovered to Lo level, SCP starts again. If RCS level reaches Hi level, then a signal to stop the safety injection pumps are transmitted. The design corresponding level sensors, signal transmitters and logic devices are included in the alternative. In addition, the hot leg level should be set for generating the relevant signals, e.g. Lo-Lo, Lo, Hi level (figure 2).

3.3.3. Diversified room cooling for safety systems (alternative 3)

The essential chillers provide chilled water to the regional cubicle coolers that remove heat from safety class equipment room ensuring equipment survivability under the accidents. APR+ design adopts four trains of essential chilled water (ECW) system and each train has 100% heat removal capacity. The removed heat is transferred to the component cooling water (CCW) system in the heat exchangers of the condenser, and another heat removal cycle, essential service water (ESW) system is connected to CCW system. Therefore, the loss of ultimate heat sink or common caused failures of CCW/ESW components result in the function loss of the ECW system. Eventually, these all series events lead to the function loss of safety systems and components that require heat removal during the accident.

The basic event, WOCHKQ4 in MCS represents CCF event of four essential chillers. In order to break the common caused failure mode, the third alternative proposes the diversification in the design of ECW and its supporting systems. First, two additional ECWs are introduced, which use different types or different manufacturers from the original pumps, chillers. And next, the added ECWs are introduced the heat sink as air-cooled type while the original ECWs use water-cooled type. Through these two diversified designs, risk associated with LPSD operation mode would be significantly reduced.

Figure 3: Design alternatives for reducing LPSD Risk
(Additional chilled water systems using air condensing)

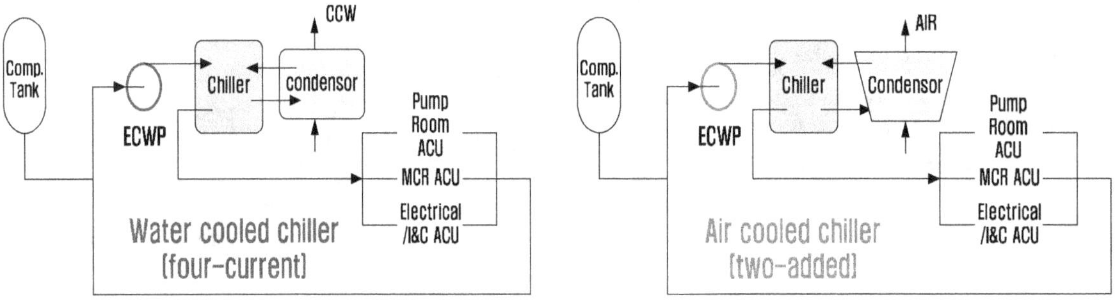

3.4. Sensitivity Analysis for Design Alternatives

Sensitivity analysis is used to evaluate the effectiveness of the proposed design alternatives with the associated parameters. The detailed modeling and risk analysis will be performed in the fifth step. The case studies for sensitivity analysis are performed changing the associated risk parameters with the factor of 0%, 10%, 50%, and 100%. The changes of core damage frequencies are measured depending on the factored values and the effectiveness of the design alternatives are interpreted based on the result of the sensitivity analysis. The combination effect of the first and second alternatives is evaluated as well.

First, the risk parameters are designated for the sensitivity analysis. In the first alternative, the initiating event of over drain during mid-loop operation, POS 5 and POS11 is selected as the representative risk parameter. And the second alternative chooses the basic events for the operator actions associated with feed and bleed operation during accident mitigation. The common caused failure event of essential chillers is designated as the representative risk parameter.

Second, the effectiveness is measured for the case of 0%, 10%, 50%, and 100% sensitivities. In terms of the first alternative, when the risk parameter is factored to 10%, the total LPSD CDF is reduced to 25%. The selected parameters are mainly effective for POS 5 and POS 11, mid-loop operation. For the second alternative, automated design of safety injection, when the risk parameter is factored to 10%, the total LPSD CDF is reduced to 34%. The selected parameters are associated with operator action for feed & bleed. For more efficient work process, the operational state and applicable cutsets are limited to 5&6 and the top 100. The top 100 cutsets cover almost 90% of the total LPSD CDF so, the simplified application would be sufficient to identify the effectiveness of the alternatives and obtain the associated risk insights. In terms of the combination alternative (1&2), when the risk parameters are factored to 10%, the total LPSD CDF is reduced to 43% from its original value. For the third alternative, when the probability value for the essential chiller CCF is factored to 10%, the total LPSD CDF is reduced to 14%. Although the risk parameter may not represent the whole design alternatives and the exact value can be evaluated with detailed design and modeling, the effectiveness for the alternatives can be estimated roughly with simple parameter sensitivity analysis.

In this paper, the maximum and minimum of the effectiveness are predicted by the result of 0% and 10% sensitivity cases. When a design alternative is implemented with very effective manners, the associated frequency or probability parameter would be reduced to very low value, e.g. $10^{-7} \sim 10^{-8}$ or even close to zero with optimistic perspectives. However, when the design alternative is poorly implemented, the gain from associated frequency or probability parameter would not be more than 10% with pessimistic perspective. These are the basis of qualitative assessment for the design alternatives and associated engineering judgments are included.

Table 2: Risk Sensitivity Analysis for the Design Alternatives

Alternative	Description	Risk Parameter	Sensitivity	Value	Total CDF (/yr)	CDF Reduction	Effectiveness (qualitative)
1	Prevention to Loss of Shutdown Cooling	%SOP05 %SOP11	100%	1.77E-03	2.76E-06	0%	High
			50%	8.87E-04	2.37E-06	14%	
			10%	1.77E-04	2.06E-06	25%	
			0%	0.00E+00	1.98E-06	28%	
2	Automated Safety Injection	HR-FB-** (POS 5&11)	100%	–	2.76E-06	0%	High+
			50%	–	2.35E-06	15%	
			10%	–	1.81E-06	34%	
			0%	–	1.70E-06	38%	
1+2	Alternative 1 + Alternative 2	%SOP05 %SOP11 HR-FB-** (POS 5&11)	100%	–	2.76E-06	0%	High++
			50%	–	2.05E-06	26%	
			10%	–	1.58E-06	43%	
			0%	–	1.52E-06	45%	
3	Diversify Room Cooling for Safety System	WOCHWQ4	100%	2.70E-05	2.76E-06	0%	Medium
			50%	1.35E-05	2.55E-06	8%	
			10%	2.70E-06	2.38E-06	14%	
			0%	0.00E+00	2.34E-06	15%	

3.5. Priority Determination

The result of the sensitivity parameter shows that the most effective alternatives are the case of combination 1&2. And next followings are the second alternative, the first alternative, and the third alternative. In the process of priority determination, not only the effectiveness of risk reduction but also other negative effects should be considered with more comprehensive manners. For instance, when the second alternative is implemented, there are possibilities for the faulted level sensing or unstable RCS level that triggers unexpected safety injection resulting in the radioactive contamination for the workers who are installing nozzle dam during mid-loop operation. As for the third alternative, although the design itself is simple to implementation, the added equipment and systems requires new space so that general arraignment should be changed and relevant additional seismic analysis should be performed. Moreover, if the design status is almost complete, the other associated design changes should be considered, e.g. HVAC configuration, new drawings for piping. On the more comprehensive perspective with risk evaluation, design, licensing, industrial safety, and constructability, the implementation of the first design alternative would be the best solution for APR+ design and the alternative plan proceeds to the fifth step.

4. CONCLUSION

This paper proposes three design alternatives to reduce low power shutdown risk for APR+ design. In order to determine the best alternative, risk parameters associated with each alternative have been identified and the corresponding conceptual designs are performed. The sensitivity analysis has been performed to measure the effectiveness of the proposed alternatives. Accordingly, comprehensive evaluations considering the negative effect of the design alternatives including design, licensing, and industrial safety has been done in the priority determination step. A best resolution has been determined and progressed to the next step.

As for the future work, the fifth step will be implemented including the detailed design and risk evaluation for the selected best alternative. Finally the completed work will be delivered to the construction phase of APR+ design.

References

[1] USNRC GENERIC LETTER NO. 88-17, Loss of Decay Heat Removal, October 17, 1988.
[2] NUMARC 91-06, "Guideline for Industry Actions to Assess Shutdown Management," December 1991.
[3] KHNP, "APR+ Probabilistic Safety Assessment Technical Report," June, 2011.
[4] USNRC NUREG/CR-6144, "Evaluation of Potential Severe Accidents during Low Power and Shutdown Operations at Surry, Unit 1," October, 1995.
[5] KHNP, "APR1400 Probabilistic Safety Assessment for Low Power and Shutdown Operations," January, 2012.
[6] KHNP, "APR+ Final Safety Analysis Report," January, 2012.
[7] J. Oh, et al. "Risk Assessment and Safety Improvement for Low Power Shutdown Operation in Advanced Nuclear Power", *Proceedings of KSME Reliability*, 2014, p5.
[8] Mohammad Modarres, "Reliability Engineering and Risk Analysis," Marcel Dekker, 1999, New York.

An Implementation Strategy of Low Power Shutdown PSA for KHNP NPPs

Jang-Hwan Na, Seok-Won Hwang, Ho-Jun Jeon

Central Research Institute of Korea Hydro & Nuclear Power Co.,Ltd, 1312-70 Beon-gil, Yuseong-daero, Yuseong-Gu, Daejeon, Korea, 305-343
Tel. +82-42-870-5642, Fax. +82-42-870-5659, E-mail: swhwang@khnp.co.kr

Abstract: Rightly after the Fukushima accidents, the Korean Regulatory Agency with the support from a group of academic and research experts evaluated the safety of Korean nuclear power plants including plants on construction. The expert group particularly focused on any possible design vulnerabilities in view of ultimate heat sinks and power sources considering external hazards such as seismic, flood or complex initiated events. They identified several common or plant-wise improvement factors and elicited 49 post-action items as near term Fukushima accident measures. One of the measures is to develop SAMG (Severe Accident Management Guideline) during LPSD (Low Power and Shutdown) operation in addition to the existing SAMG on the full power operation. At first, KHNP (Korea Hydro & Nuclear Power) decided to develop the LPSD PSA (Probabilistic Safety Assessment) models to increase the quality of LPSD SAMG. To get a technical adequacy, KHNP decided to revise the full spectrum of existing PSA models including full power, external or level 2 PSA incorporating up-to-date reliability data and methodologies. This paper presents an implementation strategy of developing LPSD PSA models including the status of upgrading full power PSA models at the end of 2013.

Keywords: LPSD PSA, PSA Technical Adequacy, SAMG, CCF, HRA

1. INTRODUCTION

Rightly after the Fukushima accidents, the Korean Regulatory Agency with the support from a group of academic and research experts evaluated the safety of Korean nuclear power plants including plants on construction. The expert group particularly focused on any possible design vulnerabilities in view of ultimate heat sinks and power sources considering external hazards such as seismic, flood or complex initiated events. They identified several common or plant-wise improvement factors and elicited 49 post-action items as near term Fukushima accident measures.

One of the measures is to develop SAMG during LPSD operation in addition to the existing SAMG on the full power operation. At first, KHNP decided to develop the LPSD PSA models to increase the quality of LPSD SAMG. To get a technical adequacy, KHNP decided to revise the full spectrum of existing PSA models including full power, external or level 2 PSA incorporating up-to-date reliability data and methodologies.

KHNP had performed two Peer Reviews by NEI (Nuclear Energy Institute) method[1] and ASME PRA Standard[2] for two types of plants. Because the comments and insights from the peer reviews were not fully reflected on all of the PSA models yet, we try to apply the major results of the peer reviews in this project. Moreover, we standardize the methodology of CCF (Common Cause Failure) and HRA (Human Reliability Analysis) which are the most influential factors to risk measures of NPPs.

LPSD PSA models will be developed for all the operating plants based on standard outage maintenance practices and Plant Operational Status (POS). Level 2 LPSD PSA models will also be developed for two types of pilot plants, one for PWR and the other for PHWR (Pressurized Heavy Water Reactor). In addition, we decide to newly develop the models for SFP (Spent Fuel Pool) as one of lessons learned from the Fukushima accidents.

Because KHNP had the fire PSA models based on only EPRI FIVE (Fire Induced Vulnerability Evaluation)[3] methodology guided by IPEEE (Individual Plant External Event Evaluation) requirements, we decide to apply the new methodology of NUREG/CR-6850[4] to fire PSA models of a pilot plant. Through this project, we expect to improve the quality of PSA models and to set a fundament of risk management for all operating status. Also, we anticipate providing the major input data and insights to SAMG.

2. IMPLEMENTATION STRATEGY OF LOW POWER SHUTDOWN PSA

The quality of LPSD PSA models is closely related to that of full power PSA models. Figure 1 shows the development and implementation strategies of LPSD PSA based on the full power PSA models.

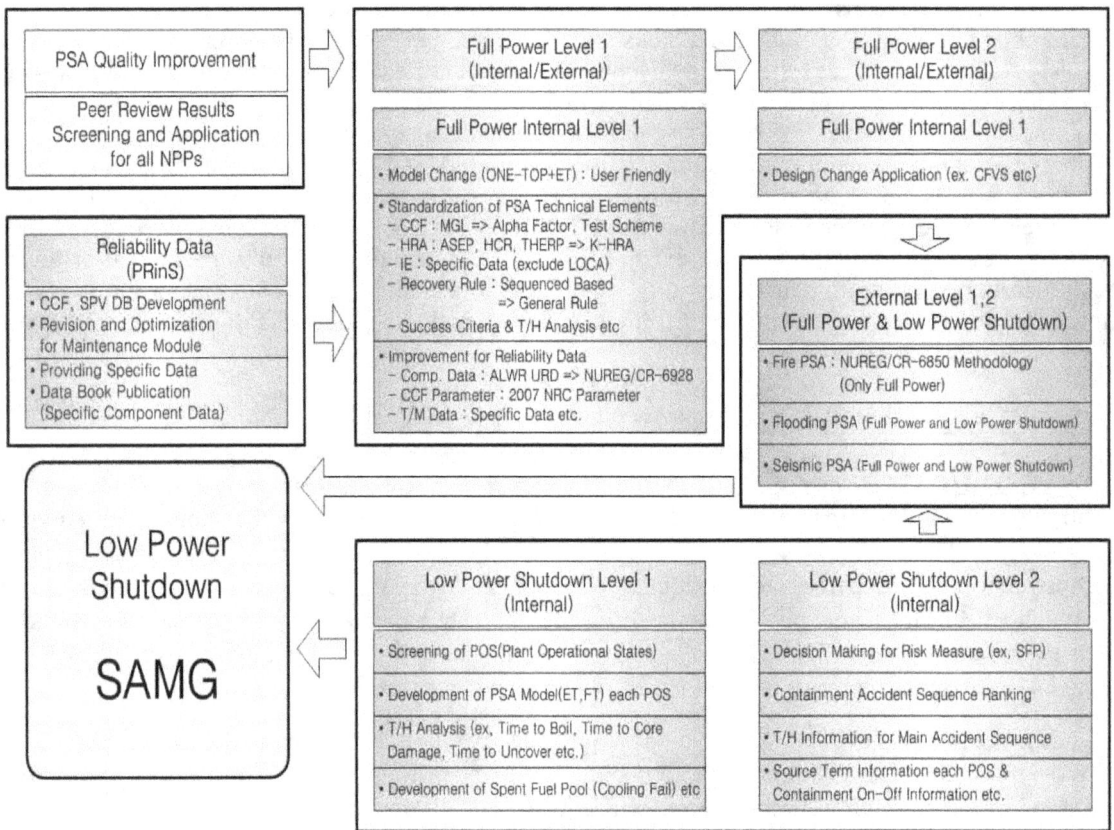

Figure 1. Implementation Strategy of Full spectrum Risk Analysis for KHNP NPPs

To ensure the quality of full power PSA models, the insights of the peer reviews are applied and the reliability data are updated with KHNP operating experiences spanning all commercial operation periods and combined with the latest generic data. The frequency of initiating events is based on country-specific empirical data by preference, and the rests applied the values provided by latest NUREG/CR-6928[5]. The generic component reliability database also used NUREG/CR-6928 instead of EPRI ALWR URD (Advanced Light Water Reactor Utility Requirement Document) [6] database. And, we perform more cases of thermal hydraulic analyses, review the existing MAAP parameter files and, verify the results of MAAP by comparing those of MARS (Multi-dimensional Analysis of Reactor Safety) [7] for main accident sequences such as feed and bleed scenarios.

As for the standardization of PSA technical elements in Figure 1, CCF and HRA have the most significant impacts on the risk measures. KHNP has used MGL (Multiple Greek Letter) parameter method for CCF analysis. However, we decided to apply Alpha Factor method because of convenience in uncertainty analysis and combining the latest international research results or data for CCF events. Also, we surveyed the test schemes of components which need modeling as CCF events to develop domestic-specific CCF Database. In case of HRA, we apply K-HRA (Korean Standard HRA)[8] methodology, which was developed and validated in 2005 based on ASEP (Accident Sequence Evaluation Program)[9] and THERP (Technique for Human Error Rate Prediction)[10] methodology. We shall develop LPSA PSA models based on the full power PSA models upgraded with the technical adequacy.

3. STANDARDIZATION AND QUALITY IMPROVEMENT OF PSA ELEMENTS

Through performing living PSA and configuration risk management, many issues related to standardization were raised for the application and management of PSA models. So, we need to standardize and manage the technical elements of PSA by applying consistent guidelines and methodologies. It is also necessary to increase the quality by applying the results and the insights from the previous peer reviews. And, we update component reliability data by using the latest operating experiences.

3.1 Application of Peer Review Findings and Model Standardization

Domestic and foreign experts performed the official peer reviews for two PSA models. KHNP has the various types of reactors such as KSNP (Korea Standard Nuclear Power), Westinghouse, Framatome, and CANDU typed reactor. Therefore, we should consider the characteristics according to reactor type when applying peer review results to other reactor types. We summarized the findings and observations from two peer reviews, 198 lists from WH type and 50 lists from KSNP. The following are the main findings and observations. The first is about using the latest reliability data or not. So, we updated IE (Initiating Event) frequencies as well as component failure data based on new generic data (NUREG/CR-6928). The second is about applying the latest references as for RCP seal modelling, SGTR modelling, and so on. It also requires that the more T/H analyses should be performed in detail to increase the quality on HRA. Considering the test schemes is recommended when estimating CCF factors. Lastly, the documents shall be reinforced such as calculating sheets and description about assumptions.

The standardization of PSA models has a close relationship with the living PSA and configuration risk management. We should maintain RIMS (RIsk Monitoring System) models as well as PSA models. PSA models are different from RIMS models in assumptions, modeling method, operating configurations to be modeled, the structures of models, and so on. Thus, integrated management is needed for smooth promotion of living PSA and RIA (Risk-Informed Application). Figure 2 shows the current status of configuration and the direction of standardization of PSA models and RIMS models.

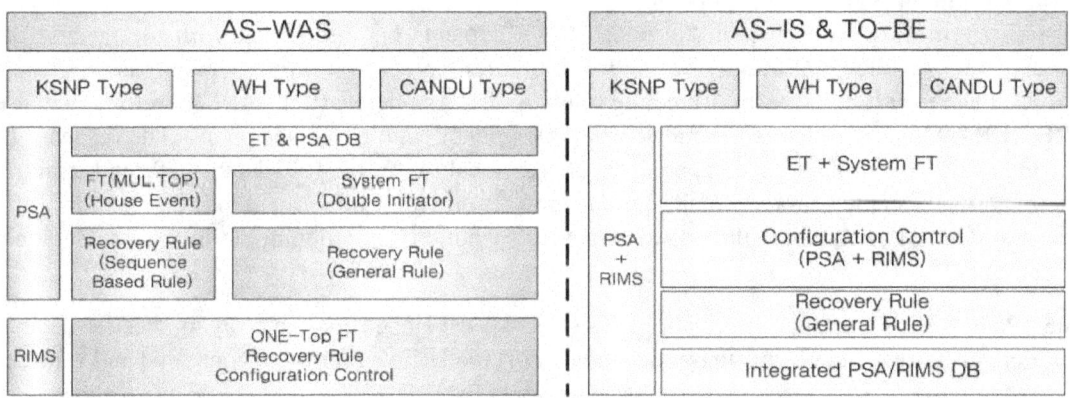

Figure 2. Integration scheme of PSA & Risk monitoring system model

3.2 Standardization of PSA Technical Elements

In this project, we emphasized the standardization of CCF, HRA, and Initiating Events. First, CCF is one of the most important factors on risk measures. We used MGL parameter method as CCF analysis assuming only staggered test scheme. Assuming staggered tests only is underestimating the risk as a factor of redundancy multiplication in CCF modeling. So, we shall consider test schemes and use Alpha Factor method which can easily analyze uncertainty and estimate parameters. In addition, we developed CCF Calculator to estimate specific CCF parameter and will develop CCF DB Module which is interchangeable with ICDE (International CCF Data Exchange) database. Figure 3 shows the user interfaces of CCF Calculator. Also, designing of database structure to estimate CCF parameter in conjunction with ERP (Enterprise Resource Planning) system is scheduled because the ERP has insufficient information of raw data and contents requested by ICDE Report Form.

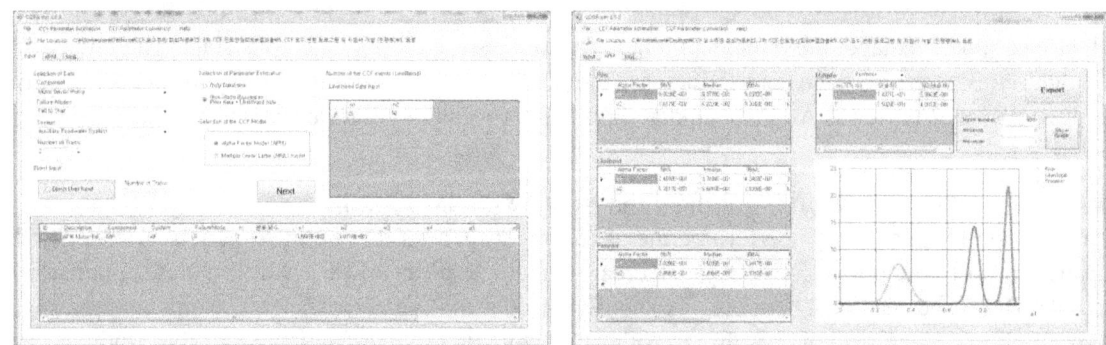

Figure 3. Calculator of CCF Parameter Estimation

Several HRA methodologies such as ASEP, THERP and HCR (Human Cognitive Reliability) are currently applied to PSA models and large difference existed by performing institutes and analysts. Technical issues are insufficiency of plant-specific analysis, lack of consistent quantification procedure, and documentation of bases. We applied standardized HRA methodology (K-HRA) consistently, because it gives detailed instruction to maintain the consistency of analysis through consensus among HRA experts in Korea. K-HRA methodology was validated through participation in various programs of NRC and OECD/NEA in order to secure the international confidences.

Initiating events analysis methodology and data are also applied to the each project-wise model. The emphasizing point is documentation on classification bases, grouping of initiating events, and FMEA (Failure Mode and Effect Analysis) on supporting systems. In case of initiating event frequencies, we used new data from NUREG/CR-6928 for LOCA (Loss of Coolant Accident) group and domestic empirical data for transient group.

3.3 Application for Newly Generic Data
Component reliability data of domestic nuclear power plants has been used with the generic data generated from ALWR URD. It does not only reflect latest component failure characteristics bus also does not match with PSA bases reflecting the "As-is, As-operated" status. Therefore, we used NUREG/CR-6928 data which is generated from the recent operating power plants. The database has following major differences as compared to conventional ALWR URD data; 1) Provides normal running, normal standby data for 9 major components 2) Provides fail to load and run data for normal standby system 3) Demand failure data is beta distribution and running failure data is gamma distribution.

PRinS (Plant Reliability Data Information System)[11], developed for the Korean specific database generation, also needed to be update by revision of PSA model. We reviewed, compared and evaluated the newly revised PSA model and database to apply NUREG/CR-6928. Also, we validated the maintenance history analysis results generated from systematic data gathering guidelines. These results were utilized to analyze the trends of component failure and to supplement design through publication of specific data book. Table 1 shows and compares the generic data of major components and latest specific data. Specific data is produced from failure history of domestic 16 PWRs. The key finding at this time is the Korean specific failure rate of DG and AFW pump is lowered, but, that of fans and chillers is increased.

Table 1. Reliability Data Trend for key Components

Component/Failure Mode	EPRI URD	KSNP	NUREG/CR-6928	New Data
DG FTR	2.40E-03	2.40E-03	8.48E-05	8.48E-05
DG FTS	1.40E-02	2.00E-02	7.43E-03	9.73E-03
AF MDP FTS	3.00E-03	7.07E-03	1.85E-03	1.87E-03
AF TDP FTS	1.50E-02	1.78E-02	9.52E-03	2.54E-03
ESW Pump FTS	2.40E-03	1.06E-02	2.23E-03	2.32E-03
HPSI Pump FTS	1.00E-03	5.66E-04	1.85E-03	1.87E-03
MOV FTS	4.00E-03	1.39E-03	1.07E-03	1.40E-03
Circuit Breaker FTS	3.00E-04	3.00E-04	2.55E-03	7.35E-04
Fan FTS	6.00E-04	6.00E-04	1.79E-03	1.48E-03

3.4 Development of LPSD PSA Model

KHNP evaluated Mid-Loop Operation of LPSD PSA[12] for two pilot plants in early 2000. After the Fukushima accidents, development of LPSD SAMG is under way and LPSD PSA should provide the plant characteristic information in developing specific LPSD SAMG. Thus we will develop internal and external LPSD Level 1 PSA model. In addition we have plans to analyze Fuel Damage Frequency due to loss of cooling in Spent Fuel Pool and to start LPSD Level 2 PSA for the first time in Korea. In LPSD Level 2 PSA, we will analyze containment accident sequence by POS, source term, and fuel building release sequences in case of loss of spent fuel pool cooling. The outputs of LPSD Level 2 PSA will be primarily utilized to develop LPSD Specific SAMG.

4. CONCLUSION

After the Fukushima accident, the significance of severe accident and PSA came to the public as well as to the industry itself. Among 50 safety-related plans, in this paper, we suggest the implementation strategies to provide major input data of LPSD SAMG through development of LPSD PSA. Also, we suggest the standardization of full-power PSA technical elements and methodologies which are the basis of development of LPSD PSA, solutions to problem of current PSA model for living PSA, application of new data. Above this, we will use NUREG/CR-6850 as a new methodology for pilot plants in case of fire risk, assess risk due to loss of Spent Fuel Pool cooling and revise the thermal-hydraulic analyses results. We established the implementation strategy for the purpose of improving safety of plants and quality of PSA.

This scheme will be an important opportunity to upgrade the PSA level in Korea. The results of LPSD PSA implementation strategy will contribute to conforming of regulatory requirement and legislation of PSA which requests the application of new methodology, RIA, PSA quality, performing optimized Living PSA through user-friendly PSA model and established long-term Roadmap.

References

[1] NEI, "Probabilistic Risk Assessment Peer Review Process Guidance," NEI-00-02, 2000.
[2] ASME, "Standard for PRA for Nuclear Power Plant Applications," ASME RA-S-2002, 2002.
[3] EPRI, "Fire-Induced Vulnerability Evaluation, EPRI TR-100370, 1992.
[4] U.S.NRC, "EPRI/NRC-RES fire PRA for nuclear power facilities," NUREG/CR-6850, 2005.
[5] U.S.NRC, "Industry-average performance for components and initiating events at U.S. commercial nuclear power plants," NUREG/CR-6928, 2007.
[6] EPRI, "Advanced Light Water Reactor Utility Requirements Document, PRA Key Assumption and Groundrules, Rev 7,(Vol II, ALWR Evolutionary Plant, Chapter 1, Appendix A)," 1995.
[7] Jeong J.J., "Development of a multi-dimensional thermal-hydraulic system code, MARS 1.3.1, Ann. Nuclear Energy 26", 1999
[8] KAERI, "Development of Standard Method for HRA of nuclear power plants – Level 1 PSA Full Power Internal HRA, " KAERI/TR-2961/2005, 2005.

[9]   A. D. Swain. "Accident Sequence Evaluation Program Human Reliability Analysis Procedure," NUREG/CR-4772, 1987.

[10]   A. D. Swain and H. E. Gutmann. "Handbook of Human Reliability Analysis With Emphasis on Nuclear Power Plant Applications", NUREG/CR-1278, 1983.

[11]   Seok-Won Hwang et al. "Development of web-based reliability data analysis algorithm model and its application", Annals of nuclear energy, Volume 37, 248-255, 2010

[12]   U.S. NRC, "Evaluation of potential severe accidents during low power and shutdown operations at surrey, unit 1", NUREG/CR-6144, 1994.

www.ingramcontent.com/pod-product-compliance
Lightning Source LLC
Chambersburg PA
CBHW080757180526
45168CB00006B/2245